Centre for Educational Research and Innovation

Understanding the Brain: The Birth of a Learning Science

OECD

ORGANISATION FOR ECONOMIC CO-OPERATION AND DEVELOPMENT

ORGANISATION FOR ECONOMIC CO-OPERATION AND DEVELOPMENT

The OECD is a unique forum where the governments of 30 democracies work together to address the economic, social and environmental challenges of globalisation. The OECD is also at the forefront of efforts to understand and to help governments respond to new developments and concerns, such as corporate governance, the information economy and the challenges of an ageing population. The Organisation provides a setting where governments can compare policy experiences, seek answers to common problems, identify good practice and work to co-ordinate domestic and international policies.

The OECD member countries are: Australia, Austria, Belgium, Canada, the Czech Republic, Denmark, Finland, France, Germany, Greece, Hungary, Iceland, Ireland, Italy, Japan, Korea, Luxembourg, Mexico, the Netherlands, New Zealand, Norway, Poland, Portugal, the Slovak Republic, Spain, Sweden, Switzerland, Turkey, the United Kingdom and the United States. The Commission of the European Communities takes part in the work of the OECD.

OECD Publishing disseminates widely the results of the Organisation's statistics gathering and research on economic, social and environmental issues, as well as the conventions, guidelines and standards agreed by its members.

> *This work is published on the responsibility of the Secretary-General of the OECD. The opinions expressed and arguments employed herein do not necessarily reflect the official views of the Organisation or of the governments of its member countries.*

Also available in French under the title:
Comprendre le cerveau : naissance d'une science de l'apprentissage

Foreword

The project on "Learning Sciences and Brain Research" was launched by the OECD's Centre for Educational Research and Innovation (CERI) in 1999. The purpose of this novel project was to encourage collaboration between learning sciences and brain research on the one hand, and researchers and policy makers on the other hand. The CERI Governing Board recognised this as a difficult and challenging task, but with a high potential pay-off. It was particularly agreed that the project had excellent potential for better understanding learning processes over the lifecycle and that a number of important ethical issues had to be addressed in this framework. Together, these potentials and concerns highlighted the need for dialogue between the different stakeholders.

Brain research is slowly but surely gaining a firm foothold into making applications in the learning field. The second phase of the project successfully initiated much cross-fertilisation across research areas and between researchers and has become internationally recognised worldwide. This has led to the instigation of many national initiatives in OECD countries to put new knowledge about the brain into educational practice. However, the number of discoveries from brain research that have been exploited by the education sector remain relatively few so far, partly because there is not enough consensus yet on the potential applications of brain research to education policies. But there are strong various reasons for fostering the pioneering brain and learning centres and promoting the creation of more bridges between the two research communities. Findings confirm about the brain's plasticity to learn anew over the individual's lifecycle, and technologies of non-invasive brain scanning and imaging are opening up totally new approaches. By bringing the two research communities closer, more value-added discoveries will certainly be made.

This book follows from the OECD report Understanding the Brain: Towards a New Learning Science *issued in 2002 (published in seven languages; most of that publication is reflected in the present one). It aims to educate readers about the brain and understand how it learns and how learning can be optimised through nurture, training and adapted teaching processes and practices. It is intended to be accessible to non-specialists and it therefore seeks to avoid exclusive language. Its content derives from the three trans-disciplinary networks set up in 2002 to focus on literacy, numeracy and lifelong learning, and a fourth focus activity on Emotions and Learning which from 2004 on ran parallel to the three networks. The project-dedicated website also served as an innovative interactive source of input to this work soliciting feedback and substantial input from educational practitioners and civil society.*

Essential financial and substantive support was provided from the beginning by:

- *the National Science Foundation (Directorate of Research, Evaluation and Communication/ Education Division, United States);*
- *the Japanese Ministry of Education, Culture, Sports, Science and Technology (MEXT) (Japan);*
- *the Department for Education and Skills (DfES) (United Kingdom);*
- *the Finnish Ministry of Education (Finland);*
- *the Spanish Ministry of Education (Spain);*
- *the Lifelong Learning Foundation (United Kingdom).*

Essential scientific, financial and/or organisational support was provided by the RIKEN Brain Science Institute (Japan); the Sackler Institute (United States); the Learning Lab Denmark (Denmark); the ZNL within Ulm University (Germany); INSERM (France); Cambridge University (United Kingdom); the Académie des Sciences (France); the City and the University of Granada (Spain); the Royal Institution (United Kingdom).

Within the OECD, the "Learning Sciences and Brain Research" project leader Bruno della Chiesa was responsible for this report, along with Cassandra Davis, Koji Miyamoto, and Keiko Momii. Substantive input was provided by Christina Hinton, Eamonn Kelly, Ulrike Rimmele and Ronit Strobel-Dahan as consultants to the project. The English version of the main report (Part I) was edited by David Istance and the French version by Bruno della Chiesa. The book was partially or completely reviewed by Jarl Bengtsson, Delphine Grandrieux, David Istance, Christina Hinton, Atsushi Iriki, Masao Ito, Jellemer Jolles, Hideaki Koizumi, Michael Posner, Ulrike Rimmele, Adriana Ruiz Esparza, Ronit Strobel-Dahan and the CERI "Brain Team".

Within the Secretariat, Jarl Bengtsson took the initiative of launching this project, and provided strategic and critical support throughout; Tom Schuller followed through the second phase of the project. Logistical support was provided by Vanessa Christoph, Emily Groves, and Carrie Tyler (in order of succession). Cassandra Davis was the project website editor.

Barbara Ischinger,
Director, Directorate for Education

Acknowledgements

On behalf of the Secretariat, Bruno della Chiesa would like to:

- Dedicate this work to Jarl Bengtsson, the brainchild behind the "Learning Sciences and Brain Research" project.

- Extend thanks and special appreciation to Eric Hamilton, Masao Ito, Eamonn Kelly, Hideaki Koizumi, Michael Posner, and Emile Servan-Schreiber for their utmost dedication to the project.

- Extend thanks to the main supporting partners of the project for their contributions (financial and/or substantial): Richard Bartholomew and team, Christopher Brookes, Eamonn Kelly, Juan Gallo and team, Eric Hamilton and team, Masayuki Inoue and team, Søren Kjær Jensen and team, Reijo Laukkanen and team, Pierre Léna and team, Francisco Lopez Ruperez, José Moratalla and team, Teiichi Sato, Sylvia Schmelkes del Valle, Hans Siggaard Jensen and team, Finbarr Sloane.

- Extend thanks to the high-flying scientifists who showed strong involvement throughout the project: Brian Butterworth, Stanislas Dehaene, Christina Hinton, Jellemer Jolles, Heikki Lyytinen, Bruce McCandliss, Ulrike Rimmele, Nuria Sebastian, Manfred Spitzer.

- Extend thanks to Hilary Barth, Antonio Battro, Daniel Berch, Leo Blomert, Elisa Bonilla, John Bruer, Tom Carr, Marie Cheour, Guy Claxton, Frank Coffield, Stanley Colcombe, Margarete Delazer, Guinevere Eden, Linnea Ehri, Michel Fayol, Uta Frith, Michael Fritz, Ram Frost, Peter Gärdenfors, Christian Gerlach, Usha Goswami, Sharon Griffin, Peter Hannon, Takao Hensch, Katrin Hille, Shu Hua, Petra Hurks, Walo Hutmacher, Atsushi Iriki, Layne Kalbfleisch, Ryuta Kawashima, Arthur Kramer, Morten Kringelbach, Stephen Kosslyn, Jan de Lange, Cindy Leaney, Geoff Masters, Michael Meaney, Michael Miller, Fred Morrison, Risto Näätänen, Kevin Ochsner, David Papo, Raja Parasuraman, Eraldo Paulesu, Ken Pugh, Denis Ralph, Ricardo Rosas, Wolfgang Schinagl, Mark Seidenberg, David Servan-Schreiber, Bennett Shaywitz, Sally Shaywitz, Elizabeth Spelke, Pio Tudela, Harry Uylings, Janet Werker, Daniel Wolpert, and Johannes Ziegler, members of the project's network of high calibre experts.

- Extend thanks for the facilities and hospitality given to allow for the productive conduct of the trandisciplinary meetings to (in order of chronological succession): the Sackler Institute, United States; the University of Granada, Spain; the RIKEN Brain Science Institute, Japan; the National Board of Education, Finland; the Royal Institution, United Kingdom; the INSERM, France; the ZNL at Psychiatric Hospital University of Ulm, Germany; the Learning Lab, Denmark; the Spanish Ministry of Education, Spain; the Académie des Sciences, France; the Research Institute for Science and Technology for Society (RISTEX) of the Japan Science and Technology Agency (JST), Japan; and the Centre for Neuroscience in Education, Cambridge University, United Kingdom.

- Extend thanks to the writers and contributors of this publication: Christopher Ball, Bharti, Frank Coffield, Mélanie Daubrosse, Gavin Doyle, Karen Evans, Kurt Fisher, Ram Frost, Christian Gerlach, Usha Goswami, Rob Harriman, Liet Hellwig, Katrin Hille, Christina Hinton, David Istance, Marc Jamous, Jellemer Jolles, Eamonn Kelly, Sandrine Kelner, Hideaki Koizumi, Morten Kringelbach, Raja Parasuraman, Odile Pavot, Michael Posner, Ulrike Rimmele, Adriana Ruiz Esparza, Nuria Sebastian, Emile Servan-Schreiber, Ronit Strobel-Dahan, Collette Tayler, Rudolf Tippelt, Johannes Ziegler.

- Extend thanks to the translators: Jean-Daniel Brèque, Isabelle Hellyar, Duane Peres, Amber Robinson, Marie Surgers.

- Extend thanks to colleagues in Human Resources for the initiation and implementation of the two Brain Awareness events held at OECD Headquarters and to colleagues from the OECD's Public Affairs and Communication Directorate for their support and understanding.

- Extend thanks to the Centre for Educational Research and Innovation (CERI) staff members who have been involved in this project, namely: Francisco Benavides, Tracey Burns, Emma Forbes, Stephen Girasuolo, Jennifer Gouby, Delphine Grandrieux, David Istance, Kurt Larsen, Sue Lindsay, Cindy Luggery-Babic, and Tom Schuller.

- Lastly thank the dedicated "Brain Team": Jarl Bengtsson, Vanessa Christoph, Cassandra Davis, Emily Groves, Koji Miyamoto, Keiko Momii, and Carrie Tyler, without whom this work would not have been possible.

Table of Contents

Part I
The Learning Brain

Part II
Collaborative Articles

Boxes

Tables

Figures

ISBN 978-92-64-02912-5
Understanding the Brain: The Birth of a Learning Science
© OECD 2007

Executive Summary

> Education is like a double-edged sword. It may be turned to dangerous uses if it is not properly handled.
>
> *Wu Ting-Fang*

After two decades of pioneering work in brain research, the education community has started to realise that "understanding the brain" can help to open new pathways to improve educational research, policies and practice. This report synthesises progress on the brain-informed approach to learning, and uses this to address key issues for the education community. It offers no glib solutions nor does it claim that brain-based learning is a panacea. It *does* provide an objective assessment of the current state of the research at the intersection of cognitive neuroscience and learning, and maps research and policy implications for the next decade.

Part I "The Learning Brain" is the main report, which is the distillation from all the analyses and events over the past seven years of the OECD/CERI "Learning Sciences and Brain Research" project. Part II "Collaborative Articles" contains three articles devoted to the "learning brain" in early childhood, adolescence and adulthood, respectively. These have been written, in each case, by three experts who have combined their experience and knowledge in synergy of the different perspectives of neuroscience and education. Annex A reproduces some insights and dialogue that have emerged from the project's interactive website, open to civil society and including notably a teachers' forum. Annex B updates the reader with developments in neuroimaging technology which have proved so fundamental to the advances discussed in this report.

The first chapter offers a novel "ABC" of the contents of the report by listing and discussing keywords in alphabetical order. This serves both to give short summaries of complex concepts and to steer the reader towards the relevant chapter(s) providing the more in-depth coverage. This is followed in the first half of the following chapter by a short but essential overview of the brain's architecture and functioning.

How the brain learns throughout life

Neuroscientists have well established that the brain has a highly robust and well-developed capacity to change in response to environmental demands, a process called *plasticity*. This involves creating and strengthening some neuronal connections and weakening or eliminating others. The degree of modification depends on the *type* of learning that takes place, with long-term learning leading to more profound modification. It also depends on the *period* of learning, with infants experiencing extraordinary growth of new synapses. But a profound message is that plasticity is a core feature of the brain throughout life.

There are optimal or "sensitive periods" during which particular types of learning are most effective, despite this lifetime plasticity. For sensory stimuli such as speech sounds, and for certain emotional and cognitive experiences such as language exposure, there are relatively tight and early sensitive periods. Other skills, such as vocabulary acquisition, do not pass through tight sensitive periods and can be learned equally well at any time over the lifespan.

Neuroimaging of adolescents now shows us that the adolescent brain is far from mature, and undergoes extensive structural changes well past puberty. Adolescence is an extremely important period in terms of emotional development partly due to a surge of hormones in the brain; the still under-developed pre-frontal cortex among teenagers may be one explanation for their unstable behaviour. We have captured this combination of emotional immaturity and high cognitive potential in the phrase "high horsepower, poor steering".

In older adults, fluency or experience with a task can reduce brain activity levels – in one sense this is greater processing efficiency. But the brain also declines the more we stop using it and with age. Studies have shown that learning can be an effective way to counteract the reduced functioning of the brain: the more there are opportunities for older and elderly people to continue learning (whether through adult education, work or social activities), the higher the chances of deferring the onset or delaying the acceleration of neurodegenerative diseases.

The importance of environment

Findings from brain research indicate how nurturing is crucial to the learning process, and are beginning to provide indication of appropriate learning environments. Many of the environmental factors conducive to improved brain functioning are everyday matters – the quality of social environment and interactions, nutrition, physical exercise, and sleep – which may seem too obvious and so easily overlooked in their impact on education. By conditioning our minds and bodies correctly, it is possible to take advantage of the brain's potential for plasticity and to facilitate the learning process. This calls for holistic approaches which recognise the close interdependence of physical and intellectual well-being and the close interplay of the emotional and cognitive.

In the centre of the brain is the set of structures known as the limbic system, historically called the "emotional brain". Evidence is now accumulating that our emotions do re-sculpt neural tissue. In situations of excessive stress or intense fear, social judgment and cognitive performance suffer through compromise to the neural processes of emotional regulation. Some stress is essential to meet challenges and can lead to better cognition and learning, but beyond a certain level it has the opposite effect. Concerning positive emotions, one of most powerful triggers that motivates people to learn is the illumination that comes with the grasp of new concepts – the brain responds very well to this. A primary goal of early education should be to ensure that children have this experience of "enlightenment" as early as possible and become aware of just how pleasurable learning can be.

Managing one's emotions is one of the key skills of being an effective learner; self-regulation is one of the most important behavioural and emotional skills that children and older people need in their social environments. Emotions direct (or disrupt) psychological processes, such as the ability to focus attention, solve problems, and support relationships. Neuroscience, drawing on cognitive psychology and child development research, starts to identify critical brain regions whose activity and development are directly related to self-control.

Language, literacy and the brain

The brain is biologically primed to acquire language right from the very start of life; the process of language acquisition needs the catalyst of experience. There is an inverse relationship between age and the effectiveness of learning many aspects of language – in general, the younger the age of exposure, the more successful the learning – and neuroscience has started to identify how the brain processes language differently among young children compared with more mature people. This understanding is relevant to education policies especially regarding foreign language instruction which often does not begin until adolescence. Adolescents and adults, of course, can also learn a language anew, but it presents greater difficulties.

The dual importance in the brain of sounds (phonetics) and of the direct processing of meaning (semantics) can inform the classic debate in teaching reading between the development of specific phonetic skills, sometimes refereed to as "syllabic instruction", and "whole language" text immersion. Understanding how both processes are at work argues for a balanced approach to literacy instruction that may target more phonetics or more "whole language" learning, depending on the morphology of the language concerned.

Much of the brain circuitry involved in reading is shared across languages but there are some differences, where specific aspects of a language call on distinct functions, such as different decoding or word recognition strategies. Within alphabetical languages, the main difference discussed in this report is the importance of the "depth" of a language's orthography: a "deep" language (which maps sounds onto letters with a wide range of variability) such as English or French contrasts with "shallow", much more "consistent" languages such as Finnish or Turkish. In these cases, particular brain structures get brought into play to support aspects of reading which are distinctive to these particular languages.

Dyslexia is widespread and occurs across cultural and socioeconomic boundaries. Atypical cortical features which have been localised in the left hemisphere in regions to the rear of the brain are commonly associated with dyslexia, which results in impairment in processing the sound elements of language. While the *linguistic* consequences of these difficulties are relatively minor (*e.g.* confusing words which sound alike), the impairment can be much more significant for *literacy* as mapping phonetic sounds to orthographic symbols is the crux of reading in alphabetic languages. Neuroscience is opening new avenues of identification and intervention.

Numeracy and the brain

Numeracy, like literacy, is created in the brain through the synergy of biology and experience. Just as certain brain structures are designed through evolution for language, there are analogous structures for the quantitative sense. And, also as with language, genetically-defined brain structures alone cannot support mathematics as they need to be co-ordinated with those supplementary neural circuits not specifically destined for this task but shaped by experience to do so. Hence, the important role of education – whether in schools, at home, or in play; and hence, the valuable role for neuroscience in helping address this educational challenge.

Although the neuroscientific research on numeracy is still in its infancy, the field has already made significant progress in the past decade. It shows that even very simple numerical operations are distributed in different parts of the brain and require the co-ordination of multiple structures. The mere representation of numbers involves a complex circuit that brings together sense of magnitude, and visual and verbal representations. Calculation calls on other complex distributed networks, varying according to the operation in question: subtraction is critically dependent on the inferior parietal circuit, while addition and multiplication engage yet others. Research on advanced mathematics is currently sparse, but it seems that it calls on at least partially distinct circuitry.

Understanding the underlying developmental pathways to mathematics from a brain perspective can help shape the design of teaching strategies. Different instructional methods lead to the creation of neural pathways that vary in effectiveness: drill learning, for instance, develops neural pathways that are less effective than those developed through strategy learning. Support is growing from neuroscience for teaching strategies which involve learning in rich detail rather than the identification of correct/incorrect responses. This is broadly consistent with formative assessment.

Though the neural underpinnings of dyscalculia – the numerical equivalent of dyslexia – are still under-researched, the discovery of biological characteristics associated with specific mathematics impairments suggests that mathematics is far from a purely cultural construction: it requires the full functioning and integrity of specific brain structures. It is likely that the deficient neural circuitry underlying dyscalculia can be addressed through targeted intervention because of the "plasticity" – the flexibility – of the neural circuitries involved in mathematics.

Dispelling "neuromyths"

Over the past few years, there has been a growing number of misconceptions circulating about the brain – "neuromyths". They are relevant to education as many have been developed as ideas about, or approaches to, how we learn. These misconceptions often have their origins in some element of sound science, which makes identifying and refuting them the more difficult. As they are incomplete, extrapolated beyond the evidence, or plain false, they need to be dispelled in order to prevent education running into a series of dead-ends.

Each "myth" or set of myths is discussed in terms of how they have emerged into popular discourse, and of why they are not sustained by neuroscientific evidence. They are grouped as follows:

- "There is no time to lose as everything important about the brain is decided by the age of three."
- "There are critical periods when certain matters *must* be taught and learnt."
- "But I read somewhere that we only use 10% of our brain anyway."
- "I'm a 'left-brain', she's a 'right-brain' person."
- "Let's face it – men and boys just have different brains from women and girls."
- "A young child's brain can only manage to learn one language at a time."
- "Improve your memory!"
- "Learn while you sleep!"

The importance and promise of this new field are not the reason to duck fundamental ethical questions which now arise.

For which purposes and for whom? It is already important to re-think the use and possible abuse of brain imaging. How to ensure, for example, that the medical information it gives is kept confidential, and not handed over to commercial organisations or indeed educational institutions? The more accurately that brain imaging allows the identification of specific, formerly "hidden", aspects of individuals, the more it needs to be asked how this should be used in education.

The use of products affecting the brain: The boundary between medical and non-medical use is not always clear, and questions arise especially about healthy individuals consuming substances that affect the brain. Should parents, for instance, have the right to give their children substances to stimulate their scholarly achievements, with inherent risks and parallels to doping in sport?

Brain meets machine: Advances are constantly being made in combining living organs with technology. The advantages of such developments are obvious for those with disabilities who are thus enabled, say, to control machines from a distance. That the same technology could be applied to control individuals' behaviour equally obviously raises profound concerns.

An overly scientific approach to education? Neurosciences can importantly inform education but if, say, "good" teachers were to be identified by verifying their impact on students' brains, this would be an entirely different scenario. It is one which runs the risk of creating an education system which is excessively scientific and highly conformist.

Though educational neuroscience is still in its early days, it will develop strategically if it is trans-disciplinary, serving both the scientific and educational communities, and international in reach. Creating a common lexicon is one critical step; another is establishing shared methodology. A reciprocal relationship should be established between educational practice and research on learning which is analogous to the relationship between medicine and biology, co-creating and sustaining a continuous, bi-directional flow to support brain-informed educational practice.

A number of institutions, networks and initiatives have already been established to show the way ahead. Vignette descriptions of several leading examples are available in this report. They include the JST-RISTEX, Japan Science and Technology's Research Institute of Science and Technology for Society; Transfer Centre for Neuroscience and Learning, Ulm, Germany; Learning Lab, Denmark; Centre for Neuroscience in Education: University of Cambridge, United Kingdom; and "Mind, Brain, and Education", Harvard Graduate School of Education, United States.

Educational neuroscience is generating valuable new knowledge to inform educational policy and practice: On many questions, neuroscience builds on the conclusions of existing knowledge and everyday observation but its important contribution is in enabling the move from correlation to causation – understanding the mechanisms behind familiar patterns – to help identify effective solutions. On other questions, neuroscience is generating new knowledge, thereby opening up new avenues.

Brain research provides important neuroscientific evidence to support the broad aim of lifelong learning: Far from supporting ageist notions that education is the province only of the young – the powerful learning capacity of young people notwithstanding – neuroscience confirms that learning is a lifelong activity and that the more it continues the more effective it is.

Neuroscience buttresses support for education's wider benefits, especially for ageing populations: Neuroscience provides powerful additional arguments on the "wider benefits" of education (beyond the purely economic that counts so highly in policy-making) as it is identifying learning interventions as a valuable part of the strategy to address the enormous and costly problems of ageing dementia in our societies.

The need for holistic approaches based on the interdependence of body and mind, the emotional and the cognitive: Far from the focus on the brain reinforcing an exclusively cognitive, performance-driven bias, it suggests the need for holistic approaches which recognise the close inter-dependence of physical and intellectual well-being, and the close interplay of the emotional and cognitive, the analytical and the creative arts.

Understanding adolescence – high horsepower, poor steering: The insights on adolescence are especially important as this is when so much takes place in an individual's educational career, with long-lasting consequences. At this time, young people have well-developed cognitive capacity (high horsepower) but emotional immaturity (poor steering). This cannot imply that important choices should simply be delayed until adulthood, but it does suggest that these choices should not definitively close doors.

Better informing the curriculum and education's phases and levels with neuroscientific insights: The message is a nuanced one: there are no "critical periods" when learning *must* take place but there are "sensitive periods" when the individual is particularly primed to engage in specific learning activities (language learning is discussed in detail). The report's message of an early strong foundation for lifetimes of learning reinforces the key role of early childhood education and basic schooling.

Ensuring neuroscience's contribution to major learning challenges, including the "3Ds": dyslexia, dyscalculia, and dementia. On dyslexia, for instance, its causes were unknown until recently. Now it is understood to result primarily from atypical features of the auditory cortex (and possibly, in some cases, of the visual cortex) and it is possible to identify these features at a very young age. Early interventions are usually more successful than later interventions, but both are possible.

More personalised assessment to improve learning, not to select and exclude: Neuroimaging potentially offers a powerful additional mechanism on which to identify individuals learning characteristics and base personalisation; but, at the same time, it may also lead to even more powerful devices for selection and exclusion than are currently available.

Key areas are identified as priorities for further educational neuroscientific research, not as an exhaustive agenda but as deriving directly from the report. This agenda for further research – covering the better scientific understanding of such matters as the optimal timing for different forms of learning, emotional development and regulation, how specific materials and environments shape learning, and the continued analysis of language and mathematics in the brain – would, if realised, be well on the way to the birth to a trans-disciplinary learning science.

This is the aspiration which concludes this report and gives it its title. It is also the report's aspiration that it will be possible to harness the burgeoning knowledge on learning to create an educational system that is both personalised to the individual and universally relevant to all.

PART I

The Learning Brain

ISBN 978-92-64-02912-5
Understanding the Brain: The Birth of a Learning Science
© OECD 2007

PART I

Introduction

> Nicht das Gehirn denkt, sondern wir denken das Gehirn.
> (The brain does not think, *we* think the brain.)
>
> *Friedrich Nietzsche*

Can neuroscience[1] truly improve education? This report suggests a complex, but nonetheless definite answer: "yes, but…" Circumstances have converged to mean that there is now a global emergence of educational neuroscience. Recent advances in the field of neuroscience have significantly increased its relevance to education. Imaging technologies enable the observation of the working brain, providing insights into perceptual, cognitive, and emotional functions of consequence for education. This trend towards the greater applicability of neuroscience for education is paralleled by an increasingly receptive society. This report summarises the state of research at the intersection of neuroscience and learning, and highlights research and policy considerations for the next decade. Scientific research findings can help all stakeholders involved in education – including learners, parents, teachers and policy makers – to better understand the processes of learning and to structure nurturing learning environments. This understanding can help education systems move in evidence-based policy decisions, inform parents about how to create a sound learning environment for their children, and help learners develop their competencies.

We by no means claim that neuroscience is a panacea – and that it will start a revolution in education, especially not immediately. The project leaders have time after time warned that "neuroscience alone is unlikely to *solve* every, if any, educational issue". The answers to many educational questions are to be found elsewhere, whether within education itself or in other social science reference disciplines or indeed in philosophy. Yet, there are certain questions for which neuroscience is particularly adept and it is already making an important educational contribution: providing new perspectives on longstanding challenges, raising new issues, confirming or dispelling age-old assumptions, or reinforcing existing practices. This report shows that a genuine trans-disciplinary approach drawing from many disciplines is required to respond to the increasingly complex questions with which our societies are confronted.[2]

---

1. The term neuroscience (sometimes appearing in the plural form, "neurosciences") is used broadly in this report to encompass all overlapping fields, including neurobiology, cognitive neuroscience, behavioural neuroscience, cognitive psychology, etc.
2. The OECD Secretariat wishes to clearly dissociate itself from any interpretation in this publication which, based on the ideas of individualistic differences in the brain and of different learning styles, would try to link certain genes to IQ and hence, could have a racist connotation towards any group or groups of people within the human community. Such interpretations should be condemned.

Neuroscience is beginning to provide a detailed account of *how* human beings – or their human brains – respond to different learning experiences and classroom environments and *why* they react in the ways they do. This understanding is important for education because so much educational policy and practice is based on only limited information. At best, quantitative and qualitative research has informed certain educational policies and practices, examining a variety of learning practices, environments and outcomes. While we have a solid knowledge base on the modes of learning associated with success or failure, we largely lack the detailed explanations for these outcomes, and much about underlying learning processes is left as a black box.

When parents, teachers and policy makers, for instance, try to identify the right timing for teaching a foreign language to children, an "informed decision" might well be based on a comparison of the experiences and performance of students who started learning a foreign language at different ages. This might conclude that teaching a foreign language from a given age gives the best outcome. Although this information would in itself be useful, it does not confirm that it is the *timing* of teaching a foreign language that really accounts for the successful outcome, nor does it show *how* foreign language learning can be most effective at this given age.

Readers will find a number of statements in this report that are either tentative (*i.e.*, based on limited evidence) or restatements of conventional principles (*i.e.* "wisdoms" from decades or even centuries of practice and research in education[3]), or even still lacking consensus within scientific communities (but we will not go into mere scientific controversies here, which would go beyond the scope of this work). Tentative conclusions are, however, honest reflections of the present state of research in this area, and serve the very useful purpose of identifying research directions for the future. Complementing what is already known with scientific evidence can be useful when it strengthens support for practices that have previously lacked a rational basis. Neuroscience can also reveal that certain existing practices are not justified in terms of the way the brain learns. Some long-standing debates within education might well be now out-of-date. But practices will be enriched by new elements that neuroscience brings to light. For all these reasons, the emergence of scientific evidence will certainly strengthen educational policy and practice.

This book was designed and written in order to allow the reader to focus on one or two chapters only, depending on his/her interests. Hence, we hope that each chapter can be read independently. The price to pay for this flexibility was of course to accept that some things are repeated in several places. Nothing is perfect. Thank you for your understanding.

---

3. The reader may well note that this report does not present an exhaustive, encyclopaedic comparison of neuroscientific findings and different learning theories. Existing knowledge does not yet permit the ties between these two sides to be systematically drawn though it should soon start to be possible. Where the links can already be made, however, we refer to them without pretending detailed elaboration.

ISBN 978-92-64-02912-5
Understanding the Brain: The Birth of a Learning Science
© OECD 2007

PART I

Chapter 1

An "ABC" of the Brain

The only good is knowledge and the only evil is ignorance.

Socrates

Not to know is bad. No to wish to know is worse.

(African Proverb)

---

*Chapter 1 provides an "ABC" of the contents of the report by listing keywords and concepts in alphabetical order covered in the chapters to follow. It begins with **A**cquistion of knowledge and **B**rain, and runs through to **V**ariability, **W**ork and **XYZ**. The reader can choose a particular topic of interest, and the corresponding description points to the relevant chapter(s) which provide more in-depth coverage of the issue. This chapter is relevant for all those who are interested in the issue of "learning sciences and brain research" including learners, parents, teachers, researchers and policy makers.*

Acquisition of knowledge

The neuroscientific approach to learning provides a hard-scientifically based theoretical framework for educational practices. This rapidly emerging field of study is slowly but surely building the foundations of a "Science of Learning".

A living being is made up of various levels of organisation. The result is that a single human process may be defined differently depending on the level used as a reference. This is true of the learning process where the definition varies depending on the perspective of the person who describes it.

The differences between cellular and behavioural definitions reflect the contrasting views of neurosciences and educational sciences. Neuroscientists consider learning as a cerebral process where the brain reacts to a stimulus, involving the perception, processing and integration of information. Educators consider this as an active process leading to the acquisition of knowledge, which in turn entails lasting, measurable and specific changes in behaviour.

Brain

Even though it plays a fundamental role, the brain remains one single part of a whole organism. An individual cannot solely be reduced to this organ as the brain is in constant interaction with other parts of the human body.

The brain is the seat of our mental faculties. It assumes vital functions by influencing heart rate, body temperature, breathing, etc., as well as performing so-called "higher" functions, such as language, reasoning and consciousness.

This organ includes two hemispheres (left and right), with each further divide into lobes (occipital, parietal, temporal and frontal) – further described in Chapter 2.

The main components of cerebral tissue are glial and nerve cells (neurons). The nerve cell is considered as the basic functional unit of the brain because of its extensive interconnectivity and because it specialises in communication. Neurons are organised in functional networks that are situated in specific parts of the brain.

Cognitive functions

Having been studied at various levels, cognitive functions benefit from a rich multidisciplinary research effort. Therefore, in a complementary way, neurosciences, cognitive neuroscience and cognitive psychology seek to understand these processes.

Cognition is defined as the set of processes enabling information processing and knowledge development. These processes are called "cognitive functions". Among these, the higher cognitive functions correspond to the human brain's most elaborate processes. They are the product of the most recent phase of the brain's evolution and are mainly centred in the cortex, which is a particularly highly developed structure in humans (see Chapter 2).

Examples of these functions are certain aspects of perception, memory and learning, but also language, reasoning, planning and decision-making.

Development

The brain is continually changing – developing – throughout life. This development is guided by both biology and experience (see Chapter 2). Genetic tendencies interact with experience to determine the structure and function of the brain at a given point in time. Because of this continuous interaction, each brain is unique.

Though there is a wide range of individual differences in brain development, the brain has age-related characteristics that can have important consequences for learning. Scientists are beginning to map out these maturational changes and to understand how biology and experience interact to guide development.

Understanding development from a scientific perspective could powerfully impact educational practice. As scientists uncover age-related changes in the brain, educators will be able to use this information to design didactics that are more age-appropriate and effective.

Emotions

Emotional components have long been neglected in institutional education. Recent contributions of neuroscientists are helping to remedy this deficiency by revealing the emotional dimension of learning (see Chapter 3).

As opposed to "affect", which is their conscious interpretation, emotions arise from cerebral processes and are necessary for the adaptation and regulation of human behaviour.

Emotions are complex reactions generally described in terms of three components: a particular mental state, a physiological change and an impulsion to act. Therefore, faced with a situation perceived as dangerous, the reactions engendered will simultaneously consist of a specific cerebral activation of the circuit devoted to fear, body reactions typical of fear (*e.g.* accelerated pulse, pallor and perspiring) and the fight-or-flight reaction.

Each emotion corresponds to a distinct functional system and has its own cerebral circuit involving structures in what we call the "limbic system" (also known as the "seat of the emotions"), as well as cortical structures, mainly the prefrontal cortex which plays a prime role in regulating emotions. Incidentally, the prefrontal cortex matures particularly late in human beings, concluding its development in the third decade of an individual's development. This means that cerebral adolescence lasts longer than was, until recently, thought, which helps to explain certain features of behaviour: the full development of the prefrontal cortex, and therefore the regulation of emotions and compensation for potential excesses of the limbic system, occur relatively late in an individual's development.

Continual exchanges make it impossible to separate the physiological, emotional and cognitive components of a particular behaviour. The strength of this interconnectivity explains the substantial impact of emotions on learning. If a positively perceived emotion is associated with learning, it will facilitate success, whereas a negatively perceived emotion will result in failure.

Functionality, neural base of learning

The neuroscientific definition of learning links this process to a biological substrate or surface. From this point of view, learning is the result of integrating all information perceived and processed. This integration takes form in structural modifications within the brain. Indeed, microscopic changes occur, enabling processed information to leave a physical "trace" of its passage.

Today, it is useful, even essential, for educators and anyone else concerned with education to gain an understanding of the scientific basis of learning processes.

Genetics

The belief that there is a simple cause and effect relationship between genetics and behaviour often persists. Imagining a linear relationship between genetic factors and behaviour is only a short step away from full-blown determinism. A gene does not activate behaviour but instead consists of a sequence of DNA containing the relevant information to produce a protein. The expression of the gene varies on the basis of numerous factors, especially environmental factors. Once a protein is synthesised in the cell, it occupies a specific place and plays a role in the functioning of this cell. In this sense, it is true that if genes affect function, they consequently mould behaviour. However, this is a complex non-linear relationship with the various levels of organisation influencing one another.

As research slowly but surely advances, belief in a frontier between the innate and the acquired is disappearing, giving way to understanding the interdependence between genetic and environmental factors in brain development.

To predict behaviour on the basis of genetics will be incomplete: any approach solely influenced by genetics is not only scientifically unfounded but also ethically questionable and politically dangerous.

"Hands on" and Holistic – learning by doing

"I hear and I forget,
I see and I remember,
I do and I understand"

Confucius

Long forgotten by educators, this quote regained prominence in the 20th century with the advent of constructivism. Contrary to theories focused on expert educators who transmit knowledge, this current advanced a new concept of learning: the construction of knowledge. Learning becomes learner-centred and relies on the development of prior knowledge, based on the experience, desires and needs of each individual.

Therefore, this theoretical upheaval has given rise to the so-called active or experiential practices of "learning-through-action". The objective is to actively involve learners in interacting with their human and material environment, based on the idea that this will lead to a more profound integration of information than perception. Action necessarily implies operationalisation – the implementation of concepts. The learner not only needs to acquire knowledge and know-how, but must also be able to render them operational in real applications. Therefore the learner becomes "active", implying a better level of learning.

Not all neuroscientific discoveries give rise to innovations in terms of didactics. However, they provide a solid theoretical basis for well-tried practices which have been consolidated through experience. These scientific insights then serve to underpin the body of empirical and intuitive knowledge already accumulated, and explain why some practices fail or succeed.

Intelligence

The concept of intelligence has always been a subject of controversy. Can a single concept account for all of the intellectual faculties of an individual? Can these faculties be separated and measured? And in particular, what do they show and predict about the cerebral functioning of an individual and about social behaviour?

The notion of intelligence evokes "skills", whether they are verbal skills, spatial skills, problem-solving skills or the very elaborate skill of dealing with complexity. However, all of these aspects neglect the concept of "potential". Yet, neurobiological research on learning and cognitive functions clearly shows that these processes undergo constant evolution and are dependent on a number of factors, particularly environmental and emotional ones. This means that a stimulating environment should offer each individual the possibility to cultivate and develop his/her skills.

From this point of view, the many attempts to quantify intelligence using tests (such as IQ measurements or others) are too static and refer to standardised and culturally (sometimes even ideologically) biased faculties.

Based on a priori assumptions, intelligent tests are restrictive and therefore problematic. Based on this "intelligence calculation" or the debatable assignation of individuals to different levels of intelligence, what should be concluded for the practices or even the choices related to career orientation?

Joy of learning

"Tell me and I forget
Teach me and I remember
Involve me and I learn"

Benjamin Franklin

This maxim restores involvement to its role as an essential condition of significant learning. Involvement can be summed up as the commitment of an individual within a given action. In this sense, it results directly from the process of motivating the individual to behave in a certain way or to pursue a particular goal. This process can be triggered by internal or external factors. This is why we speak of intrinsic motivation, which solely depends on the learner's own needs and desires, or extrinsic motivation, which takes into account external influences on the individual. Motivation is largely conditioned by self-assurance, self-esteem and by the benefits the individual may accrue in terms of a targeted behaviour or goal.

The combination of motivation and self-esteem are essential to successful learning. In order to give these factors their rightful place within learning structures, the system of tutoring is gaining ground. It offers the learner personalised support and is better adapted to his/her needs. A more personal climate for learning serves to motivate learners but should not disregard the crucial role of social interactions in all modes of learning. Personalisation should not mean the isolation of learners.

Motivation has a pivotal role in the success of learning, especially intrinsic motivation. The individual learns more easily if s/he is doing it for him/herself, with the desire to understand.

Although it is currently difficult to construct educational approaches that could go beyond "carrot and stick" systems and target this intrinsic motivation, the benefits of this approach are such that it is of paramount importance for research to orient its efforts towards this domain.

Kafka

By describing in "The Castle" the vain efforts of the protagonist to attain his objectives ("There is a goal, but no way": "Es gibt zwar ein Ziel, aber keinen Weg zum Ziel"), Franz Kafka relates the feeling of despair that an individual can feel when being confronted with a deaf and blind bureaucratic machine. Reminiscent of this work, Dino Buzzati's story "K" is a tragedy of a misunderstanding, emphasising how sad, but also dangerous it can be to understand certain realities too late…

There is abundant resistance to taking on board neuroscientific discoveries for educational policies and practices, sufficient to discourage even the most fervent advocates. The reasons may be various – simple incomprehension, mental inertia, the categorical refusal to reconsider certain "truths", through to corporate reflexes to defend acquired positions, even staunch bureaucracy. The obstacles are numerous to any trans-disciplinary effort to create a new field, or even more modestly to shed new light on educational issues. This poses a delicate problem of "knowledge management". Even if some constructive scepticism can do no harm, every innovative project finds itself in the position of "K" at one point or another, seeking to reach the Castle. Despite such difficulties, a way exists; to quote Lao Tzu: "The journey is the destination."

Moreover, neuroscience unintentionally generates a plethora of "neuromyths" founded on misunderstandings, bad interpretations, or even distortions of research results. These neuromyths, which become entrenched in the minds of the public by the media, need to be identified and dispelled. They raise many ethical questions which in democratic societies need to be addressed through political debate.

We can ask whether (in the mid-term at least), it is acceptable, in any reflection about education, not to take into consideration what is known about the learning brain. Is it ethical to ignore a field of relevant and original research that is shedding new light and fundamental understanding on education?

Language

Language is a specifically human cognitive function which is also dedicated to communication. It opens up the use of a system of symbols. When a finite number of arbitrary symbols and a set of semantic principles are combined according to rules of syntax, it is possible to generate an infinite number of statements. The resulting system is a language. Different languages use phonemes, graphemes, gestures and other symbols to represent objects, concepts, emotions, ideas and thoughts.

The actual expression of language is a function that relates at least one speaker to one listener, and the two can be interchangeable. This means language can be broken down into a direction (perception or production) and also into a mode of expression (oral or written). Oral language is acquired naturally during childhood by simple exposure to spoken language; written language, on the other hand, requires intentional instruction (see Chapter 4).

Language was one of the first functions to be shown to have a cerebral basis. In the 19th century, studies of aphasia by two scientists (Broca and Wernicke) revealed that certain areas of the brain were involved in language processing. Since then, studies have confirmed that these areas belong to the cerebral circuits involved in language (see Chapter 4).

UNDERSTANDING THE BRAIN: THE BIRTH OF A LEARNING SCIENCE – ISBN 978-92-64-02912-5 – © OECD 2007

Accumulating a large body of neuroscientific knowledge on language was possible because of neuroscience's major interest in this function. The understanding of language mechanisms and how they are learned has already had an important impact on educational policies.

Memory

During the learning process, traces are left by the processing and integration of perceived information. This is how memory is activated. Memory is a cognitive process enabling past experiences to be remembered, both in terms of acquiring new information (development phase of the trace) and remembering information (reactivation phase of this trace). The more a trace is reactivated, the more "marked" memory will be. In other words, it will be less vulnerable and less likely to be forgotten.

Memory is built on learning, and the benefits of learning persist thanks to it. These two processes have such a profound relationship that memory is subject to the same factors influencing learning. This is why memorisation of an event or of information can be improved by a strong emotional state, a special context, heightened motivation or increased attention.

Learning a lesson too often means being able to recite it. Training and testing are usually based on retrieving and therefore on memorising information often to the detriment of mastering skills and even of understanding content. Is this role given to memory skills in learning justified? This is a pivotal question in the field of education, and is beginning to attract the attention of neuroscientists.

Neuron

Organised in extensively interconnected networks, neurons have electrical and chemical properties that enable them to propagate nerve impulses (see Chapter 2, especially Figure 2.1). An electrical potential is propagated within a nerve cell and a chemical process transmits information from one cell to another. These nerve cells are consequently specialised in communication.

The electrical propagation within the cell is uni-directional. Inputs are received by the neuron's dendrites or the cell body. In response to these inputs, the neuron generates action potentials. The frequency of these potentials varies according to these inputs. Therefore, the action potentials propagate through the axon.

A zone called the *synapse* serves as a junction between two neurons. The synapse consists of three components: the axon ending, the synaptic gap and the dendrite of the postsynaptic neuron. When the action potentials reach the synapse, it releases a chemical substance called the neurotransmitter, which crosses the synaptic gap. This chemical activity is regulated by the type and amount of neurotransmitters, but also by the number of receptors involved. The amount of neurotransmitters released and the number of receptors involved are responsive to experience, which is the cellular basis of plasticity (see below). The effect on postsynaptic neurons may be excitatory or inhibitory.

Therefore, this combination of electrical and chemical activity of the neurons transmits and regulates information within the networks formed by neurons.

In order to improve the understanding of cerebral activity, various functional imaging technologies (fMRI, MEG, PET, OT, etc.) (see Annex B) are used to visualise and study the activity of the changes in blood flow induced by neuronal activities.

Studies localising cerebral networks open an important door to our understanding of learning mechanisms. The better the temporal and spatial resolution, the more precise the localisation and consequently the better our understanding of cerebral function.

Opportunity windows for learning

Certain periods in an individual's development are particularly well-suited to learning certain skills. During these key moments the brain needs certain types of stimulations in order to establish and maintain long-term development of the structures involved. These are the stages at which the individual's experience becomes an overriding factor, responsible for profound changes.

These periods are called "sensitive periods" or windows of opportunity, because they are the optimum moments for individuals to learn specific skills. They are part of natural development, but experience is needed so that a change (learning) can be effective. This process can be described as "experience-expectant" learning, such as oral language (see Chapter 4). It is not the same as "experience-dependent" learning such as written language, which can take place at any moment in an individual's lifetime.

If learning does not occur in these windows of opportunity, it does not mean it cannot occur. Learning takes place throughout a whole lifetime although outside these windows of opportunity, it takes more time and cognitive resources and it will often not be as effective.

A better understanding of sensitive periods and the learning that occurs during those periods is a crucial avenue of future research. An increasingly complete map will enable us to better match instruction to the appropriate sensitive period in educational programmes with a corresponding positive impact on the effectiveness of learning.

Plasticity

The brain is capable of learning because of its flexibility (see Chapter 2). It changes in response to stimulation from the environment. This flexibility resides in one of the intrinsic properties of the brain – its plasticity.

The mechanism operates in various ways at the level of the synaptic connections (Figure 2.1). Some synapses may be generated (synaptogenesis), others eliminated (pruning), and their effectiveness may be moulded, on the basis of the information processed and integrated by the brain.

The "traces" left by learning and memorisation are the fruit of these modifications. Plasticity is consequently a necessary condition for learning and an inherent property of the brain; it is present throughout a whole lifetime.

The concept of plasticity and its implications are vital features of the brain. Educators, policy makers and all learners will all gain from understanding why it is possible to learn over a whole lifetime and indeed brain plasticity provides a strong neuroscientific argument for "lifelong learning". Would not primary school be a good place to start teaching learners how and why they are capable of learning?

Quality existence and healthy living

Like any other organ in the human body, the brain functions best with healthy living. Recent studies have looked into the impact of nutrition and physical activity on cerebral faculties and particularly on learning. Results show that a balanced diet contributes to the development and functioning of the brain, while also preventing some behavioural and learning problems (see Chapter 3). In the same respect, regular physical activity has a positive effect on the functioning of human cognition, modifying the activity in certain regions of the brain.

Sleep is also a determining factor in brain development and function (Chapters 3 and 6). Anyone who has lacked sleep knows that cognitive functions are the first to suffer. It is during sleep that some of the processes involved in plasticity and consolidation of knowledge take place, processes that consequently play a pivotal role in memorising and learning.

Environmental factors (noise, ventilation, etc.) and physiological factors (diet, exercise, sleep, etc.) influence learning. In the short run, advances in this area should lead to concrete applications in terms of school and educationally-related practices.

Representations

Human beings are constantly perceiving, processing and integrating information, *i.e.* they learn. Individuals have their own representations, which gradually build up on the basis of their experience. This organised system translates the outside world into an individual perception. An individual's system of representation governs his/her thinking processes.

Since Plato's Cave, philosophy has pondered the question of representations. Evidently, the objective here is not to respond to the eternal questions of humanity, although it is not impossible that one day our knowledge of brain functioning is such that it will bring about new elements to these eternal philosophical debates.

Skills

The term "skills" is frequently used in English when behaviour and learning are being discussed. A given behaviour can be broken down into skills, understood as the "natural units" of behaviour.

Language, for instance, can be broken down into four "meta-skills" according to transmission or reception and the means of communication. These meta-skills are oral understanding, oral production, reading and writing. Each of these meta-skills in turn can be further broken down into more distinct skills. Oral understanding, for example, consists of some ten skills, which include short term memorisation of series of sounds, discrimination of a given language's distinctive sounds, and distinction of words and identification of grammatical categories.

Each skill corresponds to a specific class of activities. This raises questions about evaluating individual progress and the distinction between skills and knowledge. What do we expect of children? Skills or knowledge? What do we want to "measure" when we test children?

Team and social interactions

Social interactions catalyse learning. Without this sort of interaction, an individual can neither learn nor properly develop. When confronting a social context an individual's learning improves in relation to the wealth and variety of that context.

Discovery triggers the processes of using and building knowledge and skills. Dealing with others enables individuals to develop strategies and refine their reasoning. This is why social interaction is a constituent condition both for early development of cerebral structures and for the normal development of cognitive functions (see Chapter 3).

What place do schools leave for interaction between learners? The appearance of new technologies in the educational sector has had far-reaching repercussions on interactivity in learning situations. What will be the impact of these changes on learning itself?

These questions are being addressed by the rapidly emerging field of social neuroscience, which deals with social processes and behaviour.

Universality

Numerous features characterise the human kind, the development of the brain being one of them. It follows a programme recorded in the genetic heritage of each individual, and is programmed as part of a "ballet" where perfectly regulated genes are constantly nourished by experience.

One of the intrinsic properties of the brain is its plasticity (see Chapter 2). The brain continually perceives, processes and integrates information derived from personal experience, and therefore undergoes changes in the physical connections within its networks of neurons. This continual development is the result of the brain's normal operation and implies a permanent learning capacity. This means that development is a constant and universal feature of cerebral activity and that a human being can learn throughout the lifespan.

"Everyone has the right to education" (Universal Declaration of Human Rights, United Nations, 10 December 1948, Article 26). Education regulates learning so that everyone has access to the fundamentals of reading, writing and arithmetic (see Chapters 4 and 5).

International evaluations are performed to check the equality and durability of the various educational systems. Although it is difficult to "measure" acquired knowledge across cultural borderlines, such evaluations heighten awareness of the need for constant improvement in education.

Variability

Experience plays a fundamental role in individual development and the make-up of a human being, but it remains personal and subjective. Representations resulting from experience are consequently different from one person to another. Experience also plays a major role in building preferential styles, leading the learner to use particular learning strategies according to the situation.

Specific learning causes changes – transitions from one state to another. Yet, the diversity of personal experience and representations implies different conditions at the outset for each person. In addition, modifications resulting from learning vary according to learning motivations, interactions and strategies. This is why the impact of instruction differs from one person to another, and why we speak of variability.

Students in the same class, taking the same course will not learn the same things. Their representations of the concepts presented will vary as they do not all start with the same basic knowledge nor the same mode of learning. The result is that their representations will not develop in the same manner. They will all maintain traces of this learning experience, but these traces will be different and specific to each individual.

Learning experiences need to take into account individual differences, so that diversification of the curriculum to accommodate them is an increasingly important educational goal.

The question of cortical differences between men and women is frequently raised. As yet, neuroscientific data neither confirm nor disprove this conjecture.

Work

A lot of work has been done and a major task has been achieved in recent years to develop educational neuroscience, and this is helping give birth to a still larger, trans-disciplinary learning science (see Chapter 7). These achievements will seem small, however, in comparison with what is still to come from those who follow us into this field. One can hope that they will meet fewer barriers, especially as they will have to deal with a much larger knowledge base. For it happens that...

... XYZ

... the story is far from ended. This CERI project is merely the beginning of an adventure and it is now up to others to take up the baton. Many have already engaged on such ways (see Chapter 7). There is much more than these three remaining letters to write into our brain alphabet. Our knowledge of the brain is like the brain itself: a continual evolution...

ISBN 978-92-64-02912-5
Understanding the Brain: The Birth of a Learning Science
© OECD 2007

PART I

Chapter 2

How the Brain Learns throughout Life

My brain? It's my second favourite organ.

Woody Allen

This chapter presents an accessible description of the brain's architecture. It describes how the brain learns throughout life, giving an introduction to three key life phases: infancy and childhood, adolescence, and adulthood (including old age). It also discusses how cognitive decline and the dysfunctions that come with ageing can be addressed and delayed through learning. It is especially useful for readers who have had no prior exposure to neuro-scientific accounts of the brain, as the basic principles and the subsequent analyses are aimed at the layperson, aided by figures and summary tables.

Learning is a highly complex process and definitions of it vary depending on the context and the perspective. The definitions used by neuroscientists and educational researchers can be quite different, and this can pose a challenge to the dialogue between the two communities. For example, the scientist Koizumi (2003) defines learning as a "process by which the brain reacts to stimuli by making neuronal connections that act as an information processing circuit and provide information storage". In contrast, Coffield (2005), from the education research side, proposes that learning refers to "significant changes in capability, understanding, attitudes or values by individuals, groups, organisations or society"; he explicitly excludes "the acquisition of further information when it does not contribute to such changes".

This chapter does not aspire to provide a general definition of learning or even an overview of different definitions of learning. The purpose of this chapter instead, recognising that the meaning of learning can vary depending on the context, is to provide basic principles of the brain's architecture and describe what happens in the brain during different periods of life when information is being processed. It also discusses ways in which brain functioning can be ameliorated when it has started to decline or is already damaged due to ageing or sickness.

The development of new brain imaging technologies (see Annex B) has enabled the emergence of cognitive neuroscience.[1] Neuroscientists have increasingly turned to learning to apply new scientific findings as well as to frame future research questions. Certain neuroscientific findings are highly relevant to curriculum design, teaching practices and modes of literacy and numeracy learning. Cognitive neurosciences can also shed light on ways in which adult learning can help treat ageing problems such as memory loss as well as more severe chronic sickness, such as senile dementia (e.g., Alzheimer's disease).[2] The growing number of findings related to lifelong learning emerging from neuroscience is the thread linking the analyses presented in this chapter.

Basic principles of the brain's architecture

The brain consists of a vast amount of neurons and glial cells,[3] both of which constitute the basic operative units of the brain. During the period of the most rapid prenatal brain development, which occurs between 10 and 26 weeks after conception, the brain is estimated to grow at a rate of 250 000 neurons per minute. At birth, the brain contains the majority of the cells it will ever have, with estimates ranging from 15 to

1. Cognitive neuroscience concerns the scientific study of the neural mechanisms underlying cognition. Cognitive neuroscience overlaps with both neuroscience, which studies the functioning of the brain more broadly, and cognitive psychology, which focuses on the neural substrates of mental process and their behavioural manifestations.
2. Dementia is a sickness that involves deterioration of intellectual faculties, such as memory, concentration, and judgment, resulting from an organic disease or a disorder of the brain (see below).
3. Glial cells are nerve tissues of the central nervous system other than the signal-transmitting neurons; they are interspersed between neurons, providing support and insulation.

32 billions. The extent of this range reflects both that cell counting is imprecise and that the number of cells varies considerably from person to person. After birth, neural networks continue to be modified: connections among neurons are sometimes formed and reinforced, sometimes weakened and eliminated. Therefore, the brain's learning capacities are driven not only by the number of neurons, but by the richness of the connectivity between them. There is plenty of room for change given that any particular neuron is often connected with several thousands other neurons. For a long time it was assumed that such changes primarily happened in childhood because the brain is already about 90% of the adult size by the age of 6. Today, this position has needed revision through the emerging scientific evidence which indicates that the brain undergoes significant changes throughout life.

Learning and memory processes are embedded in networks of interconnecting neurons. Each neuron has three distinguishable parts: dendrites, a cell body, and an axon (see Figure 2.1). Dendrites are highly branched processes that receive chemical signals from other cells. The dendrites then relay electrical signals to the cell body. The dendrites receive from, and the axon sends stimuli to, other neurons. The cell body contains the nucleus with DNA and is the main site of protein synthesis. Electrical signals then travel along the axon, a long process covered by a fatty myelin sheath which extends out from the cell body. The axon branches into axon terminals, through which chemical signals are released to transmit the information to the dendrites of other cells. The neuron that is *sending* information is termed a presynaptic neuron and a neuron which is *receiving* information is termed a postsynaptic neuron. There is a small space – the synaptic cleft – between the axon of a presynaptic neuron and the dendrites of a postsynaptic neuron. In reality, the axon terminals of many presynaptic neurons converge on the dendrites of each postsynaptic neuron. Thus, the combined activity of many presynaptic neurons determines the net effect on each postsynaptic cell. The relative activity level at each synaptic connection regulates its strengthening or weakening and, ultimately, its existence. Taken together, this phenomenon is understood to be responsible for the structural encoding of learning and memory processes in the brain.

Figure 2.1. **A synaptic connection between two neurons**

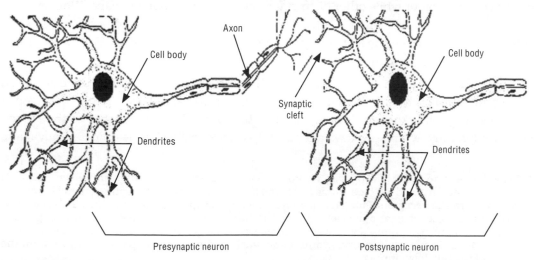

Source: Christina Hinton for OECD.

Communication between neurons is modulated by several factors. Neurons may increase their number of connections through a process called *synaptogenesis*. Alternatively, the number of synapses may decrease – "pruning". In between growth or decline, the *strength* of communication between two neurons may also be modulated by the combined effect of the amount of neurotransmitters released from the axon terminals, how quickly the neurotransmitter is removed from the synaptic cleft, and by the number of receptors the receiving neuron has on its surface. These changes amount to the strengthening or weakening of existing synaptic connections. Through these mechanisms, networks of neurons are shaped in response to experience – the brain is able to adapt to the environment.

In addition to the synaptic changes, neurons also undergo a maturing process called *myelination* in which a sheath formed by a substance called myelin wraps around the axon. Understanding this change requires consideration of what happens when neurons communicate. Communication comes about through the release of a neurotransmitter from the axon of a neuron and the axon must get a signal as to when to release the neurotransmitter. This comes through an electric impulse which travels from the body of the neuron down through the axon.[4] The axon acts like a wire and as such it can pass the current (i.e., the electric impulse) faster if it is insulated. Although most axons are not insulated at birth, they will gradually add sheaths of myelin which will act as insulation. When the axon is insulated – myelinated – the electric impulse can "jump" down the axon in the gaps between the fatty sheaths. Myelinated axons can transmit information up to 100 times faster than unmyelinated ones.

Functional organisation

The brain is highly specialised with different parts of the brain carrying out different information processing tasks – the principle of functional localisation – which holds true at almost every level of brain organisation. Each part of the brain operates different tasks and is composed of numerous inter-linked neurons. A common principle is that neurons serving the same or similar functions are connected with each other in assemblies. These assemblies are connected with other assemblies so linking a given brain area directly and indirectly with numerous other areas in complicated circuits. Brain areas are highly specialised, serving very specific sub-functions. As an example, some groups in the visual cortex code for colour while other distinct groups code for motion or shape. Whenever we "see" a given object, our brain creates a product of many specialised areas which each contributes with a given aspect of our perception. When many areas need to co-operate to provide a given function, they are referred to as *cognitive networks*.

Some functions are in place at birth. This is the case, for instance, with the brain operation that segments speech into different words (Simos and Molfese, 1997).[5] A study on French newborn babies has shown that they react to the intonation and rhythm of the French language (prosody) within five days after birth (see Chapter 4). Thus, learning already starts

---

4. This happens in the following way. Neuron A releases a neurotransmitter in the synaptic cleft between neuron A and B. Some of this neurotransmitter will pass the synaptic cleft and bind itself on to neuron B's receptors. Pumps in neuron B's membrane open so that ions outside the cell now pass into the cell whereas others leave the cell. If the influence on neuron B is strong enough – if a sufficient number of pumps are activated – the electric voltage of the cell will change in such a way that a serial reaction will occur down the axon whereby the electric impulse can travel from the body of the cell down through the axon.

5. It is probable that distinguishing words in a speech is difficult, however, as there are no spaces between the pronunciation of individual words.

taking place during the prenatal period (Pena *et al.*, 2003). Other functions are less genetically formed. The ability to read demands a complex network that involves many different areas in the brain. This network is not in place at birth, but must be formed by connecting and co-ordinating the activity of numerous specialised areas (see Chapters 4 and 5).

No two brains are alike. While every human has the same basic set of brain structures, the size of these structures and the organisation and strength of the cellular connections comprising them differ substantially from one person to the next. To begin with, each person's genetic makeup results in slightly different brain organisation as a starting point. Then, experience with the environment acts on this basic structure to bring about structural changes in the brain organisation so that different experiences can result in partially different neural networks in different people for the same cognitive process.

The structure of the brain

The human body has a line of symmetry running from the top of the head to the feet (*e.g.*, right and left eyes, hands, legs, and so forth). The brain is also divided into two major parts, the *left and right hemispheres*. The right hemisphere controls most activities on the left side of the body, and *vice versa*. Thus, a stroke suffered in the left hemisphere affects the right side of the body.

The right hemisphere has been shown to play a key role in spatial abilities and face recognition while the left hemisphere hosts crucial networks involved in language, mathematics and logic. The two hemispheres communicate through a band of up to 250 million nerve fibres called the *corpus callosum*. Therefore, even though there are some activities that appear to be dominant in one hemisphere, both hemispheres contribute to overall brain activity, each hemisphere is highly complex, and there are subsystems linking the two hemispheres. Hence, it is much too simplistic to describe any person as a "left-brain learner" or a "right-brain learner" (see discussion of neuromyths in Chapter 6).

The *cerebrum* hosts the cortex, a 2-4 mm thick multilayered sheet of cells on the surface that covers 2 000 square centimetres. The cortex is comprised of grey matter as well as white matter.[6] In order for it to fit within the skull, the cortex has many folds (gyrus or gyri) and valleys (sulcus or sulci). It hosts a large proportion of the neurons of the human brain and is mainly brought into play for higher-order functions.

The lobes

Each hemisphere is further divided into lobes (see Figure 2.2). While any complex skill depends on the co-ordinated action of neural networks across lobes, each lobe can be approximately associated with particular functions (though the following summary represents the current state of knowledge and may be modified in the future with further research). The *frontal lobe* is involved in planning and action; the *temporal lobe* plays an important role in audition, memory, and object recognition; the *parietal lobe* is involved in sensation and spatial processing; and the *occipital lobe* is essential in vision. Each lobe is further subdivided into interlocking networks of neurons specialised for very specific information processing.[7]

---

6.   Grey matter consists mainly of nerve cell bodies and dendrites; white matter consists primarily of axons that connect various areas of the brain.
7.   Any damage to these networks will disrupt the skill(s) they underlie, and each possible structural anomaly corresponds with a specific deficit.

Figure 2.2. **The major subdivisions of the cerebral cortex**

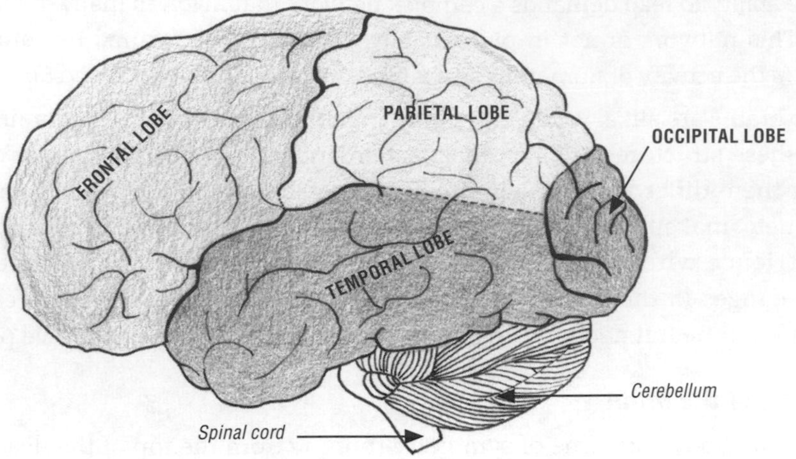

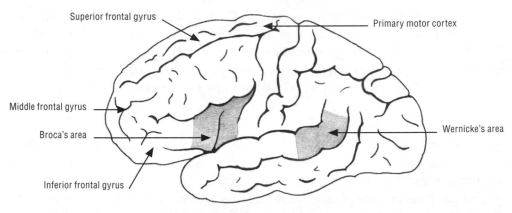

Source: Odile Pavot for the OECD.

The frontal lobe includes the *primary motor cortex*, the *superior frontal gyrus* and the *middle frontal gyrus*, and the *inferior frontal gyrus* (see Figure 2.3). The primary motor cortex is involved in the voluntary control of movements of body parts, the superior frontal gyrus in the planning and execution of behaviour, and the middle frontal gyrus in high-level executive functions and decision-making processes. *Broca's area* in the inferior frontal gyrus is associated with speech production, language processing and comprehension. Broca's area is connected to another called *Wernicke's area*, which lies at the junction of the temporal and parietal lobes in the left hemisphere, and is involved in the recognition of speech.[8]

Figure 2.3. **The frontal lobe**

Source: Odile Pavot for the OECD.

8. Damage to Broca's area can lead to Broca's aphasia, which severely limits individual's ability to form or understand complex sentences. Damage to Wernicke's area also severely impedes language processing (see Chapter 4).

The *frontal lobe* is associated with many higher-order cognitive functions including planning, judgment, memory, problem-solving, and behaviour. In general, the frontal cortex has an executive function, controlling and co-ordinating behaviour (including socially undesirable behaviour[9]). As the human brain matures to adulthood, there is a progressive myelination from the back to the front of the brain. Given that myelinated axons carry impulses faster than unmyelinated ones, brain maturity is associated with better executive functioning while demyelination (loss of myelin) is associated with diseases such as multiple sclerosis.[10]

The *parietal lobe* can be subdivided into the superior parietal lobule and inferior parietal lobule, which are separated by the intraparietal sulcus. The precuneus, the postcentral gyrus, the supramaginal gyrus and the angular gyrus are parts of the parietal lobe. Areas of the parietal lobe have been associated with mathematics learning (see Chapter 5). The parietal lobe also integrates sensory information and visuospatial processing. The angular gyrus is associated with language and cognition including the processing of metaphors and other abstractions.

The *temporal lobe* is related to auditory processing and hearing, including speech which is particularly the case in the *left temporal lobe*. It is associated with naming, comprehension, and other language functions. The *left fusiform gyrus* is part of the temporal lobe and is associated with word recognition, number recognition, face recognition and processing colour information.

The *occipital lobe* is located at the back of the brain above the cerebellum. In the interior portion of this lobe is the primary visual cortex. The occipital lobe is associated with visual processing, colour discrimination and movement discrimination.

How the brain learns over the lifetime

> Whoever ceases to be a student has never been a student.
>
> *George Iles*

Parts of the brain, including the hippocampus which plays a crucial role in learning and memory, have recently been found to generate new neurons all through life. This birth of new neurons (neurogenesis), works in tandem with the death of neurons to modify brain structure throughout the lifespan. Moreover, neurons are continually refining their connections through synapse formation (synaptogenesis), elimination (pruning),

9. Damage to or immaturity of the frontal lobe is associated with impulsivity, decreased ability to plan complex sequences of actions or persisting with a course of action without adaptability and perseveration. An example of what happens when the frontal lobe is impaired was given by David Servan-Schreiber at the New York Forum (co-organised by CERI and the Sackler Institute) on Brain Mechanisms and Early Learning (2000) by introducing the study by Antonio Damasio of a previously successful and intelligent (IQ 130, according to traditional measurement) accountant in Iowa. This accountant had part of his brain removed due to a lesion. Following the surgery while he was under medical observation, he continued to have an IQ well above average for several years. However, his social judgement became impaired so that he lost his job, failed to keep another job, got involved in a number of shadowy business ventures and eventually divorced his wife of 17 years to marry a considerably older woman.

10. Multiple sclerosis is a chronic autoimmune disease which occurs in the central nervous system. The myelin slowly disintegrates in patches throughout the brain or spinal cord (or both), and this interferes with the nerve pathways resulting in muscular weakness, loss of co-ordination, and speech and visual disturbances.

strengthening and weakening. New neurons are born and new connections are formed throughout life, and as the brain processes information from the environment, the most active connections are strengthened and the least active are weakened. Over time, inactive connections become weaker and weaker and, when all of a neuron's connections become persistently inactive, the cell itself can die. At the same time, active connections are strengthened. Through these mechanisms, the brain is tailored to fit the environment. Thus it becomes more efficient, taking account of experience in order to develop optimal architecture (Sebastian, 2004; Goswami, 2004; Koizumi, 2005). These structural changes underlie learning.

Plasticity and sensitive periods

Neuroscientists have known for some time that the brain changes significantly over the lifespan as a response to learning experiences. This flexibility of the brain to respond to environmental demands is called *plasticity*. The brain is physically modified through strengthening, weakening, and elimination of existing connections, and the growth of new ones. The degree of modification depends on the type of learning that takes place, with long-term learning leading to more profound modification.

The brain's ability to remain flexible, alert, responsive and solution-oriented is due to its lifelong capacity for plasticity. Before, it was thought that only infant brains were plastic. This was due to the extraordinary growth of new synapses paralleled with new skill acquisition. However, data uncovered over the last two decades have confirmed that *the brain retains its plasticity over the lifespan. And because plasticity underlies learning, we can learn at any stage of life albeit in somewhat different ways in the different stages* (Koizumi, 2003; OECD, 2002).

Plasticity can be classified into two types: *experience-expectant* and *experience-dependent*. Experience-expectant plasticity describes the genetically-inclined structural modification of the brain in early life and experience-dependent plasticity the structural modification of the brain as a result of exposure to complex environments over the lifespan.[11] Many researchers believe that experience-expectant plasticity characterises species-wide development: it is the natural condition of a healthy brain, a feature which allows us to learn continuously until old age.

In parallel to plasticity, learning can also be described as experience-expectant or experience-dependent. *Experience-expectant learning* takes place when the brain encounters the relevant experience, ideally at an optimal stage termed a "sensitive period". Sensitive periods are the times in which a particular biological event is likely to occur best.[12] Scientists have documented sensitive periods for certain types of sensory stimuli such as vision and speech sounds, and for certain emotional and cognitive experiences, such as language exposure. However, there are many mental skills, such as vocabulary acquisition and the ability to see colour, which do not appear to pass through tight sensitive periods. These can be considered as *experience-dependent learning* that takes place over the lifespan.

The different types of plasticity play a different role in different stages of life. The following section takes the three different stages of life, namely early childhood, adolescence, and adulthood (including ageing adults), and describes the distinctive characteristics of the learning process in each stage. Part II also discusses these key stages.

---

11. Myelination is also considered a process of experience-dependent plasticity (Stevens and Fields, 2000).
12. It should be emphasised that sensitive periods need to be considered as "windows of opportunities" rather than as times that, if missed, the opportunity will be totally lost.

Childhood (approximately 3-10 years)

> The direction in which education starts a man will determine his future life.
>
> *Plato*

Early childhood education and care has attracted enormous attention over the past decade. This has been partly driven by research indicating the importance of quality early experiences to children's short-term cognitive, social and emotional development, as well as to their long-term success in school and later life. The equitable access to quality pre-school education and care has been recognised as key to laying the foundations of lifelong learning for all children and supporting the broad educational and social needs of families. In most OECD countries, the tendency is to give all children at least two years of free public provision of education before the start of compulsory schooling; governments are thus seeking to improve staff training and working conditions and also to develop appropriate pedagogical frameworks for young children (OECD, 2001). Neuroscience will not be able to provide solutions to all the challenges facing early childhood education and care but neuroscientific findings can be expected to provide useful insights for informed decision-making in this field.

Very young children are able to develop sophisticated understandings of the phenomena around them – they are "active learners" (US National Research Council, 1999). Even at the moment of birth, the child's brain is not a *tabula rasa*. Children develop theories about the world extremely early and revise them in light of their experience. The domains of early learning include linguistics, psychology, biology and physics as well as how language, people, animals, plants and objects work. Early education needs to take good account of both the distinctive mind and individual conceptualisation of young children and this will help to identify the preferred modes of learning, *e.g.* through play.[13]

Infants have a competence for numbers. Research has indicated that very young infants, in the first months of life, already attend to the number of objects in their environment (McCrink and Wynn, 2004). There is also evidence that infants can operate with numbers (Dehaene, 1997). They develop mathematical skills through interaction with the environment and by building upon their initial number sense (further explored in Chapter 5). The educational question is then how best to build upon the already-existing competence of children. Is there an optimal timing and are there any preferred modes of learning?

---

13. According to Alison Gopnik (at the New York Forum co-organised by CERI and the Sackler Institute on "Brain Mechanisms and Early Learning", 2000), infants come equipped to learn language. But they also learn about how people around them think, feel, and how this is related to their own thinking and feeling. Children learn everyday psychology. They also learn everyday physics (how objects move and how to interact with them), and everyday biology (how living things, plants, and animals work). They master these complex domains before any official schooling takes place. It would be interesting to see whether school practices could build directly on the knowledge children have gained in their earliest environments. For instance early school could teach everyday psychology. In the case of physics and biology, schools could start to teach children from their natural conceptions (and misconceptions) about reality in order to achieve a more profound understanding of the scientific concepts that describe it. Schools could capitalise more on play, spontaneous exploration, prediction, and feedback, which are so potent in spontaneous home learning. Schools should be providing even the youngest children with the chance to be scientists and not just tell them about science.

There has long been a general belief among the non-specialists that from birth to 3 years of age, children are the most receptive to learning (Bruer, 1999).[14] On this view, if children have not been exposed fully and completely to various stimuli, they will not be able to recuperate the benefits of early stimulus later on in life. However, even for the skills for which sensitive periods exist, the capacity to learn will not be lost even after the sensitive period. While there is no scientific evidence that over-stimulating a normal, healthy infant has any beneficial effect, there is evidence that it may be a waste of time (Sebastian, 2004). The findings on which these arguments are based relate to very basic functioning such as vision; it would not be appropriate to apply this directly to the learning of cognitive skills. For more comprehensive understanding of how the experience during early childhood affects later development, a large cohort study would be required.[15]

Sensitive periods do nevertheless exist in certain areas of learning such as language acquisition (see Chapter 4). This does not imply that it is impossible to learn a foreign language after a certain age and studies have shown that the effectiveness of learning depends on the aspect of language in question. Neville (OECD, 2000) has noted that second language learning involves both comprehension and production calling for the mastery of different processes. Two of these – grammar and semantic processing – rely on different neural systems within the brain. Grammar processing relies more on frontal regions of the left hemisphere, whereas semantic processing (e.g., vocabulary learning) activates the posterior lateral regions of both the left and right hemispheres. The later that grammar is learned, the more active is the brain in the learning process.[16] Instead of processing grammatical information only with the left hemisphere, late learners process the same information with both hemispheres. This indicates that delaying exposure to language leads the brain to use a different strategy when processing grammar. Confirmatory studies have additionally shown that the subjects with this bilateral activation in the brain had significantly more difficulty in using grammar correctly – the bilateral activation indicates greater difficulty in learning. Thus, the earlier the child is exposed to the grammar of a foreign language, the easier and faster it is mastered. Semantic learning, however, continues throughout life and is not constrained in time.

Another example of sensitive periods is during the acquisition of speech sounds. Studies show that young infants in the first few months of their lives are capable of discriminating the subtle but relevant differences between similar-sounding consonants and between similar sounding vowels, for both native and foreign languages. Newborn babies can learn to discriminate difficult speech-sounds contrasts in a couple of hours even while they are sleeping, contrary to the view that sleep is a sedentary state when such capacities as attention and learning are reduced or absent (Cheour et al., 2002a; also see Chapter 3).

14. There has always been a misunderstanding of the first three years and invalid interpretation of scientific data on synaptogenesis has created several popular misconceptions as addressed in more detail in Chapter 6. One result of the neuromyths in relation to early childhood has been the rapid growth of the "brain-based" learning material industry, for example "CDs for stimulating your baby's brain". This is a good illustration of how the accurate understanding of scientific evidence is important for informed educational practices.

15. Kozorovitsky et al. (2005) contend that early childhood experience induces structural and biochemical changes in the adult primate brain. In 2004, "brain-based cohort studies" were launched by the Research Institute of Science and Technology for Society (RISTEX) of the Japan Science and Technology Agency (JST) partly with an aim to investigate this issue (see Chapter 7, Box 7.8).

16. More brain activation often means that the brain finds that particular task more difficult to process: see for example, the lower brain activation in expert readers than novice readers in word-recognition tests.

During the first year of life, however, this capacity in relation to non-native language is narrowed down as sensitivity to the sounds of their native language grows. This decline in non-native perception occurs during the first year of life, with the sharpest decline between eight and ten months (Werker, 2002; Kuhl, 1979). This change enhances the efficiency of the brain function by adapting to the natural environment. It should be noted that it is not sufficient to just make young infants listen to foreign languages through CDs in order to maintain the sensitivity towards foreign speech sounds.[17]

The acquisition of non-native speech sounds is nevertheless possible outside the sensitive period. Cheour et al. (2002b) have shown that 3 to 6-year-old children can also learn to distinguish non-native speech sounds in natural language environment within two months without any special training. McCandliss (2000) suggests that, with short-term training, Japanese native adults can learn to distinguish the speech sounds r and l (McCandliss, 2000).[18] However, as the most important aspect of language learning is to be able to communicate which does not necessarily require an accurate distinction of speech sounds, it is an open question whether it is necessary to invest time in training to distinguish foreign speech sounds, bearing in mind the level of accuracy required in different situations.

Adolescence (approximately 10-20 years)

> The foundation of every state is the education of its youth.
>
> *Diogenes Laertius*

Before brain imaging technologies became available, it was widely believed among scientists, including psychologists, that the brain was largely a finished product by the age of 12. One reason for this belief is that the actual size of the brain grows very little over the childhood years. By the time a child reaches the age of 6, the brain is already 90-95% of its adult size. In spite of its size, the adolescent brain can be understood as "work in progress". Brain imaging has revealed that both brain volume and myelination continue to grow throughout adolescence until the young adult period (i.e., between ages 20-30). Brain imaging studies on adolescents undertaken by Jay Giedd at the United States National Institute of Mental Health show that not only is the adolescent brain far from mature, but that both grey and white matters undergo extensive structural changes well past puberty (Giedd et al., 1999; Giedd, 2004). Giedd's studies show that there is a second wave of proliferation and pruning that occurs later in childhood and that the final critical part of this second wave, affecting some of our highest mental functions, occurs in the late teens. This neural waxing and waning alters the number of synapses between neurons (Wallis et al., 2004; Giedd et al., 1999; Giedd, 2004).

---

17. Parents may then wonder what exactly is sufficient to adequately develop their children's competencies for both native and foreign languages. At present, there is not enough evidence to say anything concrete on this, and this is a matter to be further explored, including through forthcoming OECD/CERI work.
18. It is known that native speakers of Japanese have considerable difficulty distinguishing between the English sounds r and l (hence, in distinguishing, for example, "load" and "road").

There are several parts of the brain that undergo change during adolescence (see Figure 2.4). First, the *right ventral striatum*, which regulates motivating reward behaviour, faces certain changes.[19] These differences may steer the adolescent brain toward engagement in high reward, risk behaviours.[20] Second, the *corpus callosum* develops before and during puberty. Third, the *pineal gland*, which produces the hormone melatonin critical to lead the body to sleep, is understood to cue the hormones to secrete melatonin much later in the 24-hour day during adolescence than in children or adults. Fourth, the *cerebellum*, which governs posture, movement, and balance, continues to grow into late adolescence. The cerebellum also influences other parts of the brain responsible for motor actions and is involved in cognitive functions including language.[21] Finally, the *prefrontal cortex*, which is responsible for important executive functions including high-level cognition, is the last part of the brain to be pruned. This area grows during the pre-teen years and then shrinks as neural connections are pruned during adolescence. Recent studies have suggested that the way in which the prefrontal cortex is developed during adolescence may affect emotional regulation.

Figure 2.4. **The adolescent brain**

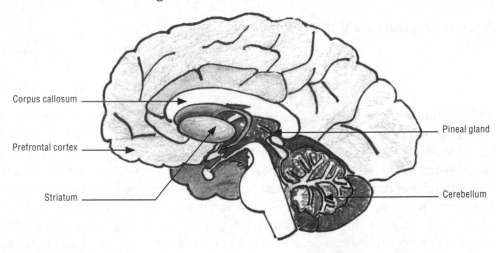

Source: Odile Pavot for the OECD.

19. A study by James M. Bjork of the National Institute on Alcohol Abuse and Alcoholism showed that this area had less activity and more errors in teenagers in comparison to adults during a reward-based gambling game (Bjork *et al.*, 2004).
20. An underdeveloped right ventral striatum is not the only factor that explains risky high reward behaviours in adolescents which will also be shaped by environmental factors such as poverty, family composition, and neighbourhood.
21. This has been established in neuroscience during the past 20 years. The cerebellum is linked with the cerebral cortex through a cerebrocerebellar communication loop. By helping the cerebral cortex by its subtle computational capability, the cerebellum contributes to all skills, whether motor or mental (such as in thought). Autism and schizophrenic symptoms such as delusion or hallucination are now explained, at least partly, by dysfunction of the cerebellum. Catherine Limperopoulos of Harvard University has demonstrated that the development of the cerebrum and cerebellum are related to each other. When there is injury to the cerebrum on one side, the cerebellum on the other side also fails to grow to a normal size. When injury occurs in one cerebellar hemisphere the opposite side of the cerebrum is smaller than normal. Limperopoulos and her colleagues suggest that in addition to motor problems, children born with cerebellar injuries have problems with higher cognitive processes such as communication, social behaviour and visual perception. The results imply the importance of cerebral-cerebellar communication in development of the brain and its disorders that would cause difficulties in mental processes for communication, social behaviour and learning (Masao Ito, 2005, upon request from the OECD Secretariat).

Gender differences in cognitive development during adolescence have also been studied in terms of speech development and lateralisation[22] of language in the human brain (Blanton et al., 2004). This showed significant age-related increases in both white and grey matters in left inferior frontal gyrus[23] in boys aged 11 years compared with girls of the same age, and this overall area was observed as larger in boys. Both boys and girls showed asymmetry in development with the right side growing faster, but each in slightly different areas of the prefrontal cortex.

Adolescence is a time of profound mental change, which affects the emotional constitution – social awareness, character, and tendencies towards the development of mental illness. It is a period when the individual is especially open to learning and to social developments, and it is also a time when anti-social behaviour can emerge.[24] Adolescence is a crucial period in terms of emotional development partly due to a surge of hormones in the brain. Sex hormones play an important part in intense teenage emotions and have recently been found to be active in the emotional centre of the brain (i.e., the limbic system, see Figure 3.1). These hormones directly influence serotonin and other neurochemicals which regulate mood, and contribute to the known thrill-seeking behaviour of teenagers. (A study on risk assessment using a driving simulation game with adolescents and adults showed that the adolescents took more risks when playing in groups with their peers although both sets made safe choices when playing alone [Steinberg, 2004].) According to psychologist Laurence Temple, "the parts of the brain responsible for things like sensation seeking are getting turned on in big ways around the time of puberty, but the parts for exercising judgment are still maturing throughout the course of adolescence. It is like turning on the engine of a car without a skilled driver at the wheel" (Wallis et al., 2004). The still under-developed pre-frontal cortex among teenagers could thus play a role in the higher incidence of unstable behaviour during adolescence.[25]

The latest findings on immaturity of adolescent's decision-making functions in the brain could have direct implications for policy with regards to this target age group. For example, "sorting" often takes place in lower secondary schools (tracking, streaming, or even selection) which may not be compatible with the under-developed state of the adolescent brain.[26] Reflection may be needed on the laws which set the minimum driving age. According to the United States Insurance Institute for Highway Safety, teenagers are four times as likely as older drivers to be involved in a crash and three times as likely to die in one. In light of this finding, some States are considering revising bills by expanding training requirements and restricting passenger numbers and cell phone use for certain teenage drivers.[27]

22. Localisation of function on either the right or left sides of the brain.
23. The inferior frontal gyrus plays an important role in language processing, speech development, and higher level cognition.
24. In the teenage years, the incidence of anti-social behaviour has been measured to increase ten-fold (Moffitt, 1993).
25. Teenagers' underdeveloped pre-cortex necessitates their use of an alternative area of the brain, namely the amygdala (see Chapter 3 for more details). An experiment using fMRI whereby children and adults were asked to identify emotions displayed on faces in photographs, showed results that the adolescents rely heavily on the amygdala whereas adults rely less on the amygdala but more on the frontal lobe (Baird et al., 1999).
26. Moreover, result from the OECD-PISA studies suggest ineffectiveness of early classroom sorting policies. Neuroscience may further shed light on this conclusion by suggesting how the adolescent's emotional brain interacts with the type of classroom environment.
27. However, older drivers on average have more driving experience as compared with young drivers, which may also contribute to the differences in the crash rate.

Adolescence is also a time when major mental illnesses such as depression, schizophrenia[28] and bipolar disorder occur, which may contribute to the high rates of teenager suicide. Kashani and Sherman (1988) conducted epidemiologic studies in the United States to reveal the incidence of depression[29] to be 0.9% in preschool-aged children, 1.9% in school-aged children and 4.7% in adolescents. There appears to be a greater prevalence of depression among adolescent girls with a female to male ratio of 2:1 which may be due to the fact that females are more socially oriented, more dependent on positive social relationships, and more vulnerable to a loss of social relationships than are boys (Allgood-Merten, Lewinsohn and Hops, 1990). This would increase their vulnerability to the interpersonal stresses that are common among teenagers. There is also evidence that the methods girls use to cope with stress may entail less denial and more focused and repetitive thinking about the stressful event (Nolen-Hoeksema and Girgus, 1994). The higher prevalence of depression among girls, therefore, could be a result of greater vulnerability, combined with different coping mechanisms from those of boys.

More brain research on individual differences in the critical adolescent years, especially those related to gender, could help understand different brain developmental trajectories and the way individuals cope with environmental stimulation. It could help explain the fact that the incidence of suicide attempts reaches a peak during the mid-adolescent years, with mortality from suicide, which increases steadily through the teens, the third leading cause of death at that age (Hoyert, Kochanek and Murphy, 1999). Because the risk of school failure and suicide is high among depressed adolescents, more research on the teenage brain will shed light on these mental illnesses and contribute towards early detection and prevention.

Table 2.1 summarises the discussion thus far on how the brain learns at different stages of childhood and adolescence. The table is by no means complete, in part as there are so many issues yet to be uncovered and cognitive neuroscience has not shed sufficient

Table 2.1. **Summary of how the brain learns**

	Childhood (3-10 years)	Early adolescence (10-13 years)	Adolescence (13-20 years)
Brain maturation[1]	Frontal region of the left hemisphere	Right ventral striatum Cerebellum Corpus callosum Pineal gland	Prefrontal cortex Cerebellum
Associated functions[2]	Language (grammar)	Motivating rewards Posture and movement Language Sleep	Executive functions Posture and movement
Optimal timing for learning[3]	Language (grammar, accent)[4] Music[5]	Not discussed	Not discussed

1. The brain maturation row does not provide an exhaustive list of maturational changes.
2. The associated functions row does not provide an exhaustive list of functions associated with listed brain areas.
3. There may be additional developmental sensitivities not mentioned in this table.
4. Though certain aspects of language are learned more efficiently during childhood, language can be learned throughout life. Moreover, the optimal age for learning language provided in this table is an average, and varies across individuals.
5. Though certain aspects of music are learned more efficiently during childhood, music can be learned throughout life. Moreover, the optimal age for learning music provided in this table is an average, and varies across individuals.

28. MRI studies have shown that while the average teenager loses about 15% of cortical grey matter, those who develop schizophrenia lose as much as 25% (Lipton, 2001).
29. This clinical condition ranges from simple sadness to a major depressive or bipolar disorder, with risk factors including family history and poor school performance.

light in particular on the learning process among adults and elderly who are excluded in the table. Given that such research could potentially address some of the key challenges of the ageing society, it is important to make further progress in this area.

Adulthood and the elderly

> Education is the best provision for old age.
>
> *Aristotle*

Adult learning has become increasingly important in the last decade or so as it becomes clear that societies are increasingly knowledge-based. High unemployment rates among the low-skilled and the acknowledgement of the significant role of human capital for economic growth and social development lead to demands for increased learning opportunities for adults. Nevertheless, participation rate in adult learning is not always high, and at around age 50 there tends to be a considerable drop. A challenge now is to make learning more attractive to adults using appropriate pedagogical methods (OECD, 2005).

Contrary to the once-popular assertion that the brain loses 100 000 neurons every day (or more if accompanied by smoking and drinking), new technologies have shown that there is no age dependence if one counts the total number of neurons in each area of the cerebral cortex (Terry, DeTeresa and Hansen, 1987). Age-dependence only applies to the number of "large" neurons in the cerebral cortex. These large neurons decline with the consequence of increasing the number of small neurons, so that the aggregate number remains the same. There is some decrease in neuronal circuitry as neurons get smaller, however, and the number of synapses are reduced. While reduced connectivity corresponds with reduced plasticity, it does not imply reduced cognitive ability. On the contrary, skill acquisition results from pruning some connections while reinforcing others. So, people continue learning throughout life.

Do older adults learn in the same way as the young? There is considerable evidence showing that older adults show less specificity or differentiation in brain recruitment while performing an array of cognitive tasks (Park *et al.*, 2001). A recent Japanese study compared language proficiency of young and older Japanese adults (Tatsumi, 2001). The subjects were asked to say aloud as many words in a given semantic and phonological category as they could within 30 seconds. The number of words the older subjects could retrieve was approximately 75% of the number retrieved by the young subjects, showing their lower word-fluency. They also had greater difficulty retrieving famous names, with their mean performance about 55% to that of young subjects. A Positron Emission Tomography (PET) activation study was carried out during the word-fluency tasks. Among the young subjects, the left anterior temporal lobe and frontal lobe were activated during the retrieval of proper names. During the retrieval of animate and inanimate names, and syllable fluency, the left infero-posterior temporal lobe and left inferior frontal lobe (i.e., Broca's area, see Figure 2.3) were activated. By contrast, the activated areas among the elderly subjects were found to be generally smaller or sometimes inactive, though certain areas which were not active among the young subjects were active among the elderly (Tatsumi, 2001).

It is premature to base conclusions on these findings which need further investigation. One interpretation of these activation patterns is that other brain areas are brought into play among the older adults in an effort to compensate for deficient word retrieval. Alternatively and in favour of the vitality of the ageing brain, fluency or experience with a task necessarily reduces activity levels so that with higher processing efficiency, these tasks can also be shuttled to different areas of the brain for processing.

Learning to delay age-related cognitive decline

Many cognitive processes in the brain decline when we stop using them, confirming the broad thrust of the lifelong learning concept. Instead of the proverbial "you can't teach an old dog new tricks" the message is instead "use it or lose it", raising the further question of how best this should be done. This is an important question since the way in which the brain is used affects the pace with which it ages. This section describes the reduced functioning of the brain that comes with age as well as possible ways in which this can be counteracted through learning.

Combating declining cognitive function

Although our brain is flexible enough to permit learning throughout life, there is a general decline in most cognitive capacities from around age 20 to 80. The everyday impression is that the decline starts much later than at 20 years, simply because it becomes more noticeable during late adulthood. Losses of executive function and long-term memory in middle-aged adults may also not be too apparent to the individual because they are offset by increases in expertise and skill (Park *et al.*, 2001). Much remains to be understood, however, about the interaction between increasing knowledge and declining executive function and memory across the lifespan so that further research is needed in this area. Not all cognitive functions decline with age in the same way. The decline has been most clearly noted in tasks such as letter comparison, pattern comparison, letter rotation, computation span, reading span, cued recall, free recall, and so on. By contrast, increases in cognitive capacities across the lifespan up to age 70 (with some declines by age 80) have also been noted. This is the case for vocabulary, for example, which manifests an increase in experience and general knowledge counterbalancing losses in other cognitive capacities (Park *et al.*, 2001; Tisserand *et al.*, 2001; 2002).

Age-related decline is due to problems with various cognitive mechanisms rather than having a single cause. It is likely that all of the different executive processes, as well as their speed, decline with age and contribute to difficulties in higher-order cognitive functions such as reasoning and memory (Park *et al.*, 2001). Studies addressing the differential decline of neurocognitive function with age show that the speed of information-processing declines already in the fourth decade of life, and applies especially to those cognitive processes which are dependent on areas and circuits within the prefrontal cortex. Thus, the so called "executive functions" are among the first to deteriorate with age, which becomes manifest in such ways as decreasing efficiency in processing of new information, increased forgetting, lack of attention and concentration, and decreased learning potential. The effect of age differs within the prefrontal cortex, with the dorsolateral and medial areas being more affected than the orbital region. It is possible that this difference leads to lack of integrity of areas within the prefrontal cortex and plays a causal role in age-related cognitive decline (Tisserand *et al.*, 2001; 2002).

Age-related decline in higher-order cognitive functioning does not necessarily affect creativity. Indeed, there is evidence that creativity is largely independent of other cognitive functions.[30] A study examining the effects of ageing on creativity among Japanese adults ranging in age from 25 to 83 found no age differences in fluency, originality of thinking

---

30. Jellemer Jolles, 2nd meeting of the CERI's Lifelong Learning Network, Tokyo, 2003.

ability, productivity, and application of creative ability. However, gender differences were found on fluency and productivity with women outscoring men. These results suggest that certain creative abilities are maintained throughout the adult years.[31]

In addition to experience, "fitness" is another factor that affects the cognitive function (see Chapter 3). The idea that physical and mental fitness are related is an ancient one, expressed in Latin by the poet Juvenal as "mens sana in corpore sano" (i.e., "a healthy mind in a healthy body"). A review of the animal literature has found reasons for optimism in the enhancement of cognitive function (Anderson et al., 2000). Moreover, a recent meta-analysis of existing longitudinal data suggests a positive and robust association between enhanced fitness and cognitive vitality in humans, particularly in executive processes (e.g., management or control of mental processes). Emerging data suggest that the regions of the brain associated with executive processes and memory, such as the frontal cortex and hippocampus, show significant age-related decline, larger than in other regions. Such decline may be slowed by good physical fitness. In particular, task improvement has been shown to be positively correlated with cardio-vascular function. Specific training studies also show positive results for spatial orientation, inductive reasoning and complex task-switching activities such as driving. In general, there is growing evidence that behavioural interventions, including enhanced fitness and learning, can contribute to improvements in performance even into old age. An important question for the future is the applicability of these results outside of the laboratory.

Combating damaged brain function

Another demonstration of the brain's flexibility is how functional reorganisation takes place after serious damage. There was the case of a baby born without the cerebral cortex adequately formed. As this is the part of the brain responsible for all forms of conscious experience, including perception, emotion, thought and planning, the baby was in a vegetative state. Normally, the diagnosis is that the baby is unable to see and hear and that there is nothing to be done. The parents were convinced that the child was seeing something, however, so that at 14 months, its brain was examined using optical topography and this revealed that the primary visual area in the occipital lobe was activated. This example shows how flexible is the brain at this young age, adapting to the environment and compensating for the lost functions of the brain (Koizumi, 2004). Further understanding of this compensatory mechanism could contribute to efficient early treatments and rehabilitation following brain damage, such as the recently-launched three-year study in the United Kingdom studying 60 children aged 10 to 16 investigating the influence of brain injury on speech and learning abilities and how the brain can compensate for injuries through reorganisation (Action Medical Research, 2005).

There is also evidence that damaged adult brains also exhibit plasticity. Language capacity in the dominant hemisphere lost following a stroke can re-emerge in the contra-lateral hemisphere. In some cases, language-specific activation has been restored to the dominant hemisphere about a year after injury and some studies have even shown that the reorganisation can happen within two months of injury among older people recovering the hemiparesis after a stroke (Kato et al., 2002; Koizumi, 2004).

---

31. Hideaki Koizumi, 3rd meeting of the CERI's Lifelong Learning Network, Tokyo, 2004.

Neurodegenerative disorders

One of the serious challenges of the ageing societies in which life expectancy has been extended to 80 years or more is age-related disorders such as Alzheimer's disease. The neurodegenerative disorder is most acutely felt in terms of cognitive function with ageing. Such illness deprives the individuals of their sense of self and deprives society of accumulated expertise and wisdom. With the ageing of populations, this problem will increase.

While research into the processes of ageing is constantly progressing, there is an accumulated body of research on the ageing brain centred on disease models, partly motivated by the massive cost of neurodegenerative illnesses for society. This is a growing problem for it was estimated that there were about 18 million people worldwide with Alzheimer's disease in 2001, set to nearly double by 2025 to 34 million. Since Alzheimer's disease is a chronic and progressive neurodegenerative disorder, the cost of caring for these patients is enormous (World Health Organisation, 2001).[32] There is real hope for early diagnosis and appropriate interventions to defer the onset or delay the acceleration of neurodegenerative diseases in old age. Lifelong learning promises a particularly effective set of strategies for combating senility and conditions such as Alzheimer's disease. The focus on degeneration also provides important insights into the normally functioning brain.

Alzheimer's disease is responsible for irreversible brain damage. The symptoms of this disease usually start in late adulthood and involve marked defects in cognitive function, memory, language, and perceptual abilities. The pathology associated with Alzheimer's disease is the formation of plaques in the brain. These changes are particularly evident in the hippocampus, a part of the "limbic system", also crucially involved in short-term memory (and in sending to the cortex new material to be stored in long-term memory). As there are not yet reliable methods for detecting Alzheimer's disease, its early onset may be better diagnosed either behaviourally or through genetic testing. Behavioural diagnosis of early onset is difficult since too little is known about the cognitive changes associated with normal ageing: declining cognitive functions with older age overlap with and are similar to those of the preclinical symptoms of Alzheimer's disease.

Some believe that it would be profitable to direct research resources to the study of attention if the onset of Alzheimer's disease is to be detected early and this for at least two reasons. First, attentional functions are impaired even in mildly affected cases, providing possible valuable early warning. Second, a major dysfunction with Alzheimer's disease is in memory, which can often be addressed by the study of attention (see Box 3.2). The neural systems mediating attentional functions are relatively well understood, having been much studied. Two aspects of spatial selective attention – attentional shifting and spatial scaling – are markedly impaired in the early stages of Alzheimer's disease. Tasks that assess these functions can thus offer useful early diagnosis. Studies involving event-related brain potential (ERP), Positron Emission Tomography (PET), and functional Magnetic Resonance Imaging (fMRI) indicate that attentional tasks indeed provide sensitive behavioural tests of early dysfunction.

Another approach to the early detection of Alzheimer's disease is to identify adults without apparent symptoms who are at genetic risk of developing it. Recent studies implicate the inheritance of the apolipoprotein E (APOE) gene in the development of

---

32. In the United States it was estimated that in the year 2000, the direct and total (direct and indirect) national cost was approximately USD 536 billion and USD 1.75 trillion respectively. Similar detailed costings are not available for other countries (World Health Organisation, 2001).

Alzheimer's disease.[33] Compared to those without an e4 allele, e4 carriers exhibit spatial attention deficits that are qualitatively similar to those shown by clinically-diagnosed Alzheimer's disease patients. Such deficits include increased attentional disengagement and the reduced ability to scale spatial attention. These deficits can appear in otherwise healthy adults without symptoms as early as their fifties.

Both the behavioural and genetic indicators can lead to the development and testing of new markers for predicting severe cognitive decline in older adults. Equipped with improved diagnostic evidence, pharmacological and behavioural treatments and interventions for enhancing cognitive function in adults can be developed. The benefits of alertness and vigilance training (attentional cueing) have been shown to reduce the symptoms of Alzheimer's disease by reducing attention deficits and enhancing learning in both healthy adults and Alzheimer's disease patients. Such interventions hold promise because the fine structure of synaptic connections in the brain is not under direct genetic control but is shaped and reshaped throughout the lifespan by experience.

Alzheimer's disease can also lead to depression.[34] Depression is an illness associated with a host of symptoms including lack of energy, concentration and interest; insomnia; loss of appetite; and the incapacity to experience pleasure (anhedonia). Identifying the causes of depression in older adults is more complicated, compared with that found in younger people, and is thus more difficult to treat. As with other age-related disorders, senile depression imposes major health and societal burdens; currently, depression in the elderly is the second most frequent mental disease following dementia. A major difference between depression in the different age groups is the lower genetic contribution to the disease in the elderly. In addition to the organic causes noted above, depression in older people can be traced in some cases to being deprived of social roles, the loss of important and close people, and the decline in economic, physical and psychological capacity.

There has recently been a promising development in overcoming neurodegenerative disorders in Japan. The Learning Therapy (see Box 2.1) developed and experimented in Fukuoka, Japan is one possible avenue in counteracting declining brain functions.

While many of the cognitive functions decline with age, it is important to recognise that not all functions are lost with age. The more there are opportunities for older and elderly people to continue learning and use their knowledge, the better will be the outcome in deferring the onset or delaying the acceleration of neurodegenerative diseases. Encouraging adult learning is thus important as it may otherwise stop among older people. Society should also work to see that old people are not suddenly deprived of their social roles, employment, and their sense of self worth, for example through providing ways for the contributions of the "third-age" cohort to be celebrated and used. Encouraging them to provide guidance to younger people to mutual benefit can also address depression among this age group (Table 2.2 summarises the main issues addressed in this section).

33. The APOE gene is inherited as one of three alleles, e2, e3, and e4, with the e4 allele associated with greater risk of developing Alzheimer's disease (Greenwood *et al.*, 2000).
34. Other disorders such as Parkinson's disease and strokes can also lead to depression.

Box 2.1. **Learning Therapy (Japan)**

Enormous investments have been made in order to better understand and combat neurodegenerative disorders. One of the successful intervention methods developed for senile dementia is the Learning Therapy from Japan. It is an intervention method based on knowledge from brain science in the functioning of the prefrontal cortex, functional brain imaging, and characteristics of the prefrontal cortex. It aims at "improving mental functions of the learner – prefrontal cognition, communication, personal needs independence, and so on – through study, as the learner and the instructor communicate, using together materials which centre on reading aloud and calculation" (Kawashima *et al.*, 2005). This work suggests that elderly people who have developed senile dementia could continue learning, provided that they are given the right material and environment.

Among the functions of the prefrontal cortex, the most relevant to intervention for dementia is communication, independence and short-term memory. The basic idea of the Learning Therapy is that by ameliorating these functions of the prefrontal cortex, those who have developed senile dementia would be able to participate in social activities regardless of their disease. Instead of trying to heal dementia, it focuses on allowing sufferers to participate in social activities again.

Brain-imaging technologies have enabled researchers to observe how the brains of those with senile dementia react to different tasks. After several years of research, the results show that simple tasks such as reading aloud and arithmetic calculation activate the brain widely, including the prefrontal cortex, in both left and right hemispheres. The fundamental principle of the Learning Therapy is to select and repeat every day a very simple task that activates the brain. The tasks have to be simple enough to allow people who have started developing Alzheimer's disease to work on them. Normally, as learning progresses, the amount of brain activity decreases. However, the key to the Learning Therapy is the selection of simple tasks so that the activity of the brain does not decline after repeating it several times. Two tasks are used in the Learning Therapy which meet these conditions, namely reading aloud and simple calculation.

The study of the people with dementia who follow the Learning Therapy has shown positive effects based on comparison of a group with senile dementia who had continued the Learning Therapy 20 minutes per day for 18 months, with a group that did not. Those who had followed the Learning Therapy showed increased functioning of the prefrontal cortex, whereas for the other group this declined. After 18 months, there was a clear difference between those who had studied and those who had not. Furthermore, the ability to do the necessary for daily life was maintained in the group that studied, compared with decline in the others.

Besides the choice of the tasks, appropriate conditions are also required in order for learning to take place. Very often, the environment is not appropriate and therefore it is tempting to believe that the elderly cannot learn. One of the keys to the success of the Learning Therapy is the development of the appropriate learning material and environment. Sometimes, peripheral sensory problems need to be treated in order not to isolate the brain functions, which would otherwise deteriorate (OECD/CERI "Lifelong Learning Network Meeting", Tokyo, 2005). The importance of communication through the learning process should be added: immediate feedback recognising the achievement of the learner is also an essential part of the Learning Therapy.

Table 2.2. **Declining or damaged brain functioning and possible response**

	Adults	Elderly
Part of the brain that declines	Pre-frontal cortex Hippocampus	Pre-frontal cortex Hippocampus (formation of senile plaques)
Cognitive process that declines	High-order cognitive functioning ● inefficiency in processing new information ● increased forgetfulness ● lack of attention and concentration ● decreased learning potential	Cognitive functioning ● memory ● language ● perceptual abilities ● communication
Disorders	Depression	Senile dementia (including Alzheimer's disease) Depression
How to recover, prevent, or reduce the speed of decline	Physical fitness Learning	Attentional tasks (alertness, vigilance) for those detected with Alzheimer's early Learning (including "Learning Therapy") to tackle dementia

Conclusions

Neuroscientists have well established that the brain has a highly robust and well-developed capacity to change in response to environmental demands – *plasticity* – creating and strengthening some neuronal connections and weakening or eliminating others. Plasticity is a core feature of the brain throughout life. There are optimal or "sensitive periods" during which a particular learning is most effective, despite this lifetime plasticity. For sensory stimuli such as speech sounds, and for certain emotional and cognitive experiences such as language exposure, there are relatively tight and early sensitive periods. Other skills, such as vocabulary acquisition, do not pass through sensitive periods and can be learned equally well at any time over the lifespan.

Neuroimaging of adolescents now shows us that the adolescent brain is far from mature, and undergoes extensive structural changes well past puberty. Adolescence is an extremely important period in terms of emotional development partly due to a surge of hormones in the brain; the still under-developed pre-frontal cortex among teenagers may be one explanation for their unstable behaviour. In older adults, fluency or experience with a task can reduce brain activity levels – in one sense this is greater processing efficiency. But the brain also declines the more we stop using it and with age. Studies have shown that learning can be an effective way to counteract the reduced functioning of the brain: the more there are opportunities for older and elderly people to continue learning, the higher the chances of deferring the onset or delaying the acceleration of neurodegenerative diseases.

An understanding of how the brain learns and matures can inform the design of more effective and age-appropriate teaching and learning for children and adults. An understanding of ageing processes in the brain can help individuals maintain cognitive functioning throughout life. An important, scientifically-informed conception of learning is taking shape.

References

Action Medical Research (2005), "Speech and Language in Children Born Preterm", *www.action.org.uk/research_projects/grant/261/*.

Allgood-Merten, B., P.M. Lewinsohn and H. Hops (1990), "Sex Differences in Adolescent Depression", *Journal of Abnormal Psychology*, Vol. 99, No. 1, pp. 55-63.

Anderson, B.J., D.N. Rapp, D.H. Baek, D.P. McCloskey, P.S. Coburn-Litvak and J.K. Robinson (2000), "Exercise Influences Spatial Learning in the Radial Arm Maze", Physiol Behav, Vol. 70, No. 5, pp. 425-429.

Baird, A.A., S.A. Gruber, D.A. Fein, L.C. Maas, R.J. Steingard, P.F. Renshaw, B.M. Cohen and D.A. Yurgelun-Todd (1999), "Functional Magnetic Resonance Imaging of Facial Affect Recognition in Children and Adolescents", *Journal of the American Academy of Child and Adolescent Psychiatry*, Vol. 38, No. 2, pp. 195-199.

BBC News (2005), "Meditation 'Brain Training' Clues", 13 June, *http://news.bbc.co.uk/2/hi/health/4613759.stm*.

Bjork, J.M., B. Knutson, G.W. Fong, D.M. Caggiano, S.M. Bennett and D.W. Hommer (2004), "Incentive-Elicited Brain Activation in Adolescents: Similarities and Differences from Young Adults", *Journal of Neuroscience*, Vol. 24, No. 8, pp. 1793-1802.

Blanton, R.E., J.G. Levitt, J.R. Peterson, D. Fadale, M.L. Sporty, M. Lee, D. To, E.C. Mormino, P.M. Thompson, J.T. McCracken and A.W. Toga (2004), "Gender Differences in the Left Inferior Frontal Gyrus in Normal Children", *Neuroimage*, Vol. 22, No. 2, pp. 626-636.

Bruer, J.T. (1999), *The Myth of the First Three Years*, Free Press, New York.

Cheour, M., O. Martynova, R. Näätänen, R. Erkkola, M. Sillanpää, P. Kero, A. Raz, M.L. Kaipio, J. Hiltunen, O. Aaltonen, J. Savela and H. Hämäläinen (2002a), "Speech Sounds Learned by Sleeping Newborns", *Nature*, Vol. 415, No. 6872, pp. 599-600.

Cheour, M., A. Shestakova, P. Alku, R. Ceponiene and R. Naatanen (2002b), "Mismatch Negativity Shows that 3-6-year-old Children Can Learn to Discriminate Non-native Speech Sounds within Two Months", *Neuroscience Letters*, Vol. 325, No. 3, pp. 187-190.

Coffield (2005), "It takes two to tango", paper written upon request of CERI in preparation of the 4th meeting of the CERI's Lifelong Learning Network, Wako-shi, 2004.

Dehaene, S. (1997), *The Number Sense: How the Mind Creates Mathematics*, Oxford University Press, New York.

Giedd, J.N. (2004), "Structural Magnetic Resonance Imaging of the Adolescent Brain", *Annals of the New York Academy of Sciences*, Vol. 1021, pp. 77-85.

Giedd, J.N., J. Blumenthal, N.O. Jeffries, F.X. Castellanos, H. Liu, A. Zijdenbos, T. Paus, A.C. Evans and J.L. Rapoport (1999), "Brain Development during Childhood and Adolescence: A Longitudinal MRI Study", *Nature Neuroscience*, Vol. 2, pp. 861-863.

Gopnik, A. (2000), "Cognitive Development and Learning Sciences: State of the Art", presentation at the 1st CERI forum on "Brain Mechanisms and Early Learning", Sackler Institute, New York City, 17 June.

Goswami, U. (2004), "Neuroscience, Education and Special Education", *British Journal of Special Education*, Vol. 31, No. 4, pp. 175-183.

Greenwood, P.M., T. Sunderland, J. Friz and R. Parasuraman (2000), "Genetics and Visual Attention: Selective Deficits in Healthy Adult Carriers of the e4 Allele of the Apolipoprotein E gene", *Proceedings of the National Academy of Sciences*, Vol. 97, pp. 11661-11666.

Hoyert, D.L., K.D. Kochanek and S.L. Murphy (1999), *National Vital Statistics Report*, Vol. 47, No. 19, *Deaths: Final Data for 1997*, Centers for Disease Control and Prevention, National Center for Health Statistics, National Vital Statistics System, Hyattsville, MD.

Johnson, M.H. (1997), *Developmental Cognitive Neuroscience: An Introduction*, Blackwell, Oxford.

Kandel, E.R., J.H. Schwartz and T.M. Jessell (1991), *Principles of Neural Science*, Appleton and Lance, Norwalk, Connecticut, third edition.

Kashani, J.H. and D.D. Sherman (1988), "Childhood Depression: Epidemiology, Etiological Models, and Treatment Implications", *Integrated Psychiatry*, Vol. 6, pp. 1-8.

Kato, H., M. Izumiyama, H. Koizumi, A. Takahashi and Y. Itoyama (2002), "Near-infrared Spectroscopic Topography as a Tool to Monitor Motor Reorganisation after Hemiparetic Stroke: A Comparison with Functional MRI", *Stroke*, Vol. 33, No. 8, pp. 2032-2036.

Kawashima, R., K. Okita, R. Tamazaki, N. Tajima, H. Yoshida, M. Taira, K. Iwata, T. Sasaki, K. Maeyama, N. Usui and K. Sugimoto (2005), "Reading Aloud and Arithmetic Calculation Improve Frontal Function of People with Dementia", *The Journals of Gerontology Series A: Biological Sciences and Medical Sciences*, Vol. 60, pp. 380-384.

Koizumi, H. (2002), "The Scope of the Symposium 8th International Conference on Functional Mapping of the Human Brain", June 2-6, Sendai, Japan.

Koizumi, H. (2003), "Science of Learning and Education: An Approach with Brain-function Imaging", *No To Hattatsu*, Vol. 35, No. 2, pp. 126-129.

Koizumi, H. (2004), "The Concept of 'Developing the Brain': A New Natural Science for Learning and Education", *Brain and Development*, Vol. 26, No. 7, pp. 434-441.

Kozorovitsky, Y., C.G. Gross, C. Kopil, L. Battaglia, M. McBreen, A.M. Stranahan and E. Gould (2005), "Experience Induces Structural and Biochemical Changes in the Adult Primate Brain", *Proceedings of the National Academy of Sciences*, Vol. 102, No. 48, pp. 17478-17482.

Kuhl, P.K. (1979), "Speech Perception in Early Infancy: Perceptual Constancy for Spectrally Dissimilar Vowel Categories", *Journal of the Acoustical Society of America*, Vol. 66, pp. 1668-1679.

Lipton, L. (2001), "Schizophrenia: A 'Wave' of Cortical Changes", *Neuropsychiatry Reviews*, Vol. 2, No. 8, October.

Maguire, E.A., D.G. Gadian, I.S. Johnsrude, C.D. Good, J. Ashburner, R.S. Frackowiak and C.D. Frith (2000), "Navigation-related Structural Change in the Hippocampi of Taxi Drivers", *Proceedings of the National Academy of Sciences*, Vol. 97, No. 8, pp. 4398-4403.

McCandliss (2000), "Cortical Circuitry of Word Reading", presentation at the 1st CERI forum on "Brain Mechanisms and Early Learning", Sackler Institute, New York City, 17 June.

McCrink, K. and K. Wynn (2004), "Large-number Addition and Subtraction by 9-month-old in Infants", *Psychological Science*, Vol. 15, pp. 776-781.

Moffitt, T.E. (1993), "Adolescence-Limited and Life-Course-Persistent Antisocial Behaviour: A Developmental Taxonomy", *Psychological Review*, Vol. 100, No. 4, pp. 674-701.

Neville, H. (2000), "Brain Mechanisms of First and Second Language Acquisition", presentation at the 1st CERI forum on "Brain Mechanisms and Early Learning", Sackler Institute, New York City, 17 June.

New York Times (2003), "Is Buddhism Good for Your Health?", S.S. Hall, 14 September.

Nolen-Hoeksema, S. and J.S. Girgus (1994), "The Emergence of Gender Differences in Depression during Adolescence", *Psychological Bulletin*, Vol. 115, No. 3, pp. 424-443.

OECD (2000), Report on the First High Level Forum held on "Brain Mechanisms and Early Learning" at Sackler Institute, New York City, USA, 16-17 June.

OECD (2001), Report on the 2nd High Level Forum held on "Brain Mechanisms and Youth Learning" at University of Granada, Granada, Spain, 1-3 February.

OECD (2002), "Learning Sciences and Brain Research: Report of the Launching Meeting of Phase II", Royal Institution, London, 29-30 April, pp. 7-8, *www.oecd.org/dataoecd/40/36/15304667.pdf*.

OECD (2005), Report on the Third Lifelong Learning Network meeting, 20-22 January 2005, Wako-shi, Japan.

Park, D.C., T. Polk, J. Mikels, S.F. Taylor and C. Marshuetz (2001), "Cerebral Aging: Integration of Brain and Behavioral Models of Cognitive Function", *Dialogues in Clinical Neuroscience*, Vol. 3, pp. 151-165.

Pena, M., A. Maki, D. Kovacic, G. Dehaene-Lambertz, H. Koizumi, F. Bouquet and J. Mehler (2003), "Sounds and Silence: An Optical Topography Study of Language Recognition at Birth", *Proceedings of the National Academy of Sciences*, Vol. 100, No. 20, pp. 11702-11705.

Polk, T.A. and M. Farah (1995), "Late Experience Alters Vision", *Nature*, Vol. 376, No. 6542, pp. 648-649.

Sebastián Gallé Núria (2004), "A Primer on Learning: A Brief Introduction from the Neurosciences", from her paper delivered at the Social Brain Conference held in Barcelona, 17-20 July 2004, *www.ub.es/pbasic/sppb/*.

Servan-Schreiber, D. (2000), at the New York Forum on Brain Mechanisms and Early Learning at Sackler Institute, New York City, USA, 16 June 2000.

Simos, P.G. and D.L. Molfese (1997), "Electrophysiological Responses from a Temporal Order Continuum in the Newborn Infant", *Neuropsychologia*, Vol. 35, pp. 89-98.

Steinberg, L. (2004), "Risk Taking in Adolescence: What Changes, and Why?", *Annals of the New York Academy of Sciences*, Vol. 1021, pp. 51-58.

Stevens, B. and R.D. Fields (2000), "Response of Schwann Cells to Action Potentials in Development", *Science*, Vol. 287, No. 5461, pp. 2267-2271.

Tatsumi, I. (2001), "A PET Activation Study on Retrieval of Proper and Common Nouns in Young and Elderly People", Third High Level Forum, Tokyo, Japan, 26/27 April.

Terry, R.D., R. DeTeresa and L.A. Hansen (1987), "Neocortical Cell Counts in Normal Human Adult Ageing", *Annuals of Neurology*, Vol. 21, No. 6, pp. 530-539.

Tisserand, D.J., H. Bosma, M.P. van Boxtel and J. Jolles (2001), "Head Size and Cognitive Ability in Nondemented Older Adults are Related", *Neurology*, Vol. 56, No. 7, pp. 969-971.

Tisserand, D.J., J.C. Pruessner, E.J. Sanz Arigita, M.P. van Boxtel, A.C. Evans, J. Jolles and H.B. Uylings (2002), "Regional Frontal Cortical Volumes Decrease Differentially in Aging: An MRI Study to Compare Volumetric Approaches and Voxel-based Morphometry", *Neuroimage*, Vol. 17, No. 2, pp. 657-669.

US National Research Council (1999), *How People Learn: Brain, Mind, Experience, and School*, J.D. Bransford, A.L. Brown and R.R. Cocking (eds.), National Academy Press, Washington DC.

Wallis, C., K. Dell and A. Park (2004), "What Makes Teens Tick; A Flood of Hormones, Sure", *Time Magazine*, 10 May.

Werker, J.F. and R.C. Tees (2002), "Cross-language Speech Perception: Evidence for Perceptual Re-organisation during the First Year of Life", *Infant Behavior and Development*, Vol. 25, pp. 121-133.

World Health Organization (2001), *World Health Report: Mental Health, New Understanding, New Hope*, WHO, Geneva.

ISBN 978-92-64-02912-5
Understanding the Brain: The Birth of a Learning Science
© OECD 2007

PART I

Chapter 3

The Impact of Environment on the Learning Brain

A person is not born a human, but becomes one.

Erasmus

This chapter assesses evidence from brain research to understand how the learning process is mediated by various environmental factors including social environment and interactions, nutrition, physical exercise, and sleep. It also examines the key areas of emotions and motivation, linking what is known through neuroscience to educational issues. Such information is particularly important for parents and teachers who have a prime role in the learning environment of children. It is also of relevance to policy makers who help to shape and maintain a desirable learning environment.

This chapter describes the conditions under which the brain can be further nurtured, and hence contribute to learning capabilities throughout the lifecycle. Findings emerging from brain research indicate how nurturing is important to the learning process and are beginning to provide policy makers and practitioners with new information on the appropriate timing for different types of learning. Many of the relevant approaches are everyday matters such as ensuring better mental and physical environments, which can be easily overlooked by educational policy. By conditioning our minds and bodies correctly, it is possible to take advantage of the brain's potential for plasticity and to facilitate learning processes. Continuing brain research in this area will contribute to the constructive review of curricula and teacher training.

While this chapter extracts common features that facilitate the learning process, the optimal process of nurturing the brain is not necessarily the same for everyone. The brain is an active plastic organ which adapts readily to its environment throughout life. Given the differences both in genetic characteristics and in the environment people face, brains are shaped differently. There is an interactive relationship between experience and brain structure: experience causes changes in brain structure, which in turn, influence the effects that subsequent experience has on the brain. Hence, it continually undergoes a process of experience-dependent reorganisation throughout life.

So, individual learning differences arise as a result of a continual and cumulative interaction between genetic factors and their environmental contexts. The environment influences the expression of genes relevant to learning throughout the lifespan, which expression results in structural changes in the brain. These modifications then affect subsequent experience-elicited genetic expression. In this way, each individual's brain accumulates structural idiosyncrasies which mediate learning processes. This means that it is difficult to prescribe one ideal learning environment for everyone – while the sensible nurturing of the brain will benefit all, it will not necessarily be equally effective.[1] The significance of the learning environment is profound, with individual differences importantly shaped by experience and by certain types of education and training. Future research agendas need to reflect on the types of learning that have the most potent impact on the brain and behaviour. It is necessary to recognise individual learning differences and to factor them in when considering how the brain can be nurtured at different stages in life.

The relatively recent development of neuroscience challenges the Cartesian tradition, which draws a sharp distinction between mind and body (Damasio, 1994). Physical health and bodily condition do have a direct influence on mental capabilities, and *vice versa*. This link should therefore be taken into account in educational practices, as well as the environmental factors that directly influence either physical or mental capabilities.

---

1. For example, the Dutch longitudinal ageing study in Maastricht has provided evidence that there are large individual differences in the deterioration of various memory functions and that low level of education influences performance much more than age.

Box 3.1. **Nutrition**

The importance of nutrition for health and human well-being is clear. It has direct implications for physical health and particularly for how well the brain functions. We can boost learning capacity through what we eat. For example, studies show that skipping breakfast interferes with cognition and learning. However, many students start school with an inadequate breakfast or no breakfast at all.

A landmark study undertaken in the United States examined the effects of school breakfast on academic performance among 1 023 low-income students from third through fifth grades. Results showed that children who participated in the study made significantly greater gains in overall standardised test scores and showed improvements in maths, reading and vocabulary scores. In addition, rates of absence and tardiness were reduced among participants (Meyers *et al.*, 1989). In Minnesota elementary schools, a three-year Universal School Breakfast Programme pilot study showed a general increase in composite maths and reading scores, improved student behaviour, reduced morning trips to the nurse and increased student attendance and test scores (Minnesota Department of Children, Families and Learning, 1998). Another study tested 29 school children throughout the morning on four successive days with a different breakfast each day (either cereal or glucose drink or no breakfast). A series of computerised tests of attention, working memory and episodic secondary memory was conducted prior to breakfast and again 30, 90, 150 and 210 minutes later. Having the glucose drink breakfast or no breakfast was followed by declines in attention and memory, but the declines were significantly reduced following a cereal breakfast. This study demonstrates that a typical breakfast of cereal rich in complex carbohydrates can help maintain mental performance over the morning (Wesnes *et al.*, 2003).

It is thus essential to take account of these nutritional needs throughout the day and to distribute nutritional intake to take account of these needs. Beyond this, 39 vital elements are not produced by the body and therefore need to be obtained from a dietary source (OECD, 2003b).

A recent finding has confirmed the benefits of the old-fashioned bane of children's daily diet – the spoonful of cod liver oil. This staple "potion", like other fish oils, is particularly rich in highly unsaturated fatty acids (HUFA), now commonly referred to as omega-3 fatty acids. They are particularly important for hormone balance and the immune system both of which are crucial for a healthy brain. In many modern diets, fatty acids have become relatively scarce, yet they are still essential to normal brain development and function. While it is necessary not to be swept along by faddish enthusiasm for omega-3 fatty acids before more extensive brain studies have provided evidence for such claims, a randomised controlled trial of dietary supplementation with omega-3 and omega-6 fatty acids *versus* a placebo was conducted on 117 children aged 5-12 years with Developmental Dyspraxia, also known as Developmental Co-ordination Disorder (DCD). Results showed that while no effect of the active treatment on motor skills was apparent, significant improvements were found in reading, spelling and behaviour over 3 months of treatment in parallel groups. The conclusion reached was that fatty acid supplementation may offer a safe and efficient treatment option for educational and behavioural problems in children with DCD (Richardson and Montgomery, 2005).

Another study was undertaken in prisons in the United Kingdom to test whether adequate intakes of vitamins, minerals and essential fatty acids caused a reduction in antisocial behaviour, including violence. This was indeed the case and is particularly relevant for those with poor diets (Gesch *et al.*, 2002).

Although scientific evidence shows that a diet rich in essential fatty acids and eating a good breakfast contribute to good health and improved learning, the clear, messages from this research have, to date, not been widely taken on board by policy to ensure its practical application. It is thus necessary to expand studies and apply such findings to the education domain. Promoting healthy behaviour among students should be a fundamental mission of schools: providing young people with the knowledge and skills they need to become healthy and productive adults. This will improve their capacity to learn; reduce absenteeism; and improve physical fitness and mental alertness. School administrators, school board members, teachers, social workers and parents should be encouraged to seek information and resources on the importance of nutrition for children's health and their academic performance.

Social interactions

There are social influences on the brain which have a direct impact on its ability to function optimally for learning. The importance of positive social influences on physiology and behaviour has been established.

In the past two decades, infants have been increasingly recognised as seekers and providers of social interaction and communication. Even though much of early learning appears to be automatic, it requires a naturally rich and stimulating environment in which social interaction is very important (Blakemore, Winston and Frith, 2004). A study of Romanian orphanages has shown that a lack of emotional nourishment can lead to attachment disorder (O'Connor, Bredenkamp and Rutter, 1999). Another study of children reared in an extreme social environment in which they were deprived of any of the normal care found that such deprivation can produce relatively permanent changes in a child's brain chemistry, impairing the production of hormones such as oxytocin[2] that are integral to bonding and social interaction. These findings support the view that there is a crucial role for early social experience in the development of the brain systems underlying key aspects of human social behaviour (Fries *et al.*, 2005).

The relatively new field of social cognitive neuroscience has emerged to study the brain in social context, exploring the neural mechanisms underlying social cognitive processes. Scanning the brains of people in tandem, whereby one subject does nothing but observe another doing something, is opening new windows into research on neuroscience and cognition. Brain imaging has shown that the act of observing movements in someone else affects the observer's peripheral motor system in the specific muscles that are used in the movements being observed (Fadiga *et al.*, 1995). Research using the so-called "mirror system" is seeking to identify the neurophysiological activity that underlies the ability to understand the meaning of one's own and another person's actions. This class of mechanism may be fundamental to several higher-level social processes, where the actions of others are interpreted in such a way as to directly influence one's own actions. This is the case in the attribution of intentions to others and oneself, and the ability to imitate as well as to teach others (Blakemore, Winston and Frith, 2004). Another exemplary study in this field looks at empathy by scanning couples under highly controlled conditions. As one partner suffered an electric shock, the other's brain was scanned as they anticipated the partner's pain, showing that the brain regions activated by the expectation of a loved one's pain overlapped with those activated by the experience of one's own pain (Singer *et al.*, 2004).

---

2. Oxytocin plays a crucial role in social situations and the regulation of emotional behaviour. Studies in animals have shown that increased levels are stimulated by pleasing sensory experiences, such as comforting touch or smell. As levels of this hormone rise, the animals form social bonds and attachments, and these experiences are embedded in their memories.

Studies aimed at understanding and observing others' actions should benefit learning and have an impact on education. Such studies are as yet in their infancy and more are needed into what type of learning requires the interaction of others, which emotions show up most in the brain, and the role played by cultural differences. Social neuroscience should illuminate the role of the teacher, identifying which methods impact learning most profoundly, and explore how exposure to the media influences learning. Since neurobiological research has revealed that the intellectual capacities of students are influenced by their overall health, it is important for parents, educators, social workers, and policy makers to recognise the key contribution made by the overall health of students to their educational achievement. The social factors shaping the emotional context of learning must also be taken into consideration.

The regulation of emotions

There is a set of structures in the centre of the brain collectively known as the limbic system (see Figure 3.1), an important part of which are the amygdala and the hippocampus.[3] Historically called the "emotional brain", this region has connections with the frontal cortex. In situations of excessive stress or intense fear, social judgment and cognitive performance suffer through compromise to emotional regulation, including responses to reward and risk.

Figure 3.1. **Inner structure of the human brain, including the limbic system**

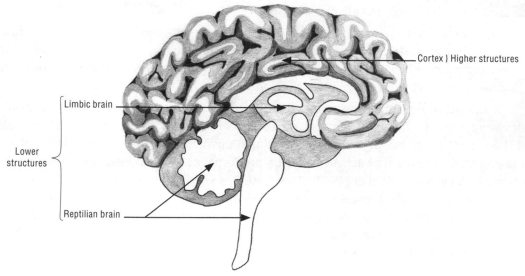

Source: Odile Pavot for the OECD.

---

3.   The word amygdala refers to a collection of neurons with the size and shape of an almond, hence the name. The word hippocampus refers to a collection of neurons with more or less the shape of a seahorse, hence the name.

Over two thousands years ago Plato declared "all learning has an emotional base", but only recently has evidence started to accumulate to show that our emotions do re-sculpt our neural tissue. Experts in the fields of neurobiology and education now view learning as a multifaceted exchange between cognitive, emotional and physiological elements. The distinction between the three is useful for analytical purposes but in reality they are intricately intertwined regarding both brain function and learning experience.[4]

Emotions are powerful and inevitable parts of life and learning. Managing one's emotions is one of the key skills of being an effective learner. Emotional regulation affects complex factors beyond the simple expression of emotion; emotions direct (or disrupt) psychological processes, such as the ability to focus attention, solve problems, and support relationships (Cole, Martin and Dennis, 2004). According to David Servan-Schreiber, "emotional competency or intelligence refers to one's ability to self-regulate that is, to restrain one's impulses and instincts, but also includes the capacity for compassion, and the ability to engage in co-operation" (OECD, 2002b).

"Emotional regulation" does not yet have a universally recognised definition:

- Thompson (1994) has proposed that it is: "… the extrinsic and intrinsic processes responsible for monitoring, evaluating, and modifying emotional reactions… to accomplish ones goals."

- Gros (2003) has defined it as the processes by which individuals influence which emotions they have, when they have them, and how they experience and express these emotions. In this definition, emotional regulation includes: coping, mood regulation, mood repair, defence, and affect regulation.

- Cole, Martin and Dennis (2004) define emotional self-regulation as the dynamic interaction of multiple behavioural, psychophysiological, attentional and affective systems which allow effective participation in the social world. This definition refers to a process distinct from emotional activation.

If emotional regulation can be measured independently of emotion itself, this should also be apparent in the brain. Neuroscientists have sought to study this by isolating the role of the prefrontal cortex in emotion processing (Ochsner et al., 2004; Eippert et al., 2006). Imaging studies show that activation of the prefrontal cortex is related to activity in the amygdala (Lewis and Stieben, 2004). These studies shed light on emotional regulation, including individual differences and developmental changes. In the past, research in this area has been primarily based on parent and teacher reports of emotional and behavioural processes in children. Brain research in this area will help to identify new ways of measuring and regulating emotions.

---

4. The following example illustrates how cognition and emotions interact: a teacher returns an exam face-down onto the desk of Ethan, a middle-school student. He flips the paper over to reveal an F staring back at him. This initiates a tango between brain regions affiliated with cognitive and emotional processes. Ethan recruits cortical structures to appraise the situation: this grade will thwart his goals to do well in the class and to convince his mother that he deserves the latest snowboard for his upcoming birthday. Almost simultaneously, his limbic (i.e., amygdala) and paralimbic (i.e., insular cortex) brain regions launch an emotional response and he begins to experience negative emotions. However, Ethan begins to cognitively regulate the situation: he depersonalises it by characterising the exam as difficult for everyone, and reinterprets it as a test that is only a partial contribution to his final grade. These regulatory strategies are reflected in both an increase in brain regions implicated in cognitive control and an attenuated response in brain areas affiliated with negative emotional response (i.e., amygdala). Regulation moderates the emotional reaction, which reinforces a cognitive appraisal that the situation is within his coping potential. This cortically-driven assessment merges with the ongoing experience and further cools the emotional reaction.

Box 3.2. **Attention as seen through the neuroscientific lens as an organ system**

Attention has always been an important topic in psychology and education because it deals with the mechanisms of subjective experience and voluntary control. Only since the advent of neuroimaging has it been possible to view attention as an organ system with its own anatomy. Imaging studies have shown that the various functions of attention such as maintaining the alert state, orienting to sensory information and resolving conflict among competing thoughts or feelings are carried out by distinct networks of neural areas. What do we mean by a network? Imaging of human task performance has shown separate brain areas must be orchestrated in even the simplest task. Each of these areas may be performing a different computation, which taken together allow performance of the task. We regard the set of activations and their connections as the network that underlies task performance.

Attentional networks are special in that their primary purpose is to influence the operation of other brain networks. Although the sites at which the influence of attentional networks can be demonstrated involve most any brain area including primary sensory, limbic system and motor cortex, the sources of these activations are much more limited.

Orienting to sensory events has been the most thoroughly studied of these networks. The convergence on the set of brain areas serving as the source of the amplification of sensory signals has been impressive. It is widely agreed that the frontal eye fields work in conjunction with superior and inferior parietal areas as the cortical nodes of the orienting network. In addition, studies have implicated some subcortical areas including the pulvinar of the thalamus and the superior colliculus. Most of the studies of this network have involved visual stimuli, but the sources of the attention influences in orienting to other modalities are much the same. Of course, the site of amplification of the sensory message is quite different for each of the sensory modalities.

Evidence to date suggests that both maintained alertness during task performance (tonic) and phasic changes induced by a warning signal involve a subcortical structure – the locus coeruleus – that is the source of the brain's norepinephrine. A great deal of evidence indicates that the tonic state depends upon an intact right cerebral hemisphere. Lesions in this hemisphere can produce profound difficulty in responding to unexpected targets. Imaging studies suggest that warning signals may have their influence more strongly on the left cerebral hemisphere.

Tasks that involve conflict between stimulus dimensions competing for control of the output often provide activation in the anterior cingulate gyrus and lateral prefrontal areas. It is thought that the conflict, induced by a stimulus, is representative of situations where different neural networks compete for control of consciousness or output. Because of this, the term "executive attention network" has been used because the network regulates the activity in other brain networks involved in thought and emotion. This network shows a strong development in childhood and its maturation is related to what in developmental psychology has been called self-regulation. Thus this network is especially critical for the ability to succeed in school.

Individual differences are invariably found in cognitive tasks involving attention. The attention network test was developed to assay the efficiency of each of the three networks. Research to date has suggested that each network has its own distinct anatomy, dominant chemical neuromodulators, and time course of development. Recent studies have explored how different alleles of specific genes contribute to network efficiency and there have also been several studies designed to explore training of the networks.

The view of attention as an organ system gives the promise of better understanding of the many forms of brain injury and pathology that involve attentional problems.

Source: Michael Posner, University of Oregon.

Emotional states induced by fear or stress directly affect learning and memory. Brain studies have illuminated how negative emotions block learning and have identified the amygdala, the hippocampus and stress hormones (glucocorticoids, epinephrine and norepinephrine), as playing a crucial role in mediating the effects of negative emotions such as fear and stress on learning and memory. Simultaneous bodily events such as an increased heart rate, perspiration, and elevated adrenaline levels also occur (Damasio, 1994; LeDoux, 2000), and in turn influence cortical activity. Some level of stress is essential for optimal adaptation to environmental challenges and can lead to better cognition and learning, but beyond this modicum it can be damaging, both physically and mentally.

Stress is provoked by strong demands on motor or cognitive systems, which are felt at the emotional level. For example, if one meets a masked gunman in the street, which is normally a dangerous stressful situation, the brain quickly registers the danger and prepares the cognitive and motor systems for survival. A set of reactions to the gunman (stressor) make up the stress response: vigilance and attention increase and the body prepares itself for fight or flight; heart rate and blood pressure rise. At the same time, digestion, growth and reproduction processes are slowed down, as these functions are not necessary for immediate survival and can be delayed without damage to the organism. This stress response is mediated by hormones, epinephrine and norepinephrine, and are secreted within milliseconds. The secretion of cortisol follows seconds later. These hormones act on the brain and thereby modulate cognition, notably by influencing learning and memory.[5] A masked gunman is thankfully very rare in the classroom. But other less extreme sources of stress can have a parallel impact – for instance, aggressive teachers, bullying students, or incomprehensible learning materials whether books or computers. If students are faced with situations that trigger fear or stress, their cognitive functions are affected.

Further investigation using psychological, neuropharmacological and neuroimaging studies is needed of the neurobiological mechanisms which underlie the effects of stress on learning and memory and the factors which can reduce and regulate it. Identifying preventive measures against harmful stress requires investigation into how people deal best with stressors and thereby maintain or even improve their cognitive performance. One known preventive measure could be physical activity (see Box 3.3). In a recent study, it was found that elite sportsmen exhibit a lower psychological stress response (less anxiety, greater calmness) as well as lower physiological (measured by cortisol level) stress response to a psycho-social stressor[6] (Rimmele et al., 2007b). Greater knowledge of neurobiological mechanisms of memory modulation by stress hormones and the amygdala and their possible physiological and cognitive manipulations would be clearly relevant to education. This is especially true in societies where the pressures to succeed are intense and negative influences such as through the media can cause serious problems with learning as well as with the emotional stability of children (if not adults!).

5. From research on levels of the stress hormone cortisol, it can be hypothesised – in accordance with findings from animal studies – that low and medium cortisol levels improve learning and enhance memory, whereas high levels of cortisol have a deleterious effect on learning and memory (McEwen and Sapolsky, 1995). This hypothesis is further supported by the finding that extremely or chronically elevated cortisol levels, as they occur in some diseases or under prolonged stress, lead to cognitive deficits and memory impairment in animals as well as in humans (McEwen and Sapolsky, 1995).
6. It is a further question whether these effects in high-level athletes are transferable to others who are not professional sportspersons.

Box 3.3. **Physical exercise**

Recently, it has been shown that the potential benefits of aerobic exercise extend beyond the cardiovascular health marker to improving brain health. There is clear evidence on this from a brain study undertaken to monitor the effects on cognitive capacities of an aerobic exercise programme in older adults. Research carried out by Arthur Kramer, who undertook a programme involving the gradual increase of periods of walking over three months, resulted in boosted performance in key areas of the brains in older adults (age 55+) subjects (Colcombe and Kramer, 2004). After six months, the subjects were scanned and given mental tests that showed an 11% improvement score, as well as changes in brain function in the middle frontal and the superior parietal regions, areas linked to keeping the brain focused on a particular task and to spatial attention. The results provide a biological basis for arguing the benefits of exercise on the brain health of older adults. Although this research explicitly targeted relatively high-functioning older adults with no known clinical syndromes, a meta-analysis of the cognitive performance of older adults found that increases in aerobic fitness were identical for both clinical and non-clinical populations.

Further research in this area might help to determine whether there is a need to increase aerobic activity in school curricula. Current research in this area is being undertaken by the Learning Lab in Denmark's Play and Learning Consortium (see Box 7.3), which has been exploring the relationship between body, mind, cognition and learning, advocating a comprehensive approach which incorporates physical activities into other teaching disciplines and not just physical training classes.

There is evidence that physical activity leads to an improvement of motor co-ordination and control (*e.g.* balance, overall motor co-ordination, specific motor skills and body awareness) with implications for addressing learning problems and attention disorders (Rudel, 1985; Nicolson, Fawcett and Dean, 1995; Roth and Winter, 1994). Several studies have also shown a clear relation between different aspects of motor and language development (Ruoho, 1990; Rintala *et al.*, 1998; Moser, 2001). Moser suggests that these arguments for the beneficial impact of exercise on cognition may also be understood as contributing to a broader argument about the need to better recognise the body and physical activity for all kinds of educational programmes, leisure time activities and for everyday life in general (Moser, 2004). At school, this is often confined to physical education classes and not sufficiently integrated into other parts of the curriculum.

Understanding the fundamental mechanisms by which exercise can affect brain function will come to inform understanding of human cognitive health, including how educational curricula might include exercise programmes to enhance learning. In the meantime, even a factor quality ventilation can make a positive difference: opening the windows from time to time and taking a pause to stretch and breathe in fresh air is beneficial to the performance of students.

Brain research, drawing on cognitive psychology and child development research, has been able to identify a critical brain region whose activity and development relate to the performance and development of self-control.[7] Self-regulation is also one of the most important behavioural and emotional skills that children need in their social environments. The capacity to control one's own impulses in order to delay gratification is also an important part of emotional regulation. The self-regulation of emotions in the social context of the educational setting is an important stepping stone in the development of a child to becoming a responsible and successful adult.[8]

A growing research base lends support to the possibility of emotional regulation as a – perhaps the – critical component of emotional competence necessary for effective interactions with others in stressful situations. How children learn to cope with stressful, negative interactions includes not only dealing with their own feeling of distress and anger, but also their reactions to the negative emotions of others; this is vital to the development and maintenance of social relationships (Eisenberg and Fabes, 1992). To date, much of the research into the development of emotional self-regulation has focused on infancy and early childhood, primarily because of the dramatic cognitive maturation which takes place during these years (Calkins, 2004). In early childhood, learning to deal with social interactions outside of the family and with peers is especially important. Social competence is important for children, and is a predictor of social and academic outcomes, such as school readiness (Carlton and Winsler, 1999). In a study of preschoolers' social competence, the emergence of emotional regulation is shown as vital to the creation and maintenance of positive relations with peers (Denham and Burton, 2003).

7. For example, one classic experiment to measure cognitive control is the "Stroop task" in which the subject is shown words that name colours, which are printed in ink that is either the same as the name (e.g. the word "red" in red ink) or different (e.g. the word "red" in blue ink). The subject is asked to say aloud the colour of the ink, which is much harder if the word names a different colour than if it names the same colour. Performance in Stroop-like tasks tends to activate a very specific region of the brain called the anterior cingulated, situated on the frontal midline just behind the orbito-frontal cortex. The anterior cingulated plays a critical role in the brain networks that are responsible for detecting errors and for regulating not only the cognitive processes (such as in the Stroop task), but also emotions, in order to achieve intentional or voluntary control of behaviour (see OECD, 2002a).

8. Neuroimaging studies have revealed that at least two-thirds of the same brain areas are activated during visual imagery and visual perception. Mental images of objects and events can engage much of the same processing that occurs during the corresponding perceptual experience. Visual imagery of aversive stimuli (e.g., burned bodies or the face of a battered person) causes skin conductance changes and changes in heart-rate. Hence, mental images affect the body. Imagery of aversive stimuli activate some brain areas more than do neutral stimuli (e.g., picture of a lamp or chair). Among the areas activated is the anterior insula, which is involved in registering the state of autonomic activity in the body. These findings suggest that people can alter their emotional state by forming specific mental images. Some researchers have claimed that such procedures can affect a host of bodily functions, including those of the endocrine and immune systems. This line of neuroimaging research could be taken further to test mental imagery techniques to overcome anxiety, create a positive learning environment or reinforce learning, which could give insights to practical applications within educational settings.

Michael Posner (OECD, 2002b) refers to the concept of "effortful control" which is the children's capacity to self-regulate their behaviour both in school and at home.[9] Effortful control can be assessed by synthesising answers from parents' questions about their child's concentration on activity (focused attention), their exercise of restraint (inhibitory control), their enjoyment of low-intensity stimulation (low-intensity pleasure), and their awareness of subtle changes in their environment.

Emotions and physiology are thus intimately connected to each other, as indeed are all the different main elements which make up the mind and the body. As a result, it should be possible to facilitate learning which is sensitive to emotions through focusing on physiological factors, though, given the complexity of the interconnections, we would not expect this to be a simple matter of cause and effect. By way of example, training heart-rate patterns to become regular has physical and psychological benefits for emotional regulation.[10]

By understanding neurofunctional mechanisms and processes, educational programmes could be devised to help develop emotional intelligence and so enhance the learning capacity of the brain. Understanding brain maturation and emotions will contribute to defining age-appropriate strategies for emotional regulation. Parents can foster a reassuring and stable emotional environment by helping their children to understand and express their emotions. Another reason to explore the brain processes through which children regulate their emotions is that it will help in the identification of emotional disorders and how to predict or prevent them.

9. A longitudinal study has illustrated the connection between delayed gratification and education. Four-year-old children were faced with the task of resisting eating one marshmallow displayed before them as they were alone in a (otherwise empty) room in order to get two marshmallows later on return of the experimenter. The time delay during which the child succeeded in resisting the impulse to eat the first marshmallow turned out to be significantly correlated with the achievement of later academic success as measured by the ability to deal with frustration, stress, task perseverance, and concentration (OECD, 2002a).

10. An emotional literacy pilot programme in Southampton, England, is using these principles to regulate the heart rate through teaching rhythmic breathing as part of a sequential set of interventions which work to stabilise the individual's physiological state in order to attain a coherent emotional state (OECD, 2003b). Other programmes work more explicitly on the problems of violence which are escalating in schools today. A non-violent communication process for example has been developed by Rosenberg (1999) and is popular in many countries, which has helped to found schools based on this teaching method. The Rosenberg methods work principally on making individuals aware of their basic needs and helping them to articulate these needs by improving communication, lack of which is seen as a key cause of violent behaviour. In China, millions of teenagers are suffering from different learning and behaviour disorders presumably linked to deficiencies in emotional development; the Chinese government seeks to encourage emotional competency so as to improve balance and quality of life and to cultivate positive social interactions (OECD, 2002c).

> ### Box 3.4. **Music**
>
> Playing music requires motor skills and co-ordination between auditory input and motor control. It also requires the brain to interpret somatosensory touch. Most musicians develop a greater ability than the average person to use both hands. One might expect that this increased co-ordination between the motor regions of the two hemispheres has an anatomical consequence and indeed, the anterior corpus callosum, which contains the band of fibres that interconnects the two motor areas, is larger in musicians than in non-musicians (Weinberger, 2004).
>
> Music also affects some of the brain's learning capacities, increasing the size of the auditory and motor cortex. One way the brain stores the learned importance of a stimulus is by devoting more brain cells to the processing of that stimulus. Studies on musicians have shown that they have additional specialisations, particularly the hyper-development of certain brain structures, but it is not yet clear how far such benefits are transferable to learning other skills. Learning retunes the brain, increasing both the responses of individual cells and the number of cells that react strongly to sounds that become important to an individual (Weinberger, 2004). A study using magnetoencephalography (MEG) in Germany has revealed that a much larger area of the auditory cortex was activated in musicians on hearing a piano tone, compared with non-musicians, and that the earlier the age at which the musician had begun to practice, the greater the enlargement (Pantev *et al.*, 1998). Research has also revealed how areas of the motor cortex corresponding specifically to the fingers of the left hand show an enhanced electrical response in violin players (Pantev, 2003). There are enlarged motor and auditory structures in musicians, indicating that prolonged periods of training can alter the underlying structure of the nervous system (Schlaug, 2003).
>
> The focus on music can provide valuable information about how the brain works. Research is uncovering how the brain has distinct circuits for perceiving, processing and playing music. Playing, listening to and creating music involve practically every cognitive function. Having already demonstrated the effect of positive emotions on learning, music might also create a positive effect to facilitate learning. This is another topic for further research, including the transferable benefits of music learning.

Motivation

> There is much pleasure to be gained from useless knowledge.
>
> *Bertrand Russell*

Motivation is crucial to successful learning and it is closely linked with understanding and emotions. Motivation can be described as the resultant force of emotional components and reflects the extent to which an organism is prepared to act physically and mentally in a focussed manner. Accordingly, motivation is intimately related to emotions as the latter constitute the brain's way of evaluating whether things should be acted upon – to be approached if pleasant and avoided if unpleasant. Therefore, it is possible to formulate the hypothesis that emotional systems create motivation.

A fundamental distinction can be drawn between *extrinsic motivation* (linked to external factors) and *intrinsic motivation* (linked to internal factors). Whereas extrinsic motivation is achieved by affecting behaviour from the outside – for example, through rewards and punishment both objective or symbolic (McGraw, 1978) – intrinsic motivation

---

Box 3.5. **Play**

Play can make a big difference in terms of motivation, whether it be story telling or something that captures the imagination in a teaching setting. Researchers have managed to motivate children successfully to do standardised tests that consist of multiple trials and which are otherwise regarded as excessively boring by the subject, by making them into games. Children are more responsive to and motivated by traditional game props such as puppets. Teachers can transform common learning barriers – oppositional behaviour, negative moods, defensive attitudes – by creating a positive learning environment through such games.

A study using near-infrared Optical Topography to "image the brain at play" showed a significant increase in cortical blood volume during puppet play as compared to the performance of a similar activity in a routine manner (Peyton et al., 2005). In a study undertaken by Nussbaum *et al.* (1999) on 300 school children using a series of games based on Gameboy, results showed a high motivation on the part of the children, both those familiar with the technology and those who had no access to it outside of the school. McFarlane, Sparrowhawk and Heald (2002), studying teachers' opinions on the limits and potential of video games, found a positive attitude from teachers to adventure games and simulations, with most teachers acknowledging that games contribute to the development of a wide variety of strategies that are extremely important to learning (Gros, 2003).

The MIT Media Laboratory invented Sam, an embodied conversational character designed to support children's language development. Sam allows children to engage in natural storytelling play with real objects, in collaboration with a virtual playmate who shares access to those real objects. This programme enhances story listening through a computer interface enabling children to tell stories that are more imaginative and have more complex narratives than when they play by themselves (Cassell *et al.*, 2000).

So, not only do games motivate, but they can help students develop their imagination and an active approach which impacts on skills, abilities and strategies. However, many schools have cut back on or eliminated recess and an increasing emphasis on testing leaves less time for recreational, playful activities. Parents also share part of the responsibility if they enrol their children in extramural academic-related activities to boost their scores, if this leaves the children little or no time for play. In a world where classrooms are increasingly consumed by a strictly academic mission, play in the classroom risks to be an exception not the rule.

---

reflects wishes to fulfil internal needs and desires. Traditional education systems focus on extrinsic motivation through punishment and reward; neuroscientific research to date has also concentrated on extrinsic motivation in the context of learning since the mechanisms of internal drives are not well understood and are presently difficult to study with neuroimaging techniques. But since much learning depends on intrinsic motivation rather than external factors, neuroscience will need to address the intrinsic motivational system.

When playing in childhood most people will have experienced the intrinsic motivation underlying effective learning. Many will have retained the ability to reach what Csikszentmihalyi (1990) describes as "flow"[11] – the state when really engaged in pursuits which afford us fundamental pleasure with no promise of external reward. Of the many

---

11. Csikszentmihalyi (1990) describes "flow" as a mental state in which a person is intrisically motivated to learn, characterised by energised focus, full involvement and optimal fulfilment.

Box 3.6. **Video games**

Surprisingly little is known about the reasons for and the effects of video games exposure on children's health, cognitive, social, and behavioural outcomes; research has not kept pace with the rapid evolution of the technology itself.

In Ulm, Germany, at the Transfer Centre for Neuroscience and Learning (Box 7.7), researchers have begun investigating the influence of the consumption of media on the psychological and physical well-being of children. Similarly, studies from the University of Rochester have found that young adults who spend a lot of time playing fast-paced video games showed better visual skills than those who did not, with video gamers displaying a pronounced improvement in their ability to pay attention to complex visual environments. They also "could keep better track of more objects simultaneously and process fast-changing visual information more efficiently" (Green and Bavelier, 2003). The current generation of computer games emphasises spatial and dynamic imagery and requires the player to simultaneously maintain attention around different parts of the screen, improving visual and attention skills.

The prevalence of violence in most popular computer games, however, cannot be ignored. Some studies have shown that an overload of emotionally-charged imagery can increase anti-social behaviour (Anderson, 2004). The emotional responses evoked by these games – danger, violence, a sense of challenge – all play a role in the highly charged emotional appeal that these games offer. Are games harmless, or are they teaching children to be more aggressive, or worse in extreme cases? Those opposed to the games point to new studies suggesting a direct link between violent games and aggressive behaviour and some researchers claim that playing a video game triggers the same violent responses in the brain as actual aggression. For example, a study from the University of Aachen, Germany, asked men to play a game which required them to kill terrorists in order to rescue hostages. During the game, the parts of the brain that deal with emotion, including the amygdala and the anterior cingulate cortex, were shut down (Weber *et al.*, 2006). The same pattern has been seen in brain scans of people during acts of imagined aggression. Birbaumer of the University of Tübingen, Germany, has proposed that someone playing violent video games regularly would have these circuits in the brain strengthened so that, confronted with a similar real-life situation, would be more primed for aggression (Motluk, 2005).

Even though the messages from research thus far are mixed on whether playing computer games is beneficial or detrimental to children, playing games does at least help them to improve their computer literacy, something vital in today's society. More studies are needed to evaluate the impact of media on children's cognitive and emotional development. Modern technology should be utilised to: simulate real experience by using the virtual; allow private practice to avoid the emotional distress of making mistakes in front of peers; and, encourage an element of play and motivation. Better understanding in this area would be useful for school policy makers and practitioners alike – beyond its recognised importance in early childhood education – and help to establish play as a "natural learning resource" capable of ameliorating many systemic problems challenging education today.

triggers that motivate people to learn, including the desire for approval and recognition, one of most (if not the most) powerful is the illumination which comes from *understanding*. The brain responds very well to this, which happens for instance during the "eurêka" moment, when the brain suddenly makes connections and sees patterns between the available information.[12] It is the most intense pleasure the brain can experience, at least in a learning context, so that it can be described as an "intellectual orgasm" (B. della Chiesa). Having had the experience, you want to have it again. A primary goal of early education should be to ensure that children have this "experience" as early as possible and so become aware of how pleasurable learning can be.

Although neuroscience is making some headway to understanding some of the motivational processes that drive learning, more work is needed in order to link it to an educational framework. Most agree that school is not always fun, and some would argue that it acts as a constraint and can in fact be de-motivating. To address this it will help to understand how the brain becomes motivated in the interaction between the inside and outside influences. The challenge is to find out how to give purpose to learning and how to encourage the internal drive to want to learn. This is an area where focused neuroscientific research could yield an important direct benefit for education.

Sleep and learning

The function of sleep has always fascinated scientists, while remaining something of a biological mystery. Fundamental questions about it remain to be clarified. From a neurophysiological point of view, it is a specific state of brain alertness.[13] Brain studies are not necessary for us to know that adequate sleep is necessary in order for people to remain alert and awake: clearly, the functions of sleep are essential to life.[14] Researchers are unanimous that while many bodily functions can recover during wakefulness, only sleep can restore cortical functions (Horne, 2000). Sleep quality is closely related to well-being, poor sleep can also have a negative impact on mood (Poelstra, 1984) and behaviour (Dahl and Puig-Antich, 1990). Latent sleep disorders can in some cases result in psychological symptoms (Reite, 1998). In adults, daytime sleepiness is related to impairments in work and social life, increased disorders, and increases risks such as of motor vehicle accidents (Ohayon *et al.*, 1997).

Studies, from the behavioural to the molecular level, suggest that sleep contributes to memory formation in humans and other mammals (Maquet, 2001). Sleep was first implicated in learning and neural plasticity with studies performed on animals, which showed a correlation between the amount of rapid eye movement (REM) sleep and performance on a learned task (Smith, 1996).[15] Recent studies in humans provided evidence for a critical involvement of slow wave sleep and associated slow EEG oscillations in the consolidation of memories and underlying neural plasticity (Huber *et al.*, 2004, Marshall *et al.*, 2006). Approaches

12. Peter Gärdenfors of the University of Lund (Sweden), during a conference in Copenhagen organised by CERI and Learning Lab Denmark in November 2004, identified the experience "Eurêka!" as corresponding to the moment of comprehension understood as "pattern recognition". An analogy with a game for children can help convey the point. The game is one in which a series of seemingly random dots are joined one by one. Progressively, and following the number order of each dot beginning with 1, the child sees the apparently chaotic maze of dots become the representation of a recognisable object. Put another way, "to understand is to transform information into knowledge" (B. della Chiesa).

13. Only in 1953 did research pioneer Nathaniel Kleitman overthrow the commonly-held belief that sleep was simply a cessation of all brain activity (Siegel, 2003).

14. Sleep rebound occurs after sleep loss, and chronic sleep deprivation (two to three weeks) ultimately causes death in rats (Miyamoto and Hensch, 2003).

15. REM sleep is the phase where dreams take place, during which the brain shows patterns of activity very similar to when awake. There is, to date, no conclusive agreement on what the function of REM sleep is. A relationship between REM sleep and procedural learning has been researched and argued by scientists, but there remains no firm correlation between the proportion of sleep spent in REM and learning ability (Nelson, 2004).

involving human functional imaging (the recording of activity from larger neuronal networks) and genetic or pharmacological manipulation of the brain have converged to support the notion that the stages of sleep (slow wave sleep and REM sleep) function in concert to reprocess recent memory traces and consolidate memory, and this across different species and different learning tasks (Stickgold, 2003). Whereas REM sleep seems to benefit particularly the consolidation of skill memories, slow wave sleep enhances particularly the consolidation of explicit declarative memories depending on the hippocampus. Numerous sleep deprivation studies support the idea that sleep contributes to the stabilisation of acquired memory. Evidence from experiments in animals and humans support the concept of an "offline" reprocessing of recent experiences during sleep that is causative for memory consolidation (Ji and Wilson, 2007; Rasch et al., 2007), and analysis of the thalamocortical system establishes the reciprocal observation that sleep is itself a plastic process affected by waking experience (Miyamoto and Hensch, 2003). One hypothesis is that sleep plays a key role in neural plasticity, i.e. in maintaining appropriate connections between neurons through reinforcing significant connections between synapses and eliminating accidental ones. It has been proposed that the entire cortex experiences neural plasticity in sleep, as it "updates" following experiences of the world, especially the previous day's events (Kavanau, 1997).

Children's sleep disturbances have been linked with numerous somatic disorders, neurological illnesses and emotional and behavioural disturbances such as hyperactivity, as well as learning difficulties (Ferber and Kryger, 1995). Sleep disturbances are highly prevalent and persistent and are among the most common complaints throughout childhood: epidemiological studies have shown that approximately one third of all children suffer from sleep problems (Simonds and Parraga, 1984; Kahn et al., 1989; Blader et al., 1997; Rona, Gulliford and Chinn, 1998). A survey of clinical paediatricians suggested that they are the fifth leading concern of parents (following illness, feeding, behavioural problems and physical abnormalities; Mindell et al., 1994).

Although there are common sleeping disorders at all ages (Wiggs and Stores, 2001), there are also age-specific patterns as with the changes which happen during adolescence. A questionnaire-based survey of sleeping habits of 25 000 people between ages 10-90 shows that children are typically early risers, but start to sleep progressively later as they enter adolescence (see Part II, Article B), reaching a maximum lateness around the age of 20, when the curve starts to decline (Abbott, 2005). Generally, individuals have increased daytime sleepiness at puberty, whether or not there are changes in total sleep duration, suggesting that the biological need for sleep does not diminish during adolescence (Carskadon et al., 1980).

Some studies tentatively suggest that sleep deprivation and sleeping problems are associated with poorer academic performance: the less they sleep, the lower their performance (Wolfson and Carskadon, 1998). Since many children suffer from chronic sleep deprivation there is a very real concern about the potentially harmful effects that this has for the developing brain. While experimental sleep deprivation studies on children are rare for ethical reasons, those carried out have investigated cognitive consequences of sleep deprivation. An early study showed that one night's complete sleep loss had similar effects in children aged 11-14 years to those previously shown in adults (Carskadon, Harvey and Dement, 1981). Partial sleep restriction was found to lead to some impaired cognitive functions. Routine performance, on the other hand, was maintained even after one full night's sleep restriction (Randazzo et al., 1998). Shorter sleep duration has been demonstrated to lead to poorer performance on short-term memory tasks (Steenari et al., 2003).

In recent years, an increasing number of studies have reported associations between children's sleep disturbances and various psychological symptoms, including depression and behavioural problems (Morrison, McGee and Stanton, 1992; Chervin *et al.*, 1997; Dagan *et al.* 1997; Corkum, Tannock and Moldofsky, 1998; Dahl, 1998; Marcotte *et al.*, 1998; Aronen *et al.*, 2000; Smedje, Broman and Hetta, 2001). Attention Deficit Hyperactivity Disorder (ADHD) is a neuropsychological disorder in which sleep disturbance frequently occurs. Several studies have reported increased rates of sleep problems among children with ADHD (Chervin *et al.*, 1997; Marcotte *et al.*, 1998; Stein, 1999; Owens *et al.*, 2000a). Children with ADHD, compared with the control children, have higher parent-reported bedtime resistance, sleep-onset problems, sleep-related anxiety, daytime sleepiness, parasomnias and shorter sleep duration (Owens *et al.*, 2000b). Certain environmental factors have also been shown to be related to sleep disturbances. For example, high quantities of television-viewing, particularly at bedtime, have adverse effects on sleep (Owens *et al.*, 2000a). Moreover bedtime resistance (Blader *et al.*, 1997; Smedje, Broman and Hetta, 1998) as well as sleeping in the same bed as parents have been correlated with sleep-onset problems (Lozoff, Wolf and Davis, 1984; Madansky and Edelbrock, 1990; Latz, Wolf and Lozoff, 1999).

Box 3.7. **Sound pressure level**

The ear plays a key role in the classroom: information continually passes through this channel. Loud noise levels affect children's school performance. A study at London Southbank University measured noise levels outside schools during lessons in 142 primary schools in London and compared these against the published results of standard national exams taken by pupils aged 7 and 11. The results showed that the noisier the environment, the worse the results, even controlling for other possible causes of poor performance, such as socioeconomic factors (Shield and Dockrell, 2004).

According to the World Health Organisation's (WHO) Guidelines for Community Noise, the critical effects of noise for schools are speech interference, impaired comprehension and message communication, and annoyance. The guidelines for classroom noise levels stipulate that, to be able to hear and understand spoken messages in a classroom, the background sound pressure level should not exceed 35 dB during teaching sessions. For children with hearing impairments, even lower sound pressure levels may be needed, and for outdoor playgrounds, the sound pressure level of the noise from external sources should not exceed 55 dB which is the same value given for outdoor residential areas in daytime.

Any discussion of deprived or reduced sleep raises the question "how much is enough?". Because individual differences in sleep requirements are large, it is impossible to give simple guidelines that would suit every person.[16] Early school starting time has been associated with increased sleep deprivation and daytime sleepiness and poorer school performance (Carskadon *et al.*, 1998). The children involved complain significantly more about being tired throughout the day and having attention and concentration difficulties than those with a later school start time (Epstein, Chillag and Lavie, 1998). It remains debatable whether school start times are too early: more studies would be needed

---

16. Existing evidence indicates that "helping" people to increase sleep time with long-term use of sleeping pills produces no clear-cut health benefit and may actually shorten the lifespan (Siegel, 2003; see also Chapter 7).

before any firm conclusions could be made about their effects, together with experiments to better understand the relationship between sleep and learning (such as memory consolidation). There is also evidence that having a nap after learning a task appears to improve performance so that the common expression "let's sleep on it" is not an example of a proverbial neuromyth (see Chapter 6). Robert Stickgold (2003) performed studies on a group of students at Harvard University, showing that performance on a complex task requiring a great deal of attention and concentration could be restored to the levels observed early in the experiment by subjects taking a nap of between 30-60 minutes.

As sleep deprivation appears to be prevalent among children, more studies to screen for sleep disturbances and more experimental studies would help to ascertain their association with psychological symptoms and diminished cognitive performance. Longitudinal studies would also be able to shed light on the evolution of sleep needs at different stages of life and how much is needed at these stages in order to maintain a healthy learning brain and emotional well-being. Further studies on the mechanisms and evolution of sleep will usefully focus on what is repaired in the brain during sleep, which learning processes benefit most from it, and just how much sleep is necessary. The effectiveness of school lessons, workplace training sessions or conferences could increase if their scheduling and planning took account of the scientific findings on sleep. Educators might, for example, schedule lessons for adolescents later in the day and advise students on the benefits of recapping lessons after a night's sleep. Parents could play a valuable role in helping to nurture their children's brains by ensuring that they get enough sleep and, before bedtime, avoiding activities that hype up the brain, such as computer games.

Conclusions

Findings from brain research indicate how nurturing is crucial to the learning process, and are beginning to provide indication of appropriate learning environments. Many of the environmental factors conducive to improved brain functioning are everyday matters – the quality of social environment and interactions, nutrition, physical exercise, and sleep – which may seem too obvious and so easily overlooked in their impact on education. They call for holistic approaches which recognise the close interdependence of physical and intellectual well-being and the close interplay of the emotional and cognitive.

Evidence is now accumulating that our emotions re-sculpt neural tissue. In situations of excessive stress or intense fear, social judgment and cognitive performance suffer through compromise to the neural processes of emotional regulation. Some stress is essential to meet challenges but beyond a certain level it has the opposite effect. Concerning positive emotions, one of most powerful triggers that motivates people to learn is the illumination that comes with the grasp of new concepts – the brain responds very well to this. A primary goal of early education should be to ensure that children have this experience of "enlightenment" as early as possible and become aware of just how pleasurable learning can be. Managing one's emotions is key to being an effective learner; self-regulation is one of the most important skills that children and older people need in their social environments.

References

Abbott, A. (2005), "Physiology: An End to Adolescence", *Nature*, Vol. 433, No. 7021, pp. 27.

Anderson, C. (2004), "Violence in the Media: Its Effects on Children", An edited transcript of a seminar presented in Melbourne, Australia, Young Media Australia, Glenelg, South Australia and the Victorian Parenting Centre, Melbourne, Victoria, 11 September.

Aronen, E.T., E.J. Paavonen, M. Fjällberg, M. Soininen and J. Törrönen (2000), "Sleep and Psychiatric Symptoms in School-age Children", *Journal of the American Academy of Child and Adolescent Psychiatry*, Vol. 39, pp. 502-508.

Blader, J.C., H.S. Koplewicz, H. Abikoff and C. Foley (1997), "Sleep Problems of Elementary School Children: A Community Survey", *Archives of Pediatrics and Adolescent Medicine*, Vol. 151, pp. 473-480.

Blakemore, S.J., J. Winston and U. Frith (2004), "Social Neuroscience: Where Are We Heading?", *TRENDS in Cognitive Sciences*, Vol. 8, No. 5, pp. 216-222.

Calkins, S.D. (2004), "Temperament and Emotional Regulation: Multiple Models of Early Development", in Mario Beauregard (ed.), *Consciousness, Emotional Self-Regulation and the Brain. Advances in Consciousness Research 54*, John Benjamins Publishing Company, Amsterdam, pp. 35-39.

Carlton, M.P. and A. Winsler (1999), "School Readiness: The Need for a Paradigm Shift", *School Psychology Review*, Vol. 28, No. 3, pp. 338-352.

Carskadon, M.A. and C. Acebo (2002), "Regulation of Sleepiness in Adolescence: Update, Insights, and Speculation", *Sleep*, Vol. 25, pp. 606-614.

Carskadon, M.A., K. Harvey and W.C. Dement (1981), "Sleep Loss in Young Adolescents", *Sleep*, Vol. 4, pp. 299-312.

Carskadon, M.A., K. Harvey, P. Duke, T.F. Anders, I.F. Litt and W.C. Dement (1980), "Pubertal Changes in Daytime Sleepiness", *Sleep*, Vol. 2, pp. 453-460.

Carskadon, M.A., A.R. Wolfson, C. Acebo, O. Tzischinsky and R. Seifer (1998), "Adolescent Sleep Patterns, Circadian Timing, and Sleepiness at a Transition to Early School Days", *Sleep*, Vol. 21, pp. 871-881.

Cassell, J., M. Ananny, A. Basu, T. Bickmore, P. Chong, D. Mellis, K. Ryokai, J. Smith, H. Vilhjálmsson and H. Yan (2000), "Shared Reality: Physical Collaboration with a Virtual Peer?", in *Proceedings of the ACM SIGCHI Conference on Human Factors in Computing Systems (CHI)*, Amsterdam, 4-9 April, pp. 259-260.

Chervin, R.D., J.E. Dillon, C. Bassetti, D.A. Ganoczy and K.J. Pituch (1997), "Symptoms of Sleep Disorders, Inattention, and Hyperactivity in Children", *Sleep*, Vol. 20, pp. 1185-1192.

Colcombe, S.J., K.I. Erickson, N. Raz, A.G. Webb, N.J. Cohen, E. McAuley and A.F. Kramer (2003), "Aerobic Fitness Reduces Brain Tissue Loss in Aging Humans", *Journal of Gerontology*, Vol. 58A, No. 2, pp. 176-180.

Colcombe, S.J. and A.F. Kramer (2004), "Fitness Effects on the Cognitive Function of Older Adults: A Meta-analytic Study", *Psychological Science*, Vol. 14, No. 2, pp. 125-130.

Colcombe, S.J., A.F. Kramer, K.I. Erickson, P. Scalf, E. McAuley, N.J. Cohen, A.G. Webb, G.J. Jerome, D.X. Marquez and S. Elavsky (2004), "Cardiovascular Fitness, Cortical Plasticity, and Aging", *Proceedings of the National Academy of Sciences*, Vol. 101, No. 9, pp. 3316-3321.

Cole, P.M., S.E Martin and T.A. Dennis (2004), "Emotion Regulation as a Scientific Construct: Methodological Challenges and Directions for Child Development Research", *Child Development*, Vol. 75, No. 2, pp. 317-333.

Corkum, P., R. Tannock and H. Moldofsky (1998), "Sleep Disturbances in Children with Attention-Deficit/Hyperactivity Disorder", *Journal of the American Academy of Child and Adolescent Psychiatry*, Vol. 37, pp. 637-646.

Csikszentmihalyi, M. (1990), *Flow: The Psychology of Optimal Experience*, Harper and Row, New York.

Dagan, Y., S. Zeevi-Luria, Y. Sever, D. Hallis, I. Yovel, A. Sadeh and E. Dolev (1997), "Sleep Quality in Children with Attention Deficit Hyperactivity Disorder: An Actigraphic Study", *Psychiatry and Clinical Neurosciences*, Vol. 51, pp. 383-386.

Dahl, R.E. (1998), "The Development and Disorders of Sleep", *Advances in Pediatrics*, Vol. 45, pp. 73-90.

Dahl, R.E. and J. Puig-Antich (1990), "Sleep Disturbances in Child and Adolescent Psychiatric Disorders", *Pediatrician*, Vol. 17, pp. 32-37.

Damasio, A.R. (1994), *Descartes' Error: Emotion, Reason, and the Human Brain*, G.P. Putnam, New York.

Denham, S.A. and R. Burton (2003), *Social and Emotional Prevention and Intervention Programming for Preschoolers*, Kluwer-Plenum, New York.

Eippert, F., R. Veit, N. Weiskopf, M. Erb, N. Birbaumer and S. Anders (2006), "Regulation of Emotional Responses Elicited by Threat-related Stimuli", *Hum Brain Mapp*, 28 November.

Eisenberg, N. and R.A. Fabes (eds.) (1992), *Emotion and Its Regulation in Early Development: New Directions for Child and Adolescent Development*, Jossey-Bass/Pfeiffer, San Francisco, CA.

Epstein, R., N. Chillag and P. Lavie (1998), "Starting Times of School: Effects on Daytime Functioning of Fifth-grade Children in Israel", *Sleep*, Vol. 21, pp. 250-256.

Fadiga, L., L. Fogassi, G. Pavesi and G. Rizzolatti (1995), "Motor Facilitation during Action Observation: A Magnetic Stimulation Study", *Journal of Neurophysiology*, Vol. 73, No. 6, pp. 2608-2611.

Ferber, R. and M. Kryger (eds.) (1995), *Principles and Practice of Sleep Medicine in the Child*, W.B. Saunders Company, Philadelphia.

Fries, A.B., T.E. Ziegler, J.R. Kurian, S. Jacoris and S.D. Pollak (2005), "Early Experience in Humans is Associated with Changes in Neuropeptides Critical for Regulating Social Behaviour", *Proceedings of the National Academy of Sciences*, Vol. 102, No. 47, pp. 17237-17240.

Gesch, C.B., S.M. Hammond, S.E. Hampson, A. Eves and M.J. Crowder (2002), "Influence of Supplementary Vitamins, Minerals and Essential Fatty Acids on the Antisocial Behaviour of Young Adult Prisoners: Randomised, Placebo-Controlled Trial", *British Journal of Psychiatry*, Vol. 181, No. 1, pp. 22-28.

Green, C. and D. Bavelier (2003), "Action Video Game Modifies Visual Selective Attention", *Nature*, Vol. 423, pp. 534-537.

Gros, B. (2003), "The Impact of Digital Games in Education", *First Monday – Peer-reviewed Journal*.

Gross, J.J. and O.P. John (2003), "Individual Differences in Two Emotion Regulation Processes: Implications for Affect, Relationships, and Well-being", *Journal of Personality and Social Psychology*, Vol. 85, pp. 348-362.

Horne, J.A. (2000), "REM Sleep – By Default?", *Neuroscience and Biobehavioral Reviews*, Vol. 24, No. 8, pp. 777-797.

Huber, R., M.F. Ghilardi, M. Massimini and G. Tononi (2004), "Local Sleep and Learning", *Nature*, Vol. 430, No. 6995, pp. 78-81.

Ji, D. and M.A. Wilson (2007), "Co-ordinated Memory Replay in the Visual Cortex and Hippocampus during Sleep", *Nature Neuroscience*, Vol. 10, No. 1, pp. 100-107.

Johnson, S. (2004), "Thinking Faster: Are the Brain's Emotional Circuits Hardwired for Speed?", *Discover*, Vol. 25, No. 5, May.

Kahn, A., C. Van de Merckt, E. Rebuffat, M.J. Mozin, M. Sottiaux, D. Blum and P. Hennart (1989), "Sleep Problems in Healthy Preadolescents", *Pediatrics*, Vol. 84, pp. 542-546.

Kavanau, J.L. (1997), "Memory, Sleep and the Evolution of Mechanisms of Synaptic Efficacy Maintenance", *Neuroscience*, Vol. 79, No. 1, pp. 7-44.

Latz, S., A.W. Wolf and B. Lozoff (1999), "Cosleeping in Context: Sleep Practices and Problems in Young Children in Japan and the United States", *Archives of Pediatrics and Adolescent Medicine*, Vol. 153, pp. 339-346.

LeDoux, J.E. (2000), "Emotion Circuits in the Brain", *Annual Review of Neuroscience*, Vol. 23, pp. 155-184.

Lewis, M.D. and J. Stieben (2004), "Emotion Regulation in the Brain: Conceptual Issues and Directions for Developmental Research", *Child Development*, Vol. 75, No. 2, March, pp. 371-376.

Lozoff, B., A.W. Wolf and N.S. Davis (1984), "Cosleeping in Urban Families with Young Children in the United States", *Pediatrics*, Vol. 74, pp. 171-182.

Madansky, D. and C. Edelbrock (1990), "Cosleeping in a Community Sample of 2- and 3-year-old Children", *Pediatrics*, Vol. 86, pp. 197-203.

Maquet, P. (2001), "The Role of Sleep in Learning and Memory", *Science*, Vol. 294, No. 5544, pp. 1048-1052.

Marcotte, A.C., P.V. Thacher, M. Butters, J. Bortz, C. Acebo and M.A. Carskadon (1998), "Parental Report of Sleep Problems in Children with Attentional and Learning Disorders", *Journal of Developmental and Behavioral Pediatrics*, Vol. 19, pp. 178-186.

Marshall, L., H. Helgadottir, M. Molle and J. Born (2006), "Boosting Slow Oscillations during Sleep Potentiates Memory", Vol. 444, No. 7119, pp. 610-613.

McEwen, B.S. and R.M. Sapolsky (1995), "Stress and Cognitive Function", *Curr Opin Neurobiol*, Vol. 5, pp. 205-216.

McFarlane, A., A. Sparrowhawk and Y. Heald (2002), *Report on the Educational Use of Games*, TEEM, Cambridge, *www.teem.org.uk/publications/teem_gamesined_full.pdf*.

McGraw, K.O. (1978), "The Detrimental Effects of Reward on Performance: A Literature Review and a Prediction Model", in M.R. Lepper and D. Greene (eds.), *The Hidden Costs of Reward: New Perspectives on the Psychology of Human Motivation*, Lawrence Erlbaum, Hillsdale, NJ, pp. 33-60.

Meyers, A.F., A.E. Sampson, M. Weitzman, M.L. Rogers and H. Kayne (1989), "School Breakfast Program and School Performance", *American Journal of Diseases of Children*, Vol. 143, No. 10, pp. 1234-1239.

Mindell, J.A., M.L. Moline, S.M. Zendell, L.W. Brown and J.M. Fry (1994), "Pediatricians and Sleep Disorders: Training and Practice", *Pediatrics*, Vol. 94, pp. 194-200.

Minnesota Department of Children, Families and Learning (1998), *School Breakfast Programs Energizing the Classroom*, Minnesota Department of Children, Families and Learning, Roseville, MN.

Miyamoto, H. and T.K. Hensch (2003), "Reciprocal Interaction of Sleep and Synaptic Plasticity", *Molecular Interventions*, Vol. 3, No. 7, pp. 404-407.

Molteni, R., A. Wu, S. Vaynman, Z. Ying, R.J. Barnard and F. Gomez-Pinilla (2004), "Exercise Reverses the Harmful Effects of Consumption of a High-fat Diet on Synaptic and Behavioral Plasticity Associated to the Action of Brain-derived Neurotrophic Factor", *Neuroscience*, Vol. 123, No. 2, pp. 429-440.

Morrison, D.N., R. McGee and W.R. Stanton (1992), "Sleep Problems in Adolescence", *Journal of the American Academy of Child and Adolescent Psychiatry*, Vol. 31, pp. 94-99.

Moser, T. (2001), "Sprechen ist Silber, Bewegen ist Gold? Zum Zusammenhang zwischen Sprache und Bewegung aus psychomotorischer und handlungstheoretischer Sicht", in J.R.Nitsch and H. Allmer (eds.), *Denken – Sprechen – Bewegen. Bericht über die 32. Tagung der Arbeitsgemeinschaft für Sportpsychologie vom 1.3. Juni 2000 in Köln*, Köln, pp. 168-174.

Moser, T. (2004), "The Significance of Physical Activity for the Psychosocial Domain: A Crash between Myths and Empirical Reality?", in P. Jørgensen and N. Vogensen (eds.), *What's Going on in the Gym? Learning, Teaching and Research in Physical Education*, University of Southern Denmark, Odense, pp. 50-71.

Motluk, A. (2005), "Do Games Prime Brain (Sic) for Violence?", *New Scientist*, Vol. 186, No. 2505, 25 June, p. 10.

Mourão-Miranda, J., E. Volchan, J. Moll, R. de Oliveira-Souza, L. Oliveira, I. Bramati, R. Gattass and L. Pessoa (2003), "Contributions of Stimulus Valence and Arousal to Visual Activation during Emotional Perception", *Neuroimage*, Vol. 20, No. 4, pp. 1955-1963.

Nelson, L. (2004), "While You Were Sleeping", *Nature*, Vol. 430, No. 7003, pp. 962-964.

Nicolson, R.I., A.J. Fawcett and P. Dean (1995), "Time Estimation Deficits in Developmental Dyslexia: Evidence of Cerebellar Involvement", *Proceedings. Biological sciences*, Vol. 259, No. 1354, pp. 43-47.

Nussbaum, M.C. (2001), *Upheavals of Thought: the Intelligence of Emotions*, Cambridge University Press, Cambridge, p. 751.

Nussbaum, M., R. Rosas, P. Rodríguez, Y. Sun and V. Valdivia (1999), "Diseño, desarrollo y evaluaciòn de video juegos portàtiles educativos y autorregulados", *Ciencia al Dìa Internacional*, Vol. 2, No. 3, pp. 1-20.

Ochsner, K.N., R.D. Ray, J.C. Cooper, E.R. Robertson, S. Chopra, J.D. Gabrieli and J.J. Gross (2004), "For Better or for Worse: Neural Systems Supporting the Cognitive down- and up-regulation of Negative Emotion", *Neuroimage*, Vol. 23(2), pp. 483-499.

O'Connor, T.G., D. Bredenkamp and M. Rutter (1999), "Attachment Disturbances and Disorders in Children Exposed to Early Severe Deprivation", *Infant Mental Health Journal*, Vol. 20, No. 10, pp. 10-29.

OECD (2002a), First High Level Forum on Brain Mechanisms and Early Learning, New York, *www.oecd.org/dataoecd/40/18/15300896.pdf*.

OECD (2002b), *Understanding the Brain: Towards a New Learning Science*, OECD, Paris.

OECD (2002c), "Learning Sciences and Brain Research: Report of the Launching Meeting of Phase II", Royal Institution, London, 29-30 April, pp. 7-8, *www.oecd.org/dataoecd/40/36/15304667.pdf*.

OECD (2003a), "A Report of the Brain Research and Learning Sciences Mini-symposium on the Design of Rehabilitation Software for Dyscalculia", INSERM Cognitive Neuroimagery Unit, Orsay, France, 20 September, *www.oecd.org/dataoecd/50/39/18268884.pdf*.

OECD (2003b), "A Report of the Brain Research and Learning Sciences Emotions and Learning Planning Symposium", Psychiatric Hospital, University of Ulm, Germany, 3 December, *www.oecd.org/dataoecd/57/49/23452767.pdf*.

Ohayon, M.M., M. Caulet, P. Philip, C. Guilleminault and R.G. Priest (1997), "How Sleep and Mental Disorders are Related to Complaints of Daytime Sleepiness", *Archives of Internal Medicine*, Vol. 157, pp. 2645-2652.

Owens, J.A., R. Maxim, C. Nobile, M. McGuinn and M. Msall (2000a), "Parental and Self-report of Sleep in Children with Attention Deficit/Hyperactivity Disorder", *Archives of Pediatrics and Adolescent Medicine*, Vol. 154, pp. 549-555.

Owens, J.A., A. Spirito and M. McGuinn (2000b), "The Children's Sleep Habits Questionnaire (CSHQ): Psychometric Properties of a Survey Instrument for School-aged Children", *Sleep*, Vol. 23, pp. 1043-1051.

Pantev, C. (2003), "Representational Cortex in Musicians", in I. Peretz and R.J. Zatorre (eds.), *The Cognitive Neuroscience of Music*, Oxford University Press, New York, pp. 382-395.

Pantev, C., R. Oostenveld, A. Engelien, B. Ross, L.E. Roberts and M. Hoke (1998), "Increased Auditory Cortical Representation in Musicians", *Nature*, Vol. 392, No. 6678, pp. 811-813.

Peyton, J.L., W.T. Bass, B.L. Burke and L.M. Frank (2005), "Novel Motor and Somatosensory Activity is Associated with Increased Cerebral Cortical Blood Volume Measured by Near-infrared Optical Topography", *Journal of Child Neurology*, Vol. 10, pp. 817-821.

Poelstra, P.A. (1984), "Relationship between Physical, Psychological, Social, and Environmental Variables and Subjective Sleep Quality", *Sleep*, Vol. 7, pp. 255-260.

Randazzo, A.C., M.J. Muehlbach, P.K. Schweitzer and J.K. Walsh (1998), "Cognitive Function Following Acute Sleep Restriction in Children Ages 10-14", *Sleep*, Vol. 21, pp. 861-868.

Rasch, B., C. Büchel, S. Gais and J. Born (2007), "Odor Cues During Slow-Wave Sleep Prompt Declarative Memory Consolidation", *Science*.

Reite, M. (1998), "Sleep Disorders Presenting as Psychiatric Disorders", *Psychiatric Clinics of North America*, Vol. 21, pp. 591-607.

Richardson, A.J. and P. Montgomery (2005), "The Oxford-Durham Study: A Randomized Controlled Trial of Dietary Supplementation with Fatty Acids in Children with Developmental Co-ordination Disorder", *Pediatrics*, Vol. 115, No. 5, pp. 1360-1366.

Rimmele, U., B. Costa Zellweger, B. Marti, R. Seiler, C. Mohiyedinni, U. Ehlert and M. Heinrichs (2007a), "Elite Sportsmen Show Lower Cortisol, Heart Rate and Psychological Responses to a Psychosocial Stressor Compared with Untrained Men", *Psychoneuroendocrinology*.

Rimmele, U. *et al.* (2007b), "Blunted Stress Reactivity of Elite Sportsmen to Mental Stress".

Rintala, P., K. Pienimäki, T. Ahonen, M. Cantell and L. Kooistra (1998), "The Effects of Psychomotor Training Programme on Motor Skill Development in Children with Developmental Language Disorders", *Human Movement Science*, Vol. 17, No. 4-5, pp. 721-737.

Rona, R.J., L. Li, M.C. Gulliford and S. Chinn (1998), "Disturbed Sleep: Effects of Sociocultural Factors and Illness", *Archives of Disease in Childhood*, Vol. 78, pp. 20-25.

Rosenberg, M. (1999), "Non-Violent Communication: A Language of Compassion", PuddleDancer Press, Encinitas, California.

Roth, K. and R. Winter (1994), "Entwicklung Koordinativer Fähigkeiten", in J. Baur, K. Bös and R. Singer (eds.), *Motorische Entwicklung – Ein Handbuch*, Verlag Karl Hofmann, Schorndorf, pp. 191-216.

Rudel, R.G. (1985), "The Definition of Dyslexia: Language and Motor Deficits", in Frank H. Duffy (ed.), *Dyslexia: A Neuroscientific Approach to Clinical Evaluation*, Little Brown, Boston, Massachussetts, pp. 33-53.

Ruoho, K. (1990), *Zum Stellenwert der Verbosensomotorik im Konzept prophylaktischer Diagnostik der Lernfähigkeit bei finnischen Vorschulkindern im Alter von sechs Jahren*, University of Joensuu, Joensuu.

Schlaug, G. (2003), "The Brain of Musicians", in I. Peretz and R.J. Zatorre (eds.), *The Cognitive Neuroscience of Music*, Oxford University Press, Oxford, pp. 366-381.

Shield, B.M. and J.E. Dockrell (2004), "External and Internal Noise Surveys of London Primary Schools", *Journal of the Acoustical Society of America*, Vol. 115, pp. 730-738.

Siegel, J. (2003), "Why We Sleep", *Scientific American*, November.

Simonds, J.F and H. Parraga (1984), "Sleep Behaviors and Disorders in Children and Adolescents Evaluated at Psychiatric Clinics", *Journal of Developmental and Behavioral Pediatrics*, Vol. 5, pp. 6-10.

Singer, T., B. Seymour, J. O'Doherty, H. Kaube, R.J. Dolan and C.D. Frith (2004), "Empathy for Pain Involves the Affective but Not Sensory Components of Pain", *Science*, Vol. 303, No. 5661, pp. 1157-1162.

Smedje, H., J.E. Broman and J. Hetta (1998), "Sleep Disturbances in Swedish Pre-school Children and their Parents", *Nordic Journal of Psychiatry*, Vol. 52, pp. 59-67.

Smedje, H., J.E. Broman and J. Hetta (2001), "Associations between Disturbed Sleep and Behavioural Difficulties in 635 Children Aged Six to Eight Years: A Study Based on Parents' Perceptions", *European Child and Adolescent Psychiatry*, Vol. 10, pp. 1-9.

Smith, C. (1996), "Sleep States, Memory Processes and Synaptic Plasticity", *Behavioural Brain Research*, Vol. 78, No. 1, pp. 49-56.

Steenari, M.R., V. Vuontela, E.J. Paavonen, S. Carlson, M. Fjällberg and E. Aronen (2003), "Working Memory and Sleep in 6- to 13-year-old Schoolchildren", *Journal of the American Academy of Child and Adolescent Psychiatry*, Vol. 42, pp. 85-92.

Stein, M.A. (1999), "Unravelling Sleep Problems in Treated and Untreated Children with ADHD", *Journal of Child and Adolescent Psychopharmacology*, Vol. 9, pp. 157-168.

Stickgold, R. (2003), "Human Studies of Sleep and Off-Line Memory Reprocessing", in P. Maquet, C. Smith and R. Stickgold (eds.), *Sleep and Brain Plasticity*, Oxford University Press, New York, pp. 42-63.

Thompson, R.A. (1994), "Emotional Regulation: A Theme In Search Of a Definition", in N.A. Fox (ed.), *The Development of Emotion Regulation: Biological and Behavioural Considerations. Monographs of the Society for Research in Child Development*, Serial No. 240, Vol. 59, No. 2-3, University of Chicago Press, Chicago, Illinois, pp. 25-52.

Weber, R., U. Ritterfeld and K. Mathiak (2006), "Does Playing Violent Video Games Induce Aggression? Empirical Evidence of a Functional Magnetic Resonance Imaging Study", *Media Psychology*, Vol. 8, No. 1, pp. 39-60.

Weinberger, N.M. (2004), "Music and the Brain", *Scientific American*, November, pp. 67-73.

Wesnes, K.A., C. Pincock, D. Richardson, G. Helm and S. Hails (2003), "Breakfast Reduces Declines in Attention and Memory Over the Morning in Schoolchildren", *Appetite*, Vol. 41, No. 3, pp. 329-331.

Wiggs, L. and G. Stores (2001), "Sleeplessness", in G. Stores and L. Wiggs (eds.), *Sleep Disturbance in Children and Adolescents with Disorders of Development: its Significance and Management. Clinics in Developmental Medicine*, Cambridge University Press, Vol. 155, pp. 24-29.

Winter, R. and K. Roth (1994), "Entwicklung motorischer Fertigkeiten", in J. Baur, K. Bös and R. Singer (eds.), *Motorische Entwicklung – Ein Handbuch*, Verlag Karl Hofmann, Schorndorf, pp. 217-237.

Wolfson, A.R. and M.A. Carskadon (1998), "Sleep Schedules and Daytime Functioning in Adolescents", *Child Development*, Vol. 69, No. 4, pp. 875-887.

Wright, E. (2004), *Generation Kill*, Bantam Press, London.

ISBN 978-92-64-02912-5
Understanding the Brain: The Birth of a Learning Science
© OECD 2007

PART I

Chapter 4

Literacy and the Brain

Just because some of us can read and write and do a little math, that doesn't mean we deserve to conquer the Universe.

Kurt Vonnegut

This chapter describes the state of knowledge on the functioning of the brain in relation to language and reading. It helps us to address questions about when and how literacy might best be acquired and the desirable environment that supports it. Such information will be useful for those responsible for policies to enhance language and literacy education, professional educators, and indeed parents who are thinking how best to read with their children. Special attention is given to differences in languages with "deep" (such as English) and "shallow" orthographies (such as Finnish). Dyslexia is specifically discussed, and what the evidence surveyed in this chapter has to say about possible remedial strategies.

As you cast your eyes over the shapes and squiggles on this page, you are suddenly in contact with the thoughts of a person on a chilly January afternoon in Paris some time ago. This remarkable ability of words to defy the limits of time and space is of tremendous importance and enables the cumulative evolution of culture. As you read this page, you are not only in contact with the thoughts of a single person on a particular day in Paris, but indirectly, with the collective wisdom of the cultural history underpinning those thoughts (Tomasello, 1999). Without literacy as a mechanism for transmitting information across the boundaries of time and space, the capacity of human thought to build on itself would be severely constrained within the limits of memory – literacy is fundamental to human progress.

Learning to read requires the mastery of a collection of complex skills. First, the knowledge of morphology – the forms of either letters of an alphabet, syllabic symbols, or ideograms – must be acquired. Then, orthographic symbols must be understood as the labels – spelling – that can be mapped onto sounds, without which the alphabetic symbols on this page would remain arbitrary shapes. Moreover, an understanding of phonetics – mapping words to sounds – is a vital, but by itself insufficient, tool for decoding words. In alphabetic languages with deep orthographies, such as English or French,[1] grapheme-phoneme combinations are variable, with English having the highest degree of "irregular" representation among alphabetic languages, at more than a thousand possible letter combinations used to represent the 42 sounds of the language. Reading, particularly in languages with deep orthographies, therefore involves the use of supplementary strategies in addition to the phonological decoding of symbols into sounds. These strategies include using context clues, recognising whole words, and noticing partial-word analogies such as *ate* common to both "late" and "gate". Moreover, once a word has been decoded, understanding the meaning of the text requires additional skills. There is the semantic knowledge of word meanings. More than this, knowledge of syntactic rules governing the arrangements of words to show their relations to each other is also critical to meaning: Orsino loves Olivia does not mean the same thing as Olivia loves Orsino. And more even than all this, each word must be integrated with previously-read words, which requires the co-ordination of different component functions and a working memory system.

The neural circuitry underlying literacy, which calls for all these skills, is guided by the interaction and synergy between the brain and experience, and hence the applicability of a dynamic developmental framework, such as skill theory, to the understanding of literacy

---

1. Languages with "deep" orthographies are those in which sounds map onto letters with a high degree of variability. In these languages, it is often not possible to determine letter-sound associations accurately without the context of the whole word. Consider, for example, the following combination of letters in English: *ghoti*. If the *gh* is pronounced as in *laugh*, the *o* pronounced as in *women*, and the *ti* pronounced as in *nation*, *ghoti* can be read as "fish". In languages with shallow orthographies, by contrast, the correspondences between letters and sounds are close to one-to-one. In Finnish, for example, there are 23 associations that match the exact number of letters.

(Fischer, Immordino-Yang and Waber, 2007). Skill theory recognises that reading proficiency can be reached through multiple developmental pathways. Through this lens, neuroscience can enable the design of more effective and inclusive reading instruction.

Language and developmental sensitivities

The brain is biologically primed to acquire language. Chomsky (1959) proposed that the brain is equipped with a recipe for making sequences of sound into representations of meaning that is analogous to the system for translating sensory information into representations of objects. That is, the brain is designed through evolution to process certain stimuli according to universal language rules. There are indeed brain structures specialised for language: research has established the role played by the left inferior frontal gyrus and the left posterior middle gyrus (Broca's area and Wernicke's area, respectively. See Figure 2.3). Broca's area, long understood as implicated in language production, is now associated with a broader range of linguistic functions (Bookheimer, 2002). Wernicke's area is involved in semantics (Bookheimer *et al.*, 1998; Thompson-Schill *et al.*, 1999). Critically, these structures are for higher levels of processing, and therefore are not restricted to the simpler processing of incoming auditory stimuli – hearing *per se*. Visual information can also be processed linguistically, as in the case of sign language.

Though certain brain structures are biologically primed for language, the process of language acquisition needs the catalyst of experience. There are developmental sensitivities (the windows of learning opportunity referred to in Chapter 1) as language circuits are most receptive to particular experience-dependent modifications at certain stages of the individual's development. Newborns are born with an ability to discern subtle phonetic changes along a continuous range, but experience with a particular language in the first ten months renders the brain sensitive to sounds relevant to that language (Gopnik, Meltzoff and Kuhl, 1999). For example, the consonant sounds r and l occur along a continuous spectrum, and all newborns hear the sounds this way. The brains of babies immersed in an English-speaking environment, however, are gradually modified to perceive this continuous spectrogram as two distinct categories, r and l. A prototypical representation of each phoneme is developed, and incoming sounds are matched to these representations and sorted as either r or l. Babies immersed in a Japanese-speaking environment, by contrast, do not form these prototypes as this distinction is not relevant to Japanese. Instead, they form prototypes of sounds relevant to Japanese, and actually lose the ability to discriminate between r and l by ten months of age. This phenomenon occurs for varied sound distinctions across many languages (Gopnik, Meltzoff and Kuhl, 1999). Therefore, the brain is optimally suited to acquire the sound prototypes of languages to which it is exposed in the first ten months from birth.[2]

There is also a developmental sensitivity for learning the grammar of a language: the earlier a language is learned, the more efficiently the brain can master its grammar (Neville and Bruer, 2001). If the brain is exposed to a foreign language between 1 and 3 years of age, grammar is processed by the left hemisphere as in a native speaker but even delaying learning until between 4 and 6 years of age means that the brain processes grammatical information with both hemispheres. When the initial exposure occurs at the ages of 11,

---

2. But it remains possible for adults to learn sound discrimination. McClelland, Fiez and McCandliss (2002) have shown that, thanks to an exaggerated contrasted exposure, Japanese adults can learn to discriminate between the English sounds /r/ and /l/, even if this contrast is alien to Japanese language.

12 or 13 years, corresponding to the early stage of secondary schooling, brain imaging studies reveal an aberrant activation pattern. Delaying exposure to language therefore leads the brain to use a different strategy for processing grammar. This is consistent with behavioural findings that later exposure to a second language results in significant deficits in grammatical processing (Fledge and Fletcher, 1992). The pattern seems thus to be that early exposure to grammar leads to a highly effective processing strategy, in contrast with alternative, and less efficient, processing strategies associated with later exposure.

In addition, there is a sensitive period for acquiring the *accent* of a language (Neville and Bruer, 2001). This aspect of phonological processing is most effectively learned before 12 years of age. Developmental sensitivities are for very specific linguistic functions, however, and there are other aspects of phonology which do not seem even to have a sensitive period.

In sum, there is an inverse relationship between age and the effectiveness of learning many aspects of language – in general, the younger the age of exposure, the more successful the language learning. This is at odds with the education policies of numerous countries where foreign language instruction does not begin until adolescence. While further research is needed to develop a complete map of developmental sensitivities for learning various aspects of language, the implications of the current findings are clear:

The earlier foreign language instruction begins, the more efficient and effective it is likely to be.

However, for early instruction to be effective, it must be age-appropriate. It would not be useful to take rule-based methods designed for older students and insert them into early childhood classrooms. It is necessary, in other words, that early foreign language instruction is appropriately designed for young children.

Although the early learning of language is most efficient and effective, it is important to note that it is possible to learn language throughout the lifespan: adolescents and adults can also learn a foreign language, albeit with greater difficulty. Indeed, if they are immersed in a new language environment, they can learn the language "very well", though particular aspects, such as accent,[3] may never develop as completely as they would have done if the language had been learned earlier. There are also individual differences such that the degree and duration of developmental sensitivities vary from one individual to the next. Some individuals are able to master almost all aspects of a foreign language into adulthood.

Literacy in the brain

In contrast to language, there are no brain structures designed by evolution to acquire *literacy*. Experience does not trigger a set of biologically-inclined processes leading to literacy, as in the case of language. Instead, experience progressively creates the capacity for literacy in the brain through cumulative neural modifications, expressed by Pinker (1995) as "Children are wired for sound, but print is an optional accessory that must be painstakingly bolted on". Experience with the printed word gradually builds brain circuitry to support reading.

The crucial role of experience in building neural circuitry capable of supporting literacy suggests that attention needs to be given to differences in the degree to which early

---

3. For a foreigner speaking in a given language, the benefit of acquiring a "native speaker accent" is not clear anyway. As long as one can make oneself understood, what is wrong with "having a foreign accent"? But education systems too often still assume that the ultimate goal for learners is (or should be) "to reach the level of a native speaker" (which one, by the way?), even as far as phonetics is concerned.

home environments provide a foundation of pre-literacy skills. For example, Hart and Risley (2003) report that the sheer number of words that American children from disadvantaged socioeconomic backgrounds were exposed to by the age of 3 lagged behind that of non-disadvantaged children by 30 million word occurrences. Such limited exposure could be insufficient to support the development of pre-literacy skills in the brain, thereby chronically impeding later reading skills. These children may well be capable of catching up through later experience, but the reality is that they very often do not (Wolf, 2007). Therefore, of policy relevance from this work:

Initiatives aimed at ensuring that all children have sufficient opportunities to develop pre-literacy skills in early childhood are essential.

While the brain is not necessarily biologically inclined to acquire literacy, it is biologically inclined to adapt to experience. It is, for instance, endowed with language circuitry capable of processing visual input. The brain's plastic capacities of adaptability enable the stimuli coming from experience to utilise language structures when constructing the neural circuitry capable of supporting literacy. This is often expressed as literacy being built "on top of" language. In the terms of Vygotsky's classic metaphor, language structures provide scaffolding for literacy to be constructed in the brain (Vygotsky, 1978).

Since literacy is built, in part, with language circuitry, future research should investigate the possibility that developmental sensitivities for certain aspects of language acquisition influence the facility with which the different aspects of reading are acquired. If such influences were identified, this would have implications for educational policy and practice regarding the timeframe for teaching different literacy skills, and could well reinforce the importance of developing pre-literacy skills in early childhood.

Research aimed at delineating the cortical areas supporting reading is rapidly accumulating. The most comprehensive and well-supported model of reading to date is the "dual route" theory (Jobard, Crivello and Tzourio-Mazoyer, 2003). The dual route theory provides a framework for describing reading in the brain at the level of the word. As you look at the words on this page, this stimulus is first processed by the primary visual cortex. Then, pre-lexical processing occurs at the left occipito-temporal junction. The dual route theory posits that processing then follows one of two complementary pathways. The *assembled* pathway involves an intermediate step of grapho-phonological conversion – converting letters/words into sounds – which occurs in certain left temporal and frontal areas, including Broca's area. The *addressed* pathway consists of a direct transfer of information from pre-lexical processing to meaning (semantic access). Both pathways terminate in the left basal temporal area, the left interior frontal gyrus, and the left posterior middle gyrus, or Wernicke's area. The pathway involving direct access to meaning has led to the proposal of a "visual word form area" (VWFA) at the ventral junction between the occipital and temporal lobes. This area was first proposed to contain a visual lexicon or collection of words which functions to immediately identify whole words when they are seen. Recent research has suggested a modified conclusion that this region may actually consist of constellations of adjacent areas sensitive to various aspects of letter strings, such as length or order of words. The entire process from visual processing (seeing) to semantic retrieval (understanding) occurs very rapidly, all within about 600 ms.

An understanding of literacy in the brain can inform reading instruction. The dual importance in the brain of phonological processing, on the one hand, and the direct processing of semantics or meanings, on the other, can inform the classic debate between top-down and bottom-up approaches – "whole language" text immersion and the development of phonetic skills, respectively. The dual importance of both processes in the brain suggests:

A balanced approach to literacy instruction that targets both phonetics and "whole language" learning may be the most effective.[4]

To support this statement, reports by the United States' National Reading Panel (2000) and National Research Council (Snow, Burns and Griffin, 1998) confirm the educational benefits of a balanced approach to reading instruction. The more studies on reading will be relevant, the less the debates around reading instruction (teaching/learning models of literacy acquisition) will be based on ideologies, beliefs or statistical results. The discussion will become more and more anchored to scientific evidence.

Neuroscientists are only beginning to investigate reading at the level of whole sentences. Preliminary results suggest that the operations which go into sentence construction, how these operations are used to determine meanings, and the working memory systems that support these operations share common neural circuits/substrates involving both hearing and seeing (Caplan, 2004). This implies that reading sentences recruits the structures responsible for these functions in language.

Linguistically-mediated literacy development

While much of the neural circuitry underlying reading is the same across different languages, there are also some important differences. A central theme concerning the brain and reading is the way that literacy is created though the colonisation of brain structures, including those specialised for language and those best suited to serve other functions. The operations that are common to speech and printed word, such as semantics, syntax, and working memory recruit brain structures which are specialised for language and which are biologically-based and common across languages. There are biological constraints determining which brain structures are best suited to take on other functions supporting literacy. Therefore, much reading circuitry is shared across languages. Even so, literacy in different languages sometimes requires distinct functions, such as different decoding or word recognition strategies. In these cases, distinct brain structures are often brought into play to support these aspects of reading which are distinctive to these particular languages.

---

4. This statement must be qualified as brain research supporting the dual route theory of reading was conducted primarily with English speakers who had presumably followed a normative developmental pathway for learning to read. Therefore, implications of this work could be less relevant for children who learn to read in other languages or follow atypical developmental pathways. In particular, the transferability of research across languages with different levels of orthographic complexity or from alphabetic to non-alphabetic languages is questionable. Interestingly, most Anglo-Saxon research, working (unconsciously) on an extreme case, does not seem to have considered this crucial aspect. Only recently some researchers became aware of this issue.

Therefore, the dual route theory of reading, which was developed mainly based on research with English speakers, may require modification to describe reading in languages with less complex spelling and orthographic features and it is only partially relevant to non-alphabetic languages. The direct addressed route for accessing meaning without sounding words is likely to be less critical in languages with shallow orthographies, such as Italian, than in those with deep orthographies, such as English. Brain research supports the hypothesis that the routes involved differ according to the depth of the orthographical structure. The "visual word form area" (occipital-temporal VWFA) implicated in identifying word meaning based on non-phonological proprieties in English speakers appears to be less critical for Italian speakers (Paulesu *et al.*, 2001a). Indeed, preliminary results suggest that the brain of Italian native speakers employs a more efficient strategy when reading text than that of English native speakers. Remarkably, this strategy is used even when Italian native speakers read in English, suggesting that the brain circuitry underlying reading for Italian native speakers develops in a different way than that underlying reading for English native speakers.

The recent psycholinguistic "grain-sized" theory describes differences in reading strategies as a function of the orthographic complexity of a language.[5] It proposes that there is a continuum of strategies from the pure decoding of single sounds (phonemes), which have a small grain size, to mixed decoding of involving units with a larger grain size, including the beginnings of words, rhymes, syllables, up to whole words, as well as phonemes. The theory posits that the orthographic complexity of a language determines the reading strategy that develops in the brain, such that the more shallow the language, the smaller the average grain size – for example, letter sounds instead of whole words – used for decoding. This theory is relevant to behavioural data indicating that the delay in reading acquisition is roughly proportional to the degree of orthographic complexity of the language. It suggests that given instructional methods are differentially effective depending on the orthographic structure of the language, which would mean that:

The most effective balance of phonetic and "whole language" instruction will vary across different languages.

Research suggests that the forms of words in a language also influence the way literacy develops in the brain. Imaging studies reveal that Chinese native speakers employ additional areas of the brain for reading compared with English native speakers, and these areas are activated when Chinese native speakers read in English (Tan *et al.*, 2003). Specifically, Chinese native speakers engage left middle frontal and posterior parietal gyri, areas of the brain often associated with spatial information processing and the co-ordination of cognitive resources. It is likely that these areas come into play because of the spatial representation of Chinese language characters (ideograms) and their connection to a syllable-level phonological representation. While much of the neural circuitry underlying reading is shared across alphabetic and non-alphabetic languages, there are distinct structures which could correspond to the extent of reliance in any language on either the assembled or addressed processing described in dual route theory,

5. Usha Goswami and Johannes Ziegler (2005), "Learning to Read Workshop", co-organised by CERI and University of Cambridge, 29-30 September 2005, Cambridge, UK.

referred to above (Yiping, Shimin and Iversen, 2002). Together with the results on orthographic complexity (deep *vs.* shallow) and reading strategy, these findings indicate that certain aspects of literacy are created in distinctive ways in the brain depending on experience with the printed form of a particular language.

All this underscores the importance of considering reading from a developmental perspective. The neural circuitry underlying reading changes as children learn to read. For example, Pugh[6] demonstrated a shift in functional neuroanatomy underlying initial aspects of reading as the developing English reader matures from multiple temporal, frontal, and right hemisphere sites towards a more consolidated response in the left hemisphere occipito-temporal region.

Multi-variate analyses of such brain patterns examining both age and reading skill revealed that the crucial predictor was the reading skill level, suggesting that the development of literacy is guided by experience rather than merely brain maturation. Since literacy is created in the brain through gradual developmental progression, it would be most useful for teaching and learning to involve ongoing assessments which support reading development. From this, the pedagogical implication is:

Reading is most suitably assessed using formative assessment.

Formative assessment, which involves using ongoing assessment to identify and respond to students' learning needs, is highly effective in raising student achievement, increasing equity of student outcomes, and improving students' ability to learn (OECD, 2005).

As research increasingly delineates the relationships between specific experiences and their consequences for the development of reading circuitry in the brain, its interest for education will grow. For example, if early experience with print from a language with a shallow orthography is confirmed to develop more efficient reading strategies in the brain, it might be useful to explore options of building this circuitry in children who speak languages with deep orthographies. Children could, for example, first be taught to read with books containing selected words from a pool with consistent letter-sound combinations. A more radical alternative would be the actual reform of languages with "deep" orthographies so that letter-sound combinations are made more consistent.[7]

Developmental dyslexia

Although experience plays a crucial role in the development of literacy in the brain, biology also has an important part to play. Consider how biologically-based differences in language structures could affect reading given that literacy is built "on top of" these structures. Many children with access to adequate reading instruction struggle to learn to read because of biologically atypical cortical features. These children are said to have developmental dyslexia. Developmental dyslexia is a neurobiological language impairment

---

6. First joint meeting of the CERI "Literacy" and "Numeracy" Networks meeting, 30-31 January 2003, Brockton, MA, United States.
7. The fact that the most recent initiatives in terms of orthographic reforms did not fully succeed (German) or even did not succeed at all (French) does not mean it is impossible. Spanish and Turkish have shown that it is possible.

defined as a reading difficulty which does not result from global intellectual deficits or a chronic problem of motivation.[8] It has formally been defined as:

> Dyslexia is a specific learning disability that is neurobiological in origin. It is characterised by difficulties with accurate and/or fluent word recognition and by poor spelling and encoding abilities. Their difficulties typically result from a deficit in the phonological component of language that is often unexpected in relation to other cognitive abilities and the provision of effective classroom instruction (Lyon, Shaywitz and Shaywitz, 2003, p. 2).

Dyslexia is both prevalent and widespread.[9] It is the most common subtype of learning difficulty and occurs across cultural, socioeconomic, and to some extent linguistic boundaries. While the phonological deficit underlying dyslexia seems common across alphabetic languages, the degree to which it is manifest, and thus its consequence for reading, may vary as a function of the orthographic structure of a language (Paulesu *et al.*, 2001). And since reading in non-alphabetic languages calls on distinct neural circuitry compared with reading alphabetic languages, dyslexia in non-alphabetic languages may be manifest in a qualitatively different way. The implications of research on dyslexia in alphabetic languages may thus not be transferable to non-alphabetic languages.

Dyslexia is multi-faceted and has variable manifestations, but these variations notwithstanding it frequently is found with atypical cortical features localised in the left posterior parieto-temporal region and the left posterior occipito-temporal region for those native speakers in alphabetic languages (Shaywitz and Shaywitz, 2005; Shaywitz *et al.*, 2001). The functional consequence of the atypical structures is impairment in processing the sound elements of language. Children with developmental dyslexia register sound imprecisely, with

8. A first impulse upon discovering that a learning difficulty is due to a "brain problem" is to consider it beyond remediation by purely educational means. However, one can also turn this around and consider that when the breakdown of a skill into its separate information processing steps and functional modules is sufficiently understood, thanks to the tools of cognitive neuroscience, that is when efficient remediation programmes can be devised. This is precisely what Bruce McCandliss and Isabelle Beck did in the case of dyslexia, building on the intact components of reading skills in dyslexic children to come up with a new method for teaching word pronounciation. And of course, such deep understanding of how a skill is decomposed into separate cognitive processes may also help design better methods for teaching unimpaired children. Using their "Word Building Method", McCandliss and Beck showed that dyslexic children are capable of learning to read. Helping children to generalise from their reading experience enables them to transfer what they had learned about specific words to new vocabulary words. These skills involve alphabetical decoding and word building and enable reading impaired children to progressively pronounce a larger and larger amount of words. This method teaches them that with a small set of letters, a large number of words can be made. As many of school aged children have difficulty in reading, attending to this problem allows this substantial portion of learners to engage in the most fundamental linguistic exchange and lessens their potential marginalisation from society. Others, most notably Drs. Paula Tallal and Michael Merzenich, have reported similar findings with a different technique. Although these results are somewhat controversial, their method does appear to help at least some children. The key point, however, is not whether one particular available method works better than others. Rather, we note that the theoretical and methodological machinery exists to attack the problem, and progress is clearly being made. Many, like for instance Emile Servan-Schreiber, predict that the study and treatment of dyslexia will be one of the major "success stories" of cognitive neuroscience in the relatively near future.
9. It is difficult to assess the relative incidence of dyslexia across countries because its definition (mostly based on economic factors more than anything else) varies from one country to another. As recommended from 2004 (2nd meeting of the CERI "Literacy" and "Numeracy" networks, El Escorial, Spain, 3-4 March 2004), and as confirmed by Kayoko Ishii's work in 2005, a scientific definition of learning disorders, such as dyslexia or dyscalculia, would help researchers (and policy makers) to create internationally agreed-upon definitions, which would, among other things, allow for international comparisons.

difficulties in retrieving and manipulating phonemes (sounds). The *linguistic* consequences of these difficulties are relatively minor, and cover such things as experiencing difficulties with pronunciation, insensitivity to rhyme, and confusing words which sound alike. However, the consequences of the impairment for *literacy* can be much more significant as mapping phonetic sounds to orthographic symbols is the crux of reading in alphabetic languages.

The recent identification of the specific atypical cortical features responsible for the deficits in processing sounds has enabled the development of targeted interventions. Intervention studies have revealed an encouraging adaptability (plasticity) of these neural circuits. The targeted treatment can enable young individuals to develop neural circuitry in left hemisphere posterior brain systems sufficiently so as to read with accuracy and fluency (Shaywitz *et al.*, 2004). It is also possible for the dyslexic brain to construct alternative, compensatory right hemispheric circuitry, sufficient so as to enable accurate, but slow, reading (Shaywitz, 2003). There seems to be a sensitive period for developing phonetic competence in children with atypical left posterior parieto-temporal and occipito-temporal gyri, as early intervention is most effective (Lyytinen *et al.*, 2005; Shaywitz, 2003; Torgesen, 1998). These results suggest that:

Interventions targeted at developing phonological skills are often effective for helping children with dyslexia to learn to read.[10]

The early identification of dyslexia is important as early interventions are usually more successful than later interventions.

Beyond specific interventions, neuroscience can radically alter the way dyslexia is conceptualised. Now that the neurobiological basis of dyslexia has begun to be identified and confirmed as open to change, educators can design effective, targeted reading interventions and start to transform dyslexia from a disability that seriously hinders learning to an alternative developmental pathway to achieving the same end goal – the literate brain. This conception of dyslexia could have many positive consequences in the classroom, including the preservation of children's self-efficacy for literacy, which is tightly linked to achievement (Bandura, 1993).

Dyslexia is more accurately conceptualised as an alternative developmental pathway than as an insurmountable learning disability.

The value of neuroscience to help design targeted interventions for children with phonological deficits suggests the need for further research aimed at tackling other deficits resulting from atypical cortical features. Neuroscience is allowing educators to differentiate among different causes of learning problems even when outcomes appear to be similar and it can therefore be used to examine confounding or alternative neural manifestations of dyslexia, such as that associated with naming-speed deficits (Wolf, 2007). And since interventions may differ in effectiveness with age and reading experience, research should continue to develop a neurobiological understanding of the developmental trajectories of dyslexia. Future research should also investigate differences in these

---

10. Not systematically, however. The most stunning results to date, in terms of early diagnosis and remediation of dyslexia, are probably those of Heikki Lyytinen at University of Jyvaskyla (Finland). But Lyytinen and his team work on an extremely shallow language. Transferability of these results remains to be measured.

trajectories across languages, as the complexity of the orthographic structure of a language or the representation of words or letters, for example, could well influence the manifestation of dyslexia.[11]

Conclusions

Neurobiological research alters conceptions of literacy in two important ways. First, it promotes a more precise understanding of literacy. Neuroscience facilitates the delineation of the different processes involved in reading in terms of underlying neural circuits. This differentiated understanding can usefully inform the design of effective instruction. For example, the dual importance of phonological and semantic processing suggests that instruction targeting both of these processes may be most effective, at least for children who speak alphabetic languages with deep orthographic structures and follow normative developmental pathways of learning to read. The differentiated conception of literacy allows the different causes of reading difficulties to be pinpointed in the neural subcomponents of literacy, thereby increasing the probability of targeted interventions being effective.

Second, an important contribution to be made by neuroscience is in a more inclusive conception of literacy development. The neural circuitry underlying literacy consists of plastic networks that are open to change and development and are constructed over a period of time. The creation of literacy in the brain is not limited to one single pathway. As reflected by cross-linguistic research and on dyslexia, there are many possible developmental pathways to achieve the end goal of a literate brain. Environmental or biological constraints may render particular pathways more effective than others for certain children. As neuroscience uncovers relationships between specific interventions and neurobiological development, educators will be able to design instruction for different possible developmental pathways. Neuroscience can thus facilitate differentiated instruction capable of accommodating a wide range of individual differences, bringing nearer a literate society which is inclusive rather than selective, with potentially powerful consequences. The famous *Brown vs. Board of Education* United States Supreme Court case concluded that "a mind is a terrible thing to waste", and greater inclusion will provide more raw material for the cumulative evolution of culture and ultimately of human progress.

11. Though most forms of dyslexia seem to be phonologically-based, it seems that some forms may have alternative or cofounding causes.

References

Bandura, A. (1993), "Perceived Self-efficacy in Cognitive Development and Functioning", *Educational Psychologist,* Vol. 28, pp. 117-148.

Bookheimer, S. (2002), "Functional MRI of Language: New Approaches to Understanding the Cortical Organization of Semantic Processing", *Annu. Rev. Neurosci.,* Vol. 25, pp. 151–188.

Bookheimer, S.Y., T.A. Zeffiro, T. Blaxton, W.D. Gaillard, B. Malow and W.H. Theodore (1998), "Regional Cerebral Blood Flow during Auditory Responsive Naming: Evidence for Cross-Modality Neural Activation", *NeuroReport,* Vol. 9, pp. 2409-2413.

Caplan, D. (2004), "Functional Neuroimaging Studies of Written Sentence Comprehension", *Scientific Studies of Reading,* Vol. 8, No. 3, pp. 225-240.

Chomsky, N.A. (1959), "Review of B. F. Skinner's Verbal Behavior", *Language,* Vol. 38, No. 1, pp. 26-59.

Fischer, K.W., M.H. Immordino-Yang and D. Waber (2007), "Toward a Grounded Synthesis of Mind, Brain, and Education for Reading Disorders: An Introduction to the Field and this Book", in K.W. Fischer, J.H. Bernstein and M.H. Immordino-Yang (eds.), *Mind, Brain, and Education in Learning Disorders,* Cambridge University Press, MA, pp. 1-20.

Fledge, J. and K. Fletcher (1992), "Talker and Listener Effects on Degree of Perceived Foreign Accent", *Journal of the Acoustical Society of America,* Vol. 91, pp. 370-389.

Gopnik, A., A.N. Meltzoff and P.K. Kuhl (1999), *The Scientist in the Crib: What Early Learning Tells Us About the Mind,* HarperCollins Publishers Inc., NY.

Hart, B. and T.R. Risley (2003), "The Early Catastrophe: The 30 Million Word Gap", *American Educator,* Vol. 27, No. 1, pp. 4-9.

Ishii, K. (2005), "Strategies for Reading and Writing Learning Difficulties (Dyslexia)", in Science and Technology Trends, Quarterly Review No. 15, April 2005 (original in Japanese 2004).

Jobard, G., F. Crivello and N. Tzourio-Mazoyer (2003), "Evaluation of the Dual Route Theory of Reading: A Metanalysis of 35 Neuroimaging Studies", *NeuroImage,* Vol. 20, pp. 693-712.

Lyon, G.R., S.E. Shaywitz and B.A. Shaywitz (2003), "A Definition of Dyslexia", *Ann Dyslexia,* Vol. 53, pp. 1-14.

Lyytinen, H., T.K. Guttorm, T. Huttunen, J.H. Paavo, H.T. Leppänen and M. Vesterinen (2005), "Psychophysiology of Developmental Dyslexia: A Review of Findings Including Studies of Children at Risk for Dyslexia", *Journal of Neurolinguistics,* Vol. 18, No. 2, pp. 167-195.

McClelland, J.L., J.A. Fiez and B.D. McCandliss (2002), "Teaching the /r/ – /l/ Discrimination to Japanese Adults: Behavioral and Neural Aspects", *Physiology and Behavior,* Vol. 77, pp. 657-662.

National Reading Panel (2000), "Teaching Children to Read: An Evidence-based Assessment of the Scientific Research Literature on Reading and its Implications for Reading Instruction", National Institute of Child Health and Human Development, Washington DC.

Neville, H.J. and J.T. Bruer (2001), "Language Processing: How Experience Affects Brain Organization", in D.B. Bailey, Jr., J.T. Bruer, F.J. Symons and J.W. Lichtman (eds.), *Critical Thinking about Critical Periods,* Paul H. Brookes Publishing, Baltimore, MD, pp. 151-172.

OECD (2005), *Formative Assessment: Improving Learning in Secondary Classrooms,* OECD, Paris.

Paulesu, E., J. Démonet, F. Fazio, E. McCrory, V. Chanoine, N. Brunswick, S.F. Cappa, G. Cossu, M. Habib, C.D. Frith and U. Frith (2001), "Dyslexia: Cultural Diversity and Biological Unity", *Science,* Vol. 291, No. 5511, pp. 2165-2167.

Pinker, S. (1995), "The Language Instinct. How the Mind Creates Language", Harper Collins, New York.

Shaywitz, S.E. (2003), *Overcoming Dyslexia,* Random House Inc., NY.

Shaywitz, S.E. and B.A. Shaywitz (2005), "Dyslexia", *Biological Psychiatry,* Vol. 57, No. 11, pp. 1301-1309.

Shaywitz, B.A., S.E. Shaywitz, B.A. Blachman, K.R. Pugh, R.K. Fulbright, P. Skudlarski, W.E. Mencl, R.T. Constable, J.M. Holahan, K.E. Marchione, J.M. Fletcher, G.R. Lyon and J.C. Gore (2004), "Development of Left Occipito-temporal Systems for Skilled Reading in Children after a Phonologically-based Intervention", *Biol. Psychiatry,* Vol. 55, pp. 926-933.

Shaywitz, B.A., S.E. Shaywitz, K.R. Pugh, R.K. Fulbright, W.E. Mencl, R.T. Constable, P. Skudlarski, J.M. Fletcher, G. Reid and J.C. Gore (2001), "The Neurobiology of Dyslexia", *Clinical Neuroscience Research,* Vol. 1, No. 4, pp. 291-299.

Snow, G.E., M.S. Burns and P. Griffin (eds.) (1998), "Preventing Reading Difficulties in Young Children", Committee on the Prevention of Reading Difficulties in Young Children, Washington DC.

Tan, L.H., J.A. Spinks, C.M. Feng, W.T. Siok, C.A. Perfetti, J. Xiong *et al.* (2003), "Neural Systems of Second Language Reading are Shaped by Native Language", *Human Brain Mapping*, Vol. 18, pp. 158-166.

Thompson-Schill, S.L., G. Aguirre, M. D'Esposito and M.J. Farah (1999), "A Neural Basis for Category and Modality Specifics of Semantic Knowledge", *Neuropsychologia*, Vol. 37, pp. 671-676.

Tomasello, M. (1999), "The Cultural Origins of Human Cognition", Harvard University Press, MA.

Torgesen, J.K. (1998), "Catch them before they Fall: Identification and Assessment to Prevent Reading Failure in Young Children", *American Educator*, Vol. 22, pp. 32-39.

Vygotsky, L.S. (1978), "Mind and Society: The Development of Higher Mental Processes", Cambridge, MA: Havard University Press.

Wolf, M. (2007), "A Triptych of the Reading Brain: Evolution, Development, Pathology and its Interventions", in K.W. Fischer, J.H. Bernstein and M.H. Immordino-Yang (eds.), *Mind, Brain, and Education in Learning Disorders*, Cambridge University Press, MA, pp. 1-20.

Yiping, C., F. Shimin and D. Iversen (2002), "Testing for Dual Brain Processing Routes in Reading: A Direct Contrast of Chinese Character and Pinyin Reading Using fMRI", *Journal of Cognitive Neuroscience*, Vol. 14, pp. 1088-1098.

ISBN 978-92-64-02912-5
Understanding the Brain: The Birth of a Learning Science
© OECD 2007

PART I

Chapter 5

Numeracy and the Brain

> ... the science of calculation also is indispensable as far as the extraction of the square and cube roots. Algebra as far as the quadratic equation and the use of logarithms are often of value in ordinary cases: but all beyond these is but a luxury; a delicious luxury indeed; but not to be indulged in by one who is to have a profession to follow for his subsistence.
>
> *Thomas Jefferson*

This chapter describes the complex functioning of the brain when one develops numeracy, including comprehension of the concept of numbers, simple arithmetic operations, and early explorations in algebra. This is used to draw implications for mathematics instruction. It also describes the barriers to learning mathematics that have a neurological basis (called dyscalculia, the equivalent of dyslexia for mathematics). This chapter is relevant to parents, teachers and policy makers who are interested in understanding and improving numeracy and mathematic education.

Neuroscience can answer important questions relevant to mathematics education. For example, it can address the question as to whether learning higher-level mathematics impacts the brain in such a way that validates teaching non-pragmatic mathematics to the majority of the population. Currently, there is no strong evidence for or against teaching advanced mathematics, such as calculus or trigonometry, to all students. Despite this, including higher-level mathematics in the standard curriculum is the norm in OECD countries and beyond. Given the massive amount of knowledge available in modern society and the tight constraints on curriculum imposed by limited time in school, the question as to whether advanced mathematics should consume so much of the standard curriculum is an important one.[1] Future neuroscience research can help answer this question.

Currently, most neuroscience research focuses on basic mathematics. This work provides important implications for mathematics instruction. In contrast to advanced mathematics, basic mathematics is unquestionably vital for all students because knowing basic mathematics is necessary to function in modern societies, supporting such pragmatic activities as telling time, cooking, and managing money.

Creating numeracy

Like literacy, numeracy is created in the brain through the synergy of biology and experience. Just as there are brain structures that have been designed through evolution for language, there are analogous structures for the quantitative sense. As in the case of literacy, however, genetically-destined structures alone cannot support mathematics. The activities of these structures are co-ordinated with those of supplemental neural circuits that were not specifically destined for numeracy but have been shaped to fit this function by experience. The neural structures not genetically specified for numeracy are gradually customised for numerical functions, a process Dehaene (1997) terms "neuronal recycling". Though experience plays a vital role in the formation of these supplementary networks, there are also biological constraints. The supplementary networks cannot be drawn from just any area of the brain: certain neural structures can be recruited for mathematics because they are sufficiently plastic and have properties that are conducive to processing number. Therefore, mathematics involves the co-operative functioning of a collection of neural networks that includes the genetically-specific quantitative structures and the experience-dependent, biologically-compatible structures.

Since the neural circuitry underlying mathematics is shaped by both biological and environmental factors, neuroscience can inform the construction of mathematics didactics in at least two main ways. First, an understanding of biological factors can contribute to the design of mathematics instruction that is consistent with biological factors and predispositions. Second, researchers can track the neurobiological effects of various forms of

---

1. It is important to note that the question posed is not whether higher-level mathematics should be taught at all, but whether it should be taught to all students or just to a minority of students who specialise in mathematics or in adjacent disciplines.

instruction and delineate the underlying learning pathways to mathematical knowledge. Once such pathways are mapped out, educators can strategically improve instruction and develop alternative pathways that accommodate individual differences. Thus, neuroscience can enable the design of more effective and inclusive mathematics instruction.

Babies calculate

With his/her supple cheeks and wobbly head, the infant seems a docile blank slate. Indeed, it was long believed that infants are born without any quantitative abilities and discover the world through fumbling sensory exploration. Many influential developmental theories underestimate young children's numerical understandings, including Piaget's (1952) theory of cognitive development.

Recent research has revealed that the infant's brain is equipped with a quantitative sense (Ferigenson, Dehaene and Spelke, 2004; Wynn, 1998). Infants possess two core number systems that allow them to quantity (Xu, 2003). One system supports the concepts of "one", "two", and "three". Infants are able precisely to discriminate these quantities from one another and from larger quantities. Moreover, they may have an abstracted concept of these numerical quantities that is insensitive to modality as they seem to connect the "twoness" common to two sounds and two objects (Starkey, Spelke and Gelman, 1990). The other core number system is approximate. It enables infants to discriminate among larger numbers with sufficiently high ratios. Therefore, infants can, for example, distinguish between 8 and 16, but not 8 and 9.

There is even evidence that infants can perform mathematical operations with these numbers. When one object is placed behind a screen followed by a second object, they expect to see two objects when the screen is removed, suggesting that they know that one plus one should equal two (Wynn, 1992). They can also perform approximate calculations, such as computing that five plus five equals about ten (McCrink and Wynn, 2004).

Contrary to the naive conception of the infant as a fumbling blank slate, this research suggests that infants are engaged in purposeful quantitative organisation of the world. Infants seem to be evolutionarily endowed with a number sense that is used as a perceptual tool to interpret the world numerically. That is, babies are born with an intuitive inclination to use numbers to understand the world and they build upon this understanding throughout early childhood.

So, young children have a substantial foundation of numerical understandings prior to formal education. Much current mathematics pedagogy is based upon outdated theories that underestimate young children's capacities. Instruction that ignores young children's base of numerical understandings overlooks a rich source of scaffolding. This early knowledge base can be used to facilitate understanding of formal mathematical concepts. Moreover, mapping symbolic mathematics onto real-world understandings helps to forge links between procedural and conceptual knowledge, which is critical for success in mathematics (Siegler, 2003). Together, this research strongly suggests that mathematics instruction should build upon children's informal and intuitive numerical understandings.

Numeracy in the brain

Research is beginning to elucidate the neural circuitry underlying numeracy. The genetically endowed quantitative sense of infants most likely resides in the parietal lobe. Mathematics is built "on top of" genetically-specified quantitative structures in a way that is roughly analogous to how literacy is constructed using language structures. Though mathematical abilities become much more sophisticated with education, the basic underlying number-processing mechanism is maintained as there are fundamental similarities in numerical cognition across the lifespan. For example, infants, children, and adults all show an identical ratio-dependent number discrimination signature (Cantlon *et al.*, 2006). Moreover, a recent fMRI study revealed that the intraparietal sulcus is the neural substrate or surface of non-symbolic number processing in both adults and children who have not yet begun formal schooling (Cantlon *et al.*, 2006). Therefore, though cultural, linguistic, and symbolic numerical practices that come with formal education alter the network of brain regions involved in numeracy, the intraparietal sulcus may well remain the nucleus of mature mathematical networks.

The parietal cortex indeed plays a fundamental role in a variety of mathematical operations (Dehaene, 1997). Damage to this area has devastating effects on mathematical abilities. For example, patients with parietal damage cannot answer a question as simple as, which number falls between 3 and 5. However, they have no difficulty solving analogous serial tasks across other domains, such as identifying which month falls between June and August, or which musical note is between do and mi.[2] And, they are able to solve some concrete problems that they cannot solve abstractly. For example, they know that there are two hours between 9 a.m. and 11 a.m., but they are unable to subtract 9 from 11 in symbolic notation.

This pattern of results exemplifies two principles about mathematics in the brain. First, mathematics is dissociable from other cognitive domains. Second, abilities within the domain of mathematics can be dissociable from one another. The first of these principles supports the notion of a multiplicity of partially distinct intelligences. It suggests at least that deficits or talents in particular domains do not necessarily imply deficits or talents in other domains. A child may, for example, struggle with reading but have excellent mathematics abilities. Therefore, it is important that teachers provide flexible pathways to mathematical knowledge which include multiple means and methods of representation and assessment. Without such flexibility, difficulties in other domains may unnecessarily interfere with mathematics learning. Consider, for example, children with dyslexia learning mathematics. These children would have difficulties accessing mathematical knowledge from printed textbooks and would struggle to demonstrate their understanding in paper-and-pencil exams. These types of avoidable problems impede learning and mask mathematics abilities. If students are given the option of alternative means of representation and assessment, such as electronic text with text-to-speech software, children with dyslexia would not fall behind in mathematics while their reading skill is developing. This example illustrates the importance of providing flexible pathways to mathematical knowledge, which, more broadly, implies:

Mathematics teachers should provide multiple means of representation and assessment.

What is more, abilities within the domain of mathematics can be dissociable from one another: teachers cannot assume that difficulties or talents in one area of mathematics are indicative of global mathematical ability. Ability in a certain mathematical skill is not necessarily predictive of ability in another, raising questions about the validity of the criteria used when tracking children into ability groups. Since the sequencing of curricula is not informed by a knowledge of which abilities are distinct in the brain, a child may be capable of excelling at a skill classified as advanced, yet struggle with a prerequisite skill.

---

2. Do and mi correspond to C and E, respectively.

As a result, this child could be erroneously tracked into a low ability group, thereby stifling their potential. Future neuroscience research may lead to the construction of a differentiated map of mathematics in the brain. However, until this has been achieved:

The validity of the criteria used for tracking in mathematics is questionable.

Because mathematical abilities are distributed in different parts of the brain, very simple numerical operations require the co-ordination of multiple brain structures. The mere representation of numbers involves a complex circuit. The triple-code model describes three levels of number processing: magnitude, visual and verbal (Dehaene and Cohen, 1995). Magnitude representation, or abstract quantitative meaning, such as "threeness", relies on the inferior parietal circuit. Visual representation involves the inferior occipito-temporal cortex. Numerical representation, such as "3", can recruit this area bilaterally, while linguistic representation, such as "three", relies solely upon this area in the left hemisphere. Verbal representation calls on or recruits perisylvian areas in the left hemisphere. Therefore, the simple representation of number involves many different brain areas, including the inferior parietal cortex central to numeracy.

Calculation also calls on a distributed network (Dehaene, 1997). Subtraction is critically dependent on the inferior parietal circuit, while addition and multiplication engage yet other networks, including a cortico-subcortical loop involving the basal ganglia of the left hemisphere. Research on advanced mathematics is currently sparse, but it seems that higher-level operations recruit at least partially distinct circuitry. Results suggest that the neural circuits holding algebraic knowledge are largely independent of those related to mental calculation (Hittmair-Delazer, Sailer and Benke, 1995). Moreover, there are additional neural networks involved in the orchestration of mathematics, including the prefrontal cortex and the anterior cingulated cortex. Figure 5.1 provides a diagram of cerebral areas known to be linked to number processing.

Although the neural circuitry underlying mathematics is only now beginning to be uncovered, it is already clear that mathematics involves a highly dispersed network of brain structures. Even an act as simple as multiplying two digits requires the collaboration of millions of neurons distributed in many brain areas. Given that numerical knowledge

Figure 5.1. **Cerebral areas**

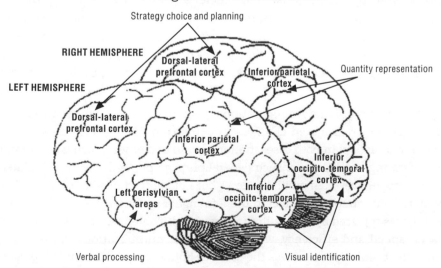

Source: Dehaene, S. and L. Cohen (1995), *Mathematical Cognition*, Vol. 1.

relies on such widely distributed brain circuitry, students need to learn to co-ordinate the activities of distinct brain regions underlying various operations and concepts, which suggests that:

One role of formal mathematics education is to bring coherence and fluidity to numerical knowledge.

This is particularly appropriate for education as mathematics is an emergent property of distributed and partially dissociable, distinct networks which are co-ordinated through experience.

Number and space

The parietal circuit critical for numeracy is also involved in the representation of space, and these two functions are intertwined (Dehaene, 1997). For example, many patients with acalculia also experience spatial difficulties, such as in distinguishing left from right (Mayer *et al.*, 1999). More generally, young children conceptualise number as spatially oriented before being introduced formally to the number line. Indeed, there may well be a biological predisposition to associate number with space. Therefore, instructional methods based upon the metaphors of number and space are formal representations of intuitive concepts, and provide concrete models for abstract concepts. Therefore, teaching tools such as the number line and concrete spatial manipulatives (*i.e.*, blocks, rods, board games, measuring tools, etc.) can reinforce and solidify children's intuitive mathematical understandings. The strong link between number and space in the brain suggests that:

Instructional methods that link number and space are powerful teaching tools.

Educational research confirms the value of such techniques. An intervention programme conducted by Griffin, Case and Siegler (1994) with a central focus on the association between number and space showed clear success. The programme made use of the number line, as well as a variety of concrete manipulatives that link number and space. The results were striking: forty 20-minute sessions propelled children who were lagging behind their peers to the top of their class.

The role of instruction

As demonstrated by Griffin, Case and Siegler's (1994) intervention programme, instruction can have powerful effects on mathematics achievement. These achievement gains likely reflect underlying neural changes for recent research indicates that learning new mathematical knowledge can dramatically alter brain activity patterns (Delazer *et al.*, 2003, 2004). The changes that occur seem to be a function of both the content and method of instruction.

Learning different mathematical operations leads to distinct changes in activation. Ischebeck *et al.* (2006) investigated changes in activation due to training with either multiplication or subtraction. In both cases, training led to decreases in activation in inferior frontal areas, indicating a reduction of such general purpose properties as working memory and executive control. However, in multiplication – but not subtraction – training also led to a shift in activation from intraparietal sulci to the left angula gyrus, suggesting that quantity-based processing was replaced by more automatic retrieval. Therefore, training in subtraction led to increased speed and efficiency, while training in multiplication resulted in novel strategies. Since the training protocol was the same in both cases, these results indicate that the neurobiological effects of learning mathematics are partially dependent upon content.

The neurobiological effects of learning mathematics are also mediated by instructional method. Delazer *et al.* (2005) found that learning by drill, which involved learning by heart to associate a specific result with two operands, was encoded in a different neural substrate than learning by strategy, which consisted of applying a sequence of arithmetic operations. Retrieval after learning by drill more strongly activated medial parietal regions extending to the left angural gyrus, while retrieval after learning by strategy was associated with the activation of the precuneus. This result demonstrates that different instructional methods can create different underlying neural pathways to the same mathematical knowledge. For example, two children may both answer that 10 plus 10 equals 20, but if one child has memorised this fact while the other is applying the strategy of double-digit addition, the children are engaging distinct neural circuitry.

These results have important implications for student assessment. Since the process by which knowledge is encoded influences its neural substrate or circuitry, dichotomous correct/incorrect measures of assessment are inadequate for assessing understanding as they cannot differentiate between, for example, knowledge which has been encoded as fact and knowledge encoded through strategy. More sensitive measures of assessment are necessary to assess underlying understanding. Stevenson and Stigler (1992) have identified an approach to mathematics assessment that more effectively assesses underlying processes. This approach, which is commonly used by Asian mathematics teachers, involves ongoing assessments which describe the process of learning in rich detail. With this assessment approach, emphasis is on the delineation of learning pathways, rather than the identification of correct or incorrect responses. Indeed, mistakes are used as opportunities to identify learning gaps and develop understanding. This type of assessment practice distinguishes between knowledge which has been encoded as a fact and knowledge encoded through strategy, suggesting that:

Process-focused assessments can provide more accurate representations of knowledge than dichotomous measures of correct/incorrect responses.

Further, Delazer *et al.* (2005) found that strategy learning resulted in greater accuracy and transferability than the drill condition. These results suggest that the neural pathway underlying drill learning is less effective than the neural pathway underlying strategy learning. While further research is needed to test this conclusion across various types of problems, the results suggest that:

Teaching by strategy leads to more robust neural encoding of mathematical information than teaching by drill.

These findings indicate that different instructional methods can lead to the creation of neural pathways that vary in effectiveness, underscoring the crucial role of instruction. Future research is needed to document the neurobiological impact of various types of instruction. It will be necessary to assess the effects of different instructional techniques on sub-populations as individual differences often play a key mediating role.

Gender and mathematics

Gender is one source of individual differences in mathematics which shows up with consistent differences in performance on particular spatial tasks. For example, males tend to outperform females on tests of mental rotation ability. It is possible that this is a biologically conditioned effect as the difference is large ($d = 0.6$ to 1.0), appears as early as testing is done, and is not decreasing across historical time (Newcombe, Mathason and Terlecki, 2002). Showing that performance is a function of biological factors does not, however, eliminate experiential factors as playing a causal role as well. Indeed, training can yield significant improvements in the mental rotation ability of both males and females, with the possible eventual elimination of sex differences (Newcombe, Mathason and Terlecki, 2002). All documented gender differences in spatial tasks seem to be responsive to training (Newcombe, Mathason and Terlecki, 2002). Therefore, it would be useful to shift the focus from the endeavour of establishing a rank ordering of abilities in their association to gender to one of identifying instructional pathways which enable all students to attain spatial abilities.

It is a common misconception that males outperform females on measures of global mathematical ability. In fact, the results are context-dependent: while males, on average, score higher on standardised mathematics assessments, females tend to outperform males in mathematics examinations in school (De Lisi and McGillicuddy-De Lisi, 2002). This pattern of results is likely due to patterned differences in use of strategy. Females tend to adhere to algorithmic methods, while males are more inclined to venture from teacher-taught algorithms and experiment with novel approaches (De Lisi and McGillicuddy-De Lisi, 2002). Therefore, males excel on standardised tests, which contain a significant portion of questions requiring non-algorithmic solutions, while females shine on classroom tests which often depend on taught algorithmic methods. These gender strategy differences underscore the importance of developing multiple pathways to mathematical knowledge to accommodate individual differences.

Barriers to learning mathematics

Some children even with sufficient instruction have difficulties with mathematics because of dyscalculia, the mathematical equivalent of dyslexia. Dyscalculia is most likely caused by an impairment of number sense – the early understandings of numerical quantities and their relations (Landerl, Bevan and Butterworth, 2004). Scientists are only beginning to investigate the neural underpinnings of dyscalculia. Recent neuro-imaging studies have revealed specific anatomical and functional features in the intraparietal sulcus of certain groups of children with dyscalculia. For example, Isaacs et al. (2001) compared the density of grey matter between two groups of adolescents who were born prematurely to a similar degree but differed in the presence or absence of dyscalculia. At the whole-brain level, those with dyscalculia showed reduced grey matter in the left intraparietal region precisely where activation is observed typically when arithmetic is being performed. Notwithstanding that further research is needed to clarify the neural underpinnings of dyscalculia, the discovery of anatomical characteristics associated with selective mathematics impairment supports the notion that mathematics does not emerge solely from a cultural construction process: it requires the full functioning and integrity of specific brain structures providing a conceptual foundation for learning.

There is a good likelihood that the deficient neural circuitry underlying dyscalculia can be addressed and rectified through targeted intervention because the mathematics circuitry seems to be plastic. Learning new number facts or strategies is able to alter brain activity (Delazer et al., 2003; Delazer et al., 2004); brain-lesioned patients with mathematics deficits can be rehabilitated; many patients have regained considerable mathematical competence as a result of intensive training specifically focused on their area of deficit (Girelli et al., 1996). These results suggest that individuals with dyscalculia could well be treated and rehabilitated. Further research is needed to identify the neural underpinnings of dyscalculia in order to design effective targeted interventions analogous to those already developed for dyslexia.

Emotion is also involved in difficulties with mathematics. Fear associated with mathematics is a relatively common experience, a condition termed math anxiety (Ashcraft, 2002). This emotional state can disrupt cognitive strategies and working memory (Ashcraft and Kirk, 2001). It is an issue for mathematics education meriting further research to identify appropriate remedies.[3]

Conclusions

Though the neuroscience of mathematics is only in its infancy, the field has made great strides in the past decade. Scientists have begun uncovering relevant biological patterns, such as the association between number and space, and linking it with the rapidly expanding field of genetics. Researchers are just beginning to explore the effects of mathematics instruction on the brain, which calls for a dynamic, developmental perspective so that the multiple underlying pathways can be charted. As with literacy, understanding the underlying developmental pathways to mathematics from a biological perspective will enable the design of differentiated instructional models that are appropriate for a wide diversity of learners.

Currently, most neuroscience research focuses on basic mathematics, providing important implications for mathematics instruction. In contrast to advanced mathematics, it is unquestionably vital for all students as being necessary to function in modern society and supporting such practical activities as telling time, cooking, and managing money.

As the effects of mathematics on the brain are better understood, it will inform the key question as to how much mathematics should be taught to all students. If, for example, learning higher-level mathematics is found to shape the brain in support of useful modes of thought, this would provide a justification for including advanced mathematics in the standard curriculum. If, on the other hand, the impacts of learning advanced mathematics are restricted to acquiring higher-level mathematics skills, it would be useful to consider whether this level of mathematics need only to be taught to learners for whom it will be useful in advancing their maths studies. In this way, neuroscience research could both provide valuable insights on how mathematics should be taught in primary school and importantly shape mathematics curriculum and instruction in the secondary school.

---

3. As Butterworth has shown, dyscalculia can, in addition to all this, have a domino-effect. Teachers, other students, parents, and even the individual with dyscalculia him/herself, may label the child as "stupid" just because of difficulties in math. Such a label will result in lower academic performance in all subjects because of its devastating effect on self-esteem.

References

Ashcraft, M.H. (2002), "Math Anxiety: Personal, Educational, and Cognitive Consequences", *Current Directions in Psychological Science*, Vol. 11, pp. 181-185.

Ashcraft, M.H. and E.P. Kirk (2001), "The Relationships among Working Memory, Math Anxiety, and Performance", *J Exp Psychol Gen*, Vol. 11, pp. 224-237.

Cantlon, J., E. Brannon, E. Carter and K. Pelphrey (2006), "Functional Imaging of Numerical Processing in Adults and 4-y-old Children", *PLoS Biology*, Vol. 4, No. 5, pp. 844-845.

Dehaene, S. (1997), *The Number Sense*, Oxford University Press, NY.

Dehaene, S. and L. Cohen (1995), *Mathematical Cognition*, Vol. 1, pp. 83-120.

Delazer, M., F. Domahs, L. Bartha, C. Brennis, A. Lochy and T. Trieb *et al.* (2003), "Learning Complex Arithmetic – An FMRI Study", *Brain Res Cogn Brain Res*, Vol. 18, No. 1, pp. 76-88.

Delazer, M., F. Domahs, L. Bartha, C. Brennis, A. Lochy and T. Trieb *et al.* (2004), "The Acquisition of Arithmetic Knowledge – An FMRI Study", *Cortex*, Vol. 40, No. 1, pp. 166-167.

Delazer, M., A. Ischebeck, F. Domahs, L. Zamarian, F. Koppelstaetter, C.M. Siedentopf, L. Kaufmann, T. Benke and S. Felber (2005), "Learning by Strategies and Learning by Drill – Evidence from an fMRI Study", *Neuroimage*, Vol. 25, pp. 838-849.

Ferigenson, L., S. Dehaene and E. Spelke (2004), "Core Systems of Number", *Trends in Cognitive Neuroscience*, Vol. 8, No. 7, pp. 1-8.

Girelli, L., M. Delazer, C. Semenza and G. Denes (1996), "The Representation of Arithmetical Facts: Evidence from Two Rehabilitation Studies", *Cortex*, Vol. 32, No. 1, pp. 49-66.

Griffin, S.A., R. Case and R.S. Siegler (1994), "Rightstart: Providing the Central Conceptual Prerequisites for First Formal Learning of Arithmetic to Students at Risk for School Failure", in K. McGilly (ed.), *Classroom Lessons: Integrating Cognitive Theory and Classroom Practice*, MIT Press, Cambridge.

Hittmair-Delazer, M., U. Sailer and T. Benke (1995), "Impaired Arithmetic Facts but Intact Conceptual Knowledge – A Single Case Study of Dyscalculia", *Cortex*, Vol. 31, pp. 139-147.

Isaacs, E.B., C.J. Edmonds, A. Lucas and D.G. Gadian (2001), "Calculation Difficulties in Children of Very Low Birthweight: A Neural Correlate", *Brain*, Vol. 124, pp. 1701-1707.

Ischebeck, A., L. Zamarian, C. Siedentopf, F. Koppelsätter, T. Benke, S. Felber and M. Delzer (2006), "How Specifically do we Learn? Imaging the Learning of Multiplication and Subtraction", *NeuroImage*, Vol. 30, pp. 1365-1375.

Landerl, K., A. Bevan and B. Butterworth (2004), *Developmental Dyscalculia and Basic Numerical Capacities: A Study of 8-9-Year-Old Students. Cognition*, Vol. 93, No. 2, pp. 99-125.

De Lisi, R. and A. McGillicuddy-De Lisi (2002), "Sex Differences in Mathematical Abilities and Achievement", in A. McGillicudddy-De Lisi and R. De Lisi (eds.), *Biology, Society, and Behavior: The Development of Sex Differences in Cognition*, Alex Publishing, London, pp. 155-181.

Mayer, E., M. Martory, A. Pegna, T. Landis, J. Delavelle and J. Annoni (1999), "A Pure Case of Gestmann Syndrome with a Subangular Lesion", *Brain*, Vol. 122, pp. 1170-1120.

McCrink, K. and K. Wynn (2004), "Large-number Addition and Subtraction by 9-Month-Old Infants", *Psychol. Sci*, Vol. 15, No. 11, pp. 776-781.

Newcombe, N., L. Mathason and M. Terlecki (2002), "Sex Differences in Mathematical Abilities and Achievement", in A. McGillicudddy-De Lisi and R. De Lisi (eds.), *Biology, Society, and Behavior: The Development of Sex Differences in Cognition*, Alex Publishing, London, pp. 155-181.

Piaget, J. (1952), *The Child's Conception of Number*, Norton, NY.

Siegler, R.S. (2003), "Implications of Cognitive Science Research for Mathematics Education", in J. Kilpatrick, W.B. Martin and D.E. Schifter (eds.), *A Research Companion to Principles and Standards for School Mathematics*, National Council of Teachers of Mathematics, Reston, VA, pp. 219-233.

Starkey, P., E.S. Spelke and R. Gelman (1990), "Numerical Abstraction by Human Infant", *Cognition*, Vol. 36, pp. 97-127.

Stevenson, H.W. and J.W. Stigler (1992), *The Learning Gap: Why our Schools are Failing and what we can Learn from Japanese and Chinese Education*, Summit Books, New York.

Wynn, K. (1992), "Addition and Subtraction by Human Infants", *Nature*, Vol. 358, pp. 749-750.

Wynn, K. (1998), "Numerical Competence in Infants", in C. Donlan (ed.), *The Development of Mathematical Skills*, Psychology Press, East Sussex, UK, pp. 1-25.

Xu, F. (2003), "Numerosity Discrimination in Infants: Evidence for Two Systems of Representations", *Cognition*, Vol. 89, No. 1, pp. 15-25.

ISBN 978-92-64-02912-5
Understanding the Brain: The Birth of a Learning Science
© OECD 2007

PART I

Chapter 6

Dispelling "Neuromyths"

When facing Truth, there are three categories of people:
Those who long for it; they are the fewest.
Those who do not care; they are the happiest.
Those who already have it; they are the most dangerous.

Anonymous

This chapter addresses some of the pitfalls that arise when erroneous or unfounded bridges get made between neuroscience and education. This is done by outlining and dispelling a number of "neuromyths". They include unfounded ideas concerning left-side and right-side thinking, the determinism of developments in infancy, gender differences, and multilingualism. This chapter is highly relevant for all those concerned about learning, and especially those who are keen to avoid faddish solutions without scientific underpinning.

What is a "neuromyth"?

Science advances through trial and error. Theories are constructed on the basis of observation which other phenomena come to confirm, modify, or refute: another theory, complementary or contradictory to the previous one, is then created, and so the process continues. This bumpy advance of science is unavoidable but it has its drawbacks. One is that hypotheses which have been invalidated nevertheless leave traces and if these have captured a wider imagination, "myths" take root. These beliefs may have been demolished by science but they prove to be stubbornly persistent and passed on through various media into the public mind.

Neuroscience is inevitably caught up in this phenomenon. Some expressions in the English language confirm this: "number sense", for example, derives from the research of a German anatomist and physiologist, Franz Joseph Gall (1758-1828). By examining the heads of convicted living criminals and dissecting the brains of deceased ones, Gall established phrenology theory: a particular talent would produce an outgrowth on the brain which pushes on the bone and distorts the skull. By feeling the head, Gall boasted that he could identify the criminal from the honest man, a "maths" person from a "literary" one. Phrenology has long been superseded, indeed discredited. To be sure, certain areas of the brain are specialised more than others with certain functions. But, contrary to the regions that Gall thought he had identified, it is instead a question of functional specialties (such as image formation, word production, tactile sensibility, etc.) and not of moral characteristics like kindness, combativeness, etc.[1]

Science itself is not solely responsible for the emergence of such myths. It is often difficult to understand all the subtleties of a study's findings, still more its protocols and methodological details. Nevertheless, human nature is often content with – even takes delight in – quick, simple, and unequivocal explanations.[2] This inevitably leads to faulty interpretations, questionable extrapolations, and, more generally, the genesis of false ideas.[3]

This chapter examines one by one the main myths belonging to brain science, with particular attention given to those most relevant to learning methods. For each myth, a historical look will explain how the idea took hold and then the current state of scientific research on the subject will be reviewed. Ironically perhaps, some myths have actually been beneficial to education in that they provided "justification" for it to diversify. But, mostly they bring unfortunate consequences and must therefore be dispelled.

1. Gall had also presupposed the existence of areas suited to languages and arithmetic.
2. The mass media critically influence opinions, and they are especially open to excessive simplification (on this, see Bourdieu, *On Television*, New Press, 1998).
3. Scientists are in no way impervious to this tendency, however. Though expected to be rigorous in their field, when addressing audiences far from their research, they are all too human and subject to subjective and emotional influences.

"There is no time to lose as everything important about the brain is decided by the age of three"

If you enter the keywords "birth to three" into a search engine on your computer, you get an impressive number of websites explaining that your child's first three years are crucial for his/her future development and that practically everything is decided at this age. You will also find numerous commercial products prepared to stimulate your young child's intelligence, before reaching this all-important threshold age.

Some physiological phenomena that take place during brain development can, indeed, lead to beliefs that the critical learning stages occur between birth and age three. But it can be easily exaggerated and distorted. It takes on mythical status when it is overused by certain policy makers, educators, toy manufacturers, and parents, who overwhelm their children with gymnastics for newborns and stimulating music in tape recorders and CD players attached above the baby's bed. What are the physiological phenomena that research has uncovered which are relevant to this belief?

The basic component of information processing in the brain is the nerve cell or neuron. A human brain contains about 100 billion neurons. Each one can be connected with thousands of others, which allows nerve information to circulate intensively and in several directions at a time. Through the connections between neurons (synapses), nerve impulses travel from one cell to another and support skill development and learning capacity. Learning is the creation of new synapses, or the strengthening or weakening of existing synapses. Compared to an adult, the number of synapses in newborns is low. After two months of growth, the synaptic density of the brain increases exponentially and exceeds that of an adult (with a peak at ten months). There is then a steady decline until age 10, when the "adult number" of synapses is reached. A relative stabilisation then occurs. The process by which synapses are produced *en masse* is called synaptogenesis. The process by which synapses decline is referred to as pruning. It is a natural mechanism, necessary for growth and development.

For a long time, science believed that the maximum number of neurons was fixed at birth; unlike most other cells, neurons were not thought to regenerate and each individual would then lose neurons regularly. In the same way, following a lesion of the brain, destroyed nerve cells would not be replaced. For the past twenty years, findings have changed this view by revealing hitherto unsuspected phenomena: new neurons appear at any point in a person's life (neurogenesis) and, in some cases at least, the number of neurons does not fluctuate throughout the lifetime.

That said, synaptogenesis is intense in the very early years of life of a human being. If learning were to be determined by the creation of new synapses – an idea with some intuitive appeal – it is a short step to deduce that it is in the early years of a child when (s)he is most capable of learning. Another version, more current in Europe, is the view that very young children must be constantly stimulated in their first two to three years in order to strengthen their learning capacities for subsequent life. In fact, these claims go well beyond the actual scientific evidence.

An experiment conducted twenty years ago may, however, have fuelled such a myth. Laboratory studies with rodents showed that synaptic density could increase when the subjects were placed in a complex environment, defined in this case as a cage with other rodents and various objects to explore. When these rats were subsequently tested on a maze learning test, they performed better and faster than other rats belonging to a control group and living in "poor" or "isolated" environments (Diamond, 2001). The conclusion was that rats living in "enriched" environments had increased synaptic density and were thus better able to perform the learning task.

The elements were in place to create a myth: a great experiment, rather easy to understand even if difficult to perform, and findings that project the expected outcome. The experiment, however, took place in the laboratory in highly artificial conditions.[4] It was conducted on rodents. Non-specialists twisted experimental data on rats, obtained with unquestionable scientific precision, and combined it with current ideas concerning human development to conclude that educational intervention, to be more effective, should be co-ordinated with synaptogenesis. Alternatively, they suggested that, "enriched environments" save synapses from pruning during infancy, or even create new synapses, and thereby contribute to greater intelligence and higher learning capacity. This is a case of using facts established in a valid study to extrapolate conclusions that go well beyond the original evidence.

The limits and lessons in this case are rather clear. There is little human neuroscientific data on the predictive relationship between synaptic density early in life and improved learning capacity. Similarly, little is available regarding the predictive relationship between the synaptic densities of children and adults. There is no direct neuroscientific evidence, for either animals or humans, linking adult synaptic density to greater learning capacity. All of this does not mean that the plasticity of the brain, and synaptogenesis in particular, might not bear some relation to learning but, on the strength of available evidence, the assumptions made in identifying such a determining role for birth-to-three development cannot be sustained.

For further reading, the reader should consult John Bruer's *The Myth of the First Three Years* (2000). He was the first systematically to contest this myth, which he presented as "rooted in our cultural beliefs about children and childhood, our fascination with the mind-brain, and our perennial need to find reassuring answers to troubling questions". Bruer goes back to the 18th century to find its origin: it was already believed that a mother's education was the most powerful force to map out the life and fate of a child; successful children were those who had interacted "well" with their family. He eliminates one by one the myths based on faulty interpretations of early synaptogenesis.

"There are critical periods when certain matters must be taught and learnt"

The influence of the intense synaptogenesis early in life on the adult brain is not yet known, but it is known that adults are less capable of learning certain things. Anyone who starts to learn a foreign language later in life, for example, will in all likelihood always have a "foreign accent"; the virtuosity of a late learner of an instrument will in all probability

---

4. In the wild, rats live in stimulating environments (docks, pipes, etc.) and have the number of synapses needed to survive; when put in an artificially impoverished environment, their brains have the synaptic density appropriate for that environment. In short, they will be just as "smart" as they need to be to live in a laboratory cage. The same reasoning could apply to human beings, but the facts remain to be proven. In this case, most people's brains are adjusted to a reasonably stimulating environment. Research has shown that even children growing up in what could be defined as an impoverished environment (a ghetto, for example) may over time come to excel in school and go on to higher education. There are simply too many factors to take into account when defining what an "enriched" environment should be for the majority of students to make any predictions about intellectual capacity so that results such as this are, in their current state, not applicable for education.

never equal that of a child practised with the same musical instruction from the age of 5. Does this mean that there are periods of life when certain tasks can no longer be learned? Or are tasks merely learned more slowly or differently at different times?

For a long time it was believed that the brain loses neurons with age, but the measures opened by new technologies have challenged this certainty. Terry and his colleagues showed that the total number of neurons in each area of the cerebral cortex is not age-dependent but only the number of "large" neurons. Nerve cells shrink, resulting in a growing number of small neurons but the aggregate number of all neurons remains the same. Certain parts of the brain, like the hippocampus, have recently been found actually to generate new neurons throughout the lifespan. The hippocampus is, among other things, involved in spatial memory and navigation processes (Burgess and O'Keefe, 1996). Research comparing London taxi drivers with random other citizens suggests a strong relationship between the relative size and activation of the hippocampus, on the one hand, and a good capacity for navigation, on the other; there is a positive correlation between the enlargement of the auditory cortex and the development of musical talent, as there is growth of motor areas of the brain following intense training of finger movements. In the latter case, changes in the neuron network configuration linked to the learning could be measured using brain imaging from the fifth day of training, *i.e.* after an extremely brief period of learning.

The processes that remodel the brain – neuron synaptogenesis, pruning, development, and modification – are grouped together under the same term: "brain plasticity". Numerous studies have shown that the brain remained plastic throughout the lifespan, in terms of numbers of both neurons and synapses. The acquisition of skills results from training and the strengthening of certain connections, but also from pruning certain others. There is a distinction that needs to be drawn between two types of synaptogenesis – the one that occurs naturally early in life and the other resulting from exposure to complex environments throughout the lifespan. Researchers refer to the first as "experience-expectant learning" and to the second as "experience-dependent learning". Grammar is learned faster and easier up to approximately age 16, while the capacity to enrich vocabulary actually improves throughout the lifespan (Neville, 2000). Grammar gives an example of *sensitive-period* learning and is experience-expectant: for learning to be done without excessive difficulty, it must ideally take place in a given lapse of time (the sensitive period). *Experience-expectant learning is thus optimal during certain periods of life.* Learning that does not depend on a sensitive period, such as the acquisition of vocabulary, is "experience-dependent": when the learning best takes place is not constrained by age or time and this type of learning can even improve as the years go by (see Chapter 2).

Are there "*critical periods*" as unique phases during which certain types of learning can only successfully take place? Can certain skills or even knowledge only be acquired during relatively short "windows of opportunity" which then close once-and-for-all at a precise stage of brain development? The concept of "critical period" dates back to experiments conducted in the 1970s by the ethologist Konrad Lorenz which are relatively well-known by the general public. He observed that fledglings on hatching became permanently attached to the prominent mobile object of the environment, usually their mother, which attachment Lorenz named as "imprinting". By taking the place of the mother, Lorenz managed to become attached to fledglings which followed him everywhere. The period that allows this attachment is very short (right after hatching); once in place, it was impossible to change the attachment object and the fledglings permanently followed the substitute instead of their mother. The term "critical period" is appropriate for such a case as an event (or its absence) during a specific period brings about an irreversible situation.[5]

No critical period for learning has yet been found for humans (though they may yet be). It is more appropriate to refer to "sensitive periods", when learning of a particular kind is easier. The scientific community acknowledges that there are sensitive periods, particularly for language learning, and has identified several of them (some in adult age). A key research question is whether the programmes of education systems match the succession of sensitive periods, and brain imaging will be able to bring new explanations concerning the biological processes linked to these periods.

Language learning provides good examples of "sensitive periods". At birth, children can distinguish all the sounds of the language, even those very different from the native language of their parents. So, for example, while Japanese adults experience difficulty in telling the difference between the r and l sounds which are perceived as identical, the very young Japanese baby is able to distinguish between them. Perception is rapidly shaped by the sound environment of the child over the course of its first twelve months, by which time (s)he can no longer pick up differences which have not been part of that environment. The ability to differentiate foreign sounds diminishes between the sixth and the twelfth months during which time the brain changes so that the child can become a very competent speaker in the native language. Since the native language sound repertoire does not require the acquisition of new sounds, but conversely the "loss" of non-perceived, non-produced ones, we can hypothesise that this process is completed by successive pruning of synapses. An important reason why it is preferable to denote this aspect of human learning in terms of "sensitive" rather than "critical" periods is that it refers to a loss not an increase in information. However described, there is no doubt that the ability to reproduce the sounds of a language (phonology, accent) and the capacity effectively to integrate grammar are optimal during childhood, while only the faculty to acquire vocabulary among linguistic competences endures equally well throughout the entire lifespan.

The work of Piaget has greatly influenced the organisation of school systems over the course of the latter decades of the 20th century. The basic Piagetian idea about development is that children experience specific periods of cognitive development so that they are not capable of learning certain skills before relatively fixed ages. This applies to reading and counting and in school systems in OECD countries, reading, writing, and arithmetic are not officially taught before ages 6 or 7. Piaget and his colleagues proposed *inter alia* that children come into this world without any preconceived ideas about numbers. But recent research on the workings of the brain has shown that children are born with an innate sense of the representation of numbers (Dehaene, 1997). It is not a matter of calling into question all of Piaget's findings and he had rightly identified the

---

5. Animal studies must in any case be considered with extreme precaution (which, by the way, Lorenz and some others seem to have forgotten at some stages). By analogy with experiments conducted on rodents, the belief has arisen that providing stimulating environments for students will increase their brain connectivity and thus produce better students. Recommendations have been suggested that teachers (and parents) should provide a colourful, interesting and sensory meaningful environment to ensure a bright child. Arguing from the data on rats about the need for "enriched environments" for children is unjustified (*e.g.*, listening to Mozart, looking at coloured mobiles), particularly considering that parallel neuroscientific studies of the affect of complex or isolated environments on the development of human brains have not been conducted.

importance of truly sensitive periods. But, children are more "gifted" at birth than researchers thought for a long time (Gopnik, Meltzoff and Kuhl, 2005). This is why Piaget's influential theories must be put into perspective by this kind of research.

"But I read somewhere that we only use 10% of our brain anyway"

It is often said that humans only use 10% (sometimes 20%) of their brain. Where did this myth come from? Some say it came from Einstein, who responded once during an interview that he only used 10% of his brain. Early research on the brain may have supported this myth. In the 1930s, Karl Lashley explored the brain using electric shocks. As many areas of the brain did not react to these shocks, Lashley concluded that these areas had no function. This is how the term "silent cortex" came into circulation. This theory is now judged to be incorrect. Dubious interpretations of the brain's functioning have also fuelled this myth.

Today, thanks to imaging techniques, the brain can be precisely described in functional areas. Each sense corresponds to one or several primary functional areas: a primary visual area, which receives information perceived by the eye; a primary auditory area, which receives information perceived by the ear, etc. Several regions are linked to the production and comprehension of language. They are sometimes described separately by physiologists, and the public which remembers these partial descriptions may gain the impression that the brain functions area by area. This would be consistent with the image that, at any one moment, only a small region of the brain is active but this is not what occurs. The primary areas are surrounded by secondary areas, so that, for example, information from images perceived by the eye is sent to the primary visual areas, and then is analysed in the secondary visual areas where the three-dimensional reconstitution of the perceived objects takes place. Information from the memory of the subject circulate in the brain to recognise objects, while semantic information from language areas comes into play so that the person can quickly name the object seen. At the same time, the brain areas that deal with posture and movement are in action under the effect of nerve signals from the entire body, allowing the person to know whether (s)he is sitting or standing, with the head turned to the right or the left, etc. Therefore, a partial, fragmented description of the areas of the brain can lead to a misinterpretation of how it works.

Another origin of the myth may be found in the fact that the brain is made up of ten glial cells for every neuron. Glial cells have a nutritional role and support nerve cells, but they do not transmit any information. In terms of transmission of nerve impulses only the neurons are recruited (or 10% of the cells comprising the brain) so that this offers a further source of misunderstanding on which the "10% myth" might come. But this vision of cell functions is simplistic: while the glial cells play a different role from that of the neurons, they are no less essential to the functioning of the whole.

Neuroscience findings now show that the brain is 100% active. In neurosurgery, when it is possible to observe the functions of the brain on patients under local anaesthetic, electric stimulations show no inactive areas, even when no movement, sensation, or emotion is being observed. No areas of the brain are completely inactive, even during sleep; if they were, it would indicate a serious functional disorder. Similarly, loss of very much less than 90% of brain tissue leads to serious consequences as no region of the brain can be damaged without causing physical or mental defects. The cases of people who have lived for years with a bullet lodged in the brain or similar trauma do not indicate "useless areas". If it is possible to completely recover from such a shock, it is the demonstration of the brain's extraordinary plasticity: neurons (or networks of neurons) have been able to replace those that were destroyed and in such cases the brain reconfigures itself to overcome the defect.

The myth is implausible for physiological reasons, too. Evolution does not allow waste and the brain, like the other organs but probably more than any other, is moulded by natural selection. It represents only 2% of the total weight of the human body but consumes 20% of available energy. With such high energy cost, evolution would not have allowed the development of an organ of which 90% is useless.

"I'm a 'left-brain', she's a 'right-brain' person"

The brain is made up of neuronal networks, it has functional areas that interact among themselves, and it is composed of a left and a right hemispheres. Each hemisphere is more specialised in certain fields than in others. Do these facts justify the strange statements to be heard in everyday life, such as: "me, I'm more left-brained" or "women have a more developed right brain"? Is there really a right brain and a left brain? A rapid overview of the origin of these terms is needed to determine whether they correspond to facts, or if it is again a matter of questionable extrapolations of scientific data. But to begin with, it needs to be underlined that the two hemispheres are not separate functional and anatomic entities: nerve structures connect them together (the corpus callosum) and many neurons have their cell nucleus in one hemisphere and extensions in the other. This alone should prompt reflection.

It has been said that the "left brain" is the seat of rational thinking, intellectual thinking, analysis, and speech. It also processes numerical information deductively or logically. It dissects the information by analysing, distinguishing, and structuring the parts of a whole, by linearly arranging the data. The left hemisphere is the best equipped to deal with tasks related to language (writing and reading), algebra, mathematic problem-solving, logical operations. Thus, it can be believed that people who are rational, intellectual, logical, and have a good analytical sense "preferentially use their 'left brain" and tend to be mathematicians, engineers, and researchers.

The "right brain" has been called the seat of intuition, emotion, non-verbal thinking, synthetic thinking, which allows representations in space, creation, and emotions. It tends to synthesise, recreates three-dimensional forms, notices similarities rather than differences, and understands complex configurations. It recognises faces and perceives spaces. From this stems the complementary myth that people who are intuitive, emotional, imaginative, and easily find their way around, "preferentially use their 'right brain' and engage in the artistic and creative professions".

This "left brain/right brain" opposition originated in the first neurophysiology research. Intellectual capacities were often then described in two classes: critical and analytical aptitudes, on the one hand, and creative and synthetic aptitudes, on the other. One of the major doctrines of neurophysiology from the 19th century associated each class to a hemisphere. In 1844, Arthur Ladbroke Wigan published *A New View of Insanity: Duality of the Mind*. He describes the two hemispheres of the brain as independent, and attributes to each one its own will and way of thinking: they usually work together but in some diseases they can work against each other. This idea caught the imagination with the publication of Robert Louis Stevenson's famous *The Strange Case of Doctor Jekyll and Mister Hyde* in 1866 which exploits the idea of a cultivated left hemisphere that opposes a primitive and emotional right hemisphere, which easily loses all control. Paul Broca, a

French neurologist, went beyond fiction to localise different roles in the two hemispheres. In the 1860s, he examined *postmortem* the brains of more than 20 patients whose language functions had been impaired. In all the brains examined, he noticed lesions in the frontal lobe of the left hemisphere, whereas the right hemisphere was still intact. He concluded that the production of spoken language had to be located in the front part of the left hemisphere. This was completed a few years later, by the German neurologist Wernicke who also examined *postmortem* brains of those who had had language development disorders; he suggested that the capacity to understand language is situated in the temporal lobe of the left hemisphere. Thus, Broca and Wernicke associated the same hemisphere of the brain, the left, to two essential components of processing language – comprehension and oral production.

Until the 1960s, methods for observing the dominant role of the left hemisphere in language use and processing (lateralisation of language) were based on studies *postmortem* of patients with brain lesions. Some neurologists nevertheless claimed that language might not be entirely a left hemisphere function in that it was impossible to conclude no role for the right hemisphere on the basis of lack of lesions there among those who had had language impairments. Lesions only on the left side could be random. The pertinence of this intuition was underlined by studies carried out on "split-brain" patients. The corpus callosum of these patients was severed in order to stop epileptic attacks from one hemisphere to the other. While the primary goal of the operation was to reduce epileptic fits, it also allowed researchers to study the role of each hemisphere on these patients. The first such studies were conducted in the 1960s and 1970s, with Medicine Nobel Prize winner Roger Sperry and his team from the California Institute of Technology playing a dominant role. They succeeded in supplying information to a single hemisphere in their "split-brain" patients and asked them to use each hand separately to identify objects without looking at them. This experimental protocol built on the fact that basic sensory and motor functions are symmetrically divided between the two hemispheres of the brain – the left hemisphere receives almost all sensory information from and controls movements to the right part of the body and *vice versa*. Sensory information from the right hand is received in the left hemisphere and that from the left hand in the right hemisphere. When patients touched an object with their right hand, they could easily name the object but not when they touched it with the left hand. Here was proof that the left hemisphere is the seat of principal language functions.

This unequal localisation of language functions created the idea of the left hemisphere as the verbal one and the right hemisphere as the non-verbal one. Since language has often been perceived as the noblest function of the human species, the left hemisphere was declared "dominant".

Other experiments with the same type of patients helped to clarify the role of the right hemisphere. A video made by Sperry and Gazzaniga about the split-brain patient W.J. gives a surprising demonstration of the superiority of the right hemisphere for spatial vision. The patient was given several dice, each with two red sides, two white sides, and two sides with alternating white and red diagonal stripes. The task of the patient was to arrange the dice according to patterns presented on cards. The beginning of the video shows W.J. quickly arranging the dice in the required pattern using his left hand (controlled, remember, by the right hemisphere). He had great difficulty, however, completing the same task using his right hand – he was slow and moved the dice indecisively. Once his left hand intervenes, he became quick and precise but when the researchers hold it back he again became indecisive. Other research by Sperry *et al.* (1969) confirmed the domination of the right hemisphere in spatial vision. This role was then confirmed by clinical case studies. Patients suffering from lesions in the right hemisphere were not able to recognise familiar faces; other patients had difficulty with spatial orientation.

Some patients with lesions in the right hemisphere have shown defects in identifying the emotional intonation of words and in recognising emotional facial expressions. Behavioural studies back up the clinical studies: speech rhythms are best perceived when the sounds are received by the left ear so that the information goes to the right hemisphere and images seen by the left visual field provoke greater emotional reaction. It was deduced from this that the right hemisphere was also specialised in the processes related to emotions.

This set of findings was ripe for spawning neuromyths. In 1970, Robert Ornstein's *The Psychology of Consciousness* hypothesised that "Westerners" use mainly the left half of their brain with a well-trained left hemisphere thanks to their focus on language and logical thinking. However, they neglect their right hemisphere and, therefore, their emotional and intuitive thinking. Ornstein associates the left hemisphere with the logical and analytical thinking of "Westerners" and the right hemisphere with emotional and intuitive "Oriental" thinking. The traditional dualism between intelligence and intuition is thus accorded a physiological origin, based on the difference between the two hemispheres of the brain. Apart from the highly questionable ethical aspect of Ornstein's ideas, they are the accumulated result of misinterpretations and distortions of available scientific findings.

Another widespread notion, without scientific foundation, stipulates that the left hemisphere tends to process quick changes and analyses the details and characteristics of the stimuli, while the right processes the simultaneous and general characteristics of the stimuli. This model remains entirely speculative. Starting from the differences between the verbal hemisphere (the left) and the non-verbal hemisphere (the right), a growing number of abstract concepts and relationships between mental functions and the two hemispheres has made their appearance on the neuromyth stage, moving further and further away from the scientific findings.

Gradually, further myths emerged in which the two hemispheres are associated not just with two ways of thinking but as revelations of two types of personality. The concepts of "left brain thinking" and "right brain thinking", together with the idea of a dominant hemisphere, led to the notion that each individual depends predominantly on one of the two hemispheres, with distinctive cognitive styles. A rational and analytical person could be characterised as "left-brained", an intuitive and emotional person as "right-brained". These cognitive styles, promoted through such media as magazines, "self-knowledge" books, and conferences, became popular and raised questions about their application in education. Is it necessary to imagine teaching methods more effectively adapted to the use of one or other of the hemispheres according to the supposed characteristics of the learner associated with that hemisphere? Do school programmes adopt teaching methods that use the entire brain or, with their focus on arithmetic and language, do they concentrate too much on the "left brain"?

The idea that western societies focus on only half of our mental capacities ("our left brain thinking") and neglect the other half ("our right brain thinking") became widespread, and some educationists and systems jumped on the bandwagon to recommend that schools change their teaching methods according to the dominant hemisphere concept. Educators like M. Hunter and E.P Torrance claimed that educational programmes were principally made for "left brains" and favour left brain-dependent activities like always sitting in class or learning algebra, instead of favouring the right hemisphere by allowing students to stretch out and learn geometry. Hence, methods were devised which sought to engage the two hemispheres, or even to emphasise activities related to the right hemisphere. Such an example is "show and tell": instead of just reading texts to the students (left hemisphere action), the teacher also shows images and graphs (right hemisphere actions). Other methods use music, metaphors, role-playing, meditation, or drawing, all to activate the synchronisation of the two hemispheres. Arguably, they have served to advance education by diversifying its methods. Nevertheless, insofar as they have borrowed on theories of the brain, they are based on scientific misinterpretation as the two halves of the brain cannot be so clearly separated.

No scientific evidence, indeed, indicates a correlation between the degree of creativity and the activity of the right hemisphere. A recent analysis of 65 studies on brain imaging and the processing of emotions concludes that such processing cannot be associated exclusively with the right hemisphere. Similarly, no scientific evidence validates the idea that analysis and logic depend on the left hemisphere or that the left hemisphere is the special seat for arithmetic or reading. Dehaene (1997) found that the two hemispheres are active when identifying Arab numerals (*e.g.* 1 or 2 or 5). Other studies show that, when the components of reading processes are analysed (*e.g.* decoding written words or recognising sounds for the higher level processes, such as reading a text), sub-systems of the two hemispheres are activated. Even a capacity associated essentially with the right hemisphere – encoding spatial relationships – proves to be within the competence of the two hemispheres but in a different way in each case. The left hemisphere is more skilful at encoding "categorical" spatial relationships (*e.g.* high/low or right/left), while the right hemisphere is more skilful at encoding metric spatial relationships (*i.e.* continuous distances). Brain imaging has shown that even in these two specific cases, areas of both hemispheres are activated and working together. A more surprising finding, perhaps, is that the dominant hemisphere for language is not necessarily connected to right- or left-handedness, as had been thought. A widespread idea is that right-handed people have their language on the left and *vice versa*, but 5% of right-handed people have the main areas related to language in the right hemisphere and nearly a third of left-handed people have them located in the left hemisphere.

Based on the latest studies, therefore, *scientists think that the hemispheres of the brain do not work separately but together for all cognitive tasks, even if there are functional asymmetries.* As a highly integrated system, it is rare that one part of the brain works individually. There are some tasks – such as recognising faces and producing speech – that are dominated by a given hemisphere, but most require that the two hemispheres work at the same time. This invalidates the "left brain" and "right brain" concepts. Even if they may have brought some benefit through supporting more diversified educational methods, classifying students or cultures according to a dominant brain hemisphere is highly dubious scientifically, potentially dangerous socially, and strongly questionable ethically. It is thus an important myth to avoid.

"Let's face it – men and boys just have different brains from women and girls"

The 2003 PISA study is only one of the latest to reveal gender-related learning and educational achievement differences. Far more questionable are the works which have appeared over recent years claiming to be inspired by scientific findings apparently to show that men and women think differently due to a different brain development. Titles like *Why Men Don't Listen and Women Can't Read Maps* have become popular reading. How much is this founded on sound research? Is there a "feminine brain" and a "masculine brain"? Should teaching styles be shaped according to gender?

There are functional and morphological differences between the male and female brain. The male brain is larger, for instance, and when it comes to language, the relevant areas of the brain are more strongly activated in females. But determining what these differences mean is extremely difficult. *No study to date has shown gender-specific processes involved in building up neuronal networks during learning*; this is another candidate for additional research.

The terms "feminine brain" and "masculine brain" refer to "ways of being" described in cognitive terms rather than to any biological reality. Baron-Cohen, who uses these expressions to describe autism and related disorders (2003), believes that men tend to be more "methodical" (ability to understand mechanical systems) and women better communicators (ability to communicate and understand others), and he suggests that autism can be understood as an extreme form of the "masculine brain". But he does not propose that men and women have radically different brains nor that autistic women have a masculine brain. He employs the terms "masculine and feminine brain" to refer to particular cognitive profiles, which is an unfortunate choice of terminology if it contributes to distorted ideas concerning the workings of the brain.

Even if it were established that, on average, a girl's brain makes her less capable of learning mathematics, would this be grounds to propose education specialised to these differences? If the goal of education were to produce intensely specialised human beings, then the question may be worth at least considering but so long as its most important role continues to be to create citizens with a basic culture, such a question loses its relevance for educational policy. Where differences can be shown to exist, they will be small and based on averages. The much more important individual variations are such as to rule out being able to know if a young girl, taken at random, will be less capable of learning a particular subject than a young boy taken at random, etc.

"A young child's brain can only manage to learn one language at a time"

Today, half the world population speaks at least two languages and multilingualism is generally considered an asset. Yet for long, many have believed that learning a new language is problematic for the native language. Superstitions on this die hard and are often based on the false representation of language in the brain. One myth is that the more one learns a new language, the more one necessarily loses the other. Another imagines two languages as occupying separate areas in the brain without contact points such that knowledge acquired in one language cannot be transferred to the other. From these ideas, it has been supposed that the simultaneous learning of two languages during infancy would create a mixture of the two languages in the brain and slow down the development of the child. The false inference is that the native language had to be learned "correctly" before beginning another one.

The myths arise from a combination of factors. Since language is important culturally and politically, these considerations colour numerous arguments, including brain research findings, to favour one "official" language to the detriment of others. Certain medical observations have played their part: cases of bi- or multilingual patients completely forgetting one language and not at all another after a head trauma helped foster the idea that languages occupied separate areas in the brain. Studies conducted at the beginning of the 20th century, which found that bilingual individuals had inferior "intelligence",[6] were carried out with faulty methodologies, being based mainly on migrant children who were often undernourished and in difficult cultural and social conditions. The protocols should have taken into account that many of these children had started learning the language of their host country around the age of 5, 6, or later, and, without a strong command of that language, they had problems learning other subjects. In short, we cannot meaningfully compare the intelligence of monolingual children from native, often well-off families with that of multilingual children from primarily underprivileged environments with limited family knowledge of the dominant language.

Recent studies have revealed overlapping language areas in the brain of people who have a strong command of more than one language.[7] This point could be twisted in favour of the myth that the brain has only "limited space" in which to store information relating to language. Other studies on bilingual subjects have shown the activation of distinct areas of a few millimetres when they described what they did that day in their native language, then in the language learned much later (Kim, 1997). The question of "language areas" in multilingual individuals has thus not yet been resolved. But from this lack of resolution, it is wrong to claim that the strong command of one's native language is weakened when a second language is learned. Abundant cases of multilingual experts are living proof that this is not so. Students who learn a foreign language at school do not get weaker in their native language but instead advance in both.[8]

"Knowledge acquired in a language is not accessible or transferable to another language" is another myth and one of the most counter-intuitive. Anyone who learns a difficult concept in one language – for example evolution – can understand it in another language. If there is incapacity to explain the concept in the second language, it is due to a lack of vocabulary not a decrease in knowledge. Experiments have found that the more knowledge is acquired in different languages, the more it is stored in areas far away from the area reserved for language: it is not only preserved in the form of words but in other forms such as images. Multilingual individuals may no longer remember in what language they learned certain things – they may forget after a while if they read a particular article or saw a particular film, for example, in French, in German, or in English.

The myth that one has first to speak well one's native language before learning a second language is counteracted by the studies showing that children who master two languages understand the structure of each language better and apply them in a more conscious way. Therefore, multilingualism helps foster other competences related to language. These positive effects are clearest when the second language is acquired early; a multilingual education does not lead to a delay in development. Sometimes, very young children may confuse languages but unless there is a defect in acquisition (such as poor differentiation of sounds), this phenomenon later disappears.

6. One must be very careful when using the word "intelligence", which has no real scientific definition anyway.
7. The conditions of the creation of such an overlapping are not resolved. One theory stipulates that the areas reserved to languages overlap when the languages are learned at a young age but that when the second or other language(s) is learned late, there is no overlapping. Another theory proposes that overlapping appears when the two languages are mastered.
8. Studies conducted in 1990 on Turkish immigrant children in the Federal Republic of Germany who follow regular schooling found that the number of mistakes made by these children diminished in both Turkish and in German.

Theories on bilingualism and multilingualism have been particularly based on cognitive theories. Future school programmes on language learning should rely on successful examples of teaching practices and be informed by research on the brain, current or future, on ages favourable to language learning (sensitive periods).

"Improve your memory!"

Memory is an essential function in learning and is also the subject of rich fantasies and distortions. "Improve your memory!" "Increase your memory capacity!" "How to get an exceptional memory fast!" cry the marketing slogans for books and pharmaceutical products. The slogans are pushed with increased insistence during examination time. Do we now know enough to understand the processes and to envisage the creation of products and methods that improve memorisation? Do we need the same forms of memory today as was called for fifty or a hundred years ago in a world of different skills and professions? Can we talk of different memories – for instance, visual, lexical, or emotional? Do learning methods use the memory in the same way they did fifty years ago? These are relevant questions in this context.

In recent years, the understanding of memory has advanced. We now know that the memory does not respond only to the type of phenomenon and it is not located in only one part of the brain. However, contrary to one popular belief, memory is not infinite and this is because information is stored in neuronal networks, the number of which is itself finite (though enormous). No-one can hope to memorise the entire *Encyclopaedia Britannica*. Research has also found that the capacity to forget is necessary for good memorisation. On this, the case of a patient followed by the neuropsychologist Alexander Luri is enlightening: the patient had a memory that seemed to be infinite but, with no capacity to forget, was incapable of finding a steady job, unless it was as a "memory champion". It seems that the forgetting rate of children is the optimal rate to build up an efficient memory (Anderson, 1990).

What about those people who have a visual, almost photographic memory, who are very good at memorising a long list of numbers drawn at random or capable of simultaneously playing several games of chess blindfolded? Researchers have come to attribute these performances to specialised ways of thinking, rather than to a specific type of visual memory. DeGroot (1965) took an interest in the great chess masters, getting them to co-operate in experiments where the layout of the chessboard was briefly shown and these excellent players had then to recreate the layout of the pieces. They succeeded at this challenge perfectly, except when the layout shown had no chance of happening during a real game of chess. The conclusion DeGroot drew was that the ability of the great players to recreate the layout of the chessboard was thus not due to visual memory, but rather to the capacity to mentally organise the information of a game that they knew extremely well. On this view, the same stimulus is perceived and understood differently depending on the depth of knowledge of the situation.

This work notwithstanding, some people do seem to possess an exceptional visual memory, which can keep an image practically intact. This is "eidetic memory". Some people can, for example, spell out an entire page written in an unknown language seen only very briefly, as if they had taken a picture of the page. The eidetic image is not formed in the brain like a picture, however – it is not a reproduction but a construction. It takes time to form it and those with this type of memory must look at the image for at least three to five seconds to be able to examine each point. Once this image is formed in the brain, the subjects are able to describe what they saw as if they were looking at what they describe. By contrast, normal subjects without eidetic memory are more hesitant in their description. It is interesting (and possibly unsettling) to know that a larger proportion of children than adults seem to possess an eidetic memory; it seems as if learning, or age, weakens this capacity (Haber and Haber, 1988). These researchers also showed that 2-15% of primary school children have an eidetic memory. Leask and his colleagues (1969) found that verbalisation while observing an image interfered with the eidetic capture of the image, thus suggesting a possible line of explanation for the loss of eidetic memory with age. Kosslyn (1980) also sought to explain this negative correlation between visual memorisation and age. According to his studies, the explanation resides in the fact that adults can encode information using words whereas children have not yet finished developing their verbal aptitudes. There is still lack of scientific evidence to confirm or contradict these explanations. Brain imaging studies on this are needed.

There are a great number of techniques to improve memory, but they tend to act on a particular type of memory only, whether through mnemonics, repetitions of the same stimulus, or the creation of concept maps (giving meaning to things that they do not necessarily have in order to learn them more easily). Joseph Novak has devoted considerable study to concept maps (see Novak, 2003) who noticed a significant increase in the ability of high school physics students to resolve problems through the use of these concept maps. This work still lacks a brain imaging study to define the cerebral areas activated during these different processes. Nevertheless, it has been observed that different areas of the brain are activated, depending to whether the subject is a novice or not in the subject concerned.[9] Neurological studies are thus still needed to understand how memory works. Considerable individual diversities exist, and the same individuals will use their memory differently throughout the lifespan depending on their age. As seen in Chapter 3, science has nevertheless confirmed the role played by physical exercise, the active use of the brain, and a well-balanced diet (including fatty acids), in developing memory and reducing the risk of degenerative diseases.

Questions relating to the use of memory in current teaching methods and, especially to the critical role played by memory in student evaluation and certification in many OECD education systems, will probably have to be reconsidered in the future in light of new neuroscientific discoveries. Many such programmes rely more on memory than comprehension. The question "Is it not better to learn to learn?" cannot be answered through neuroscience but it remains highly pertinent.

"Learn while you sleep!"

Learning while you sleep – what a fascinating, appealing idea! Quick, effortless learning is the dream of many of us. But, even the greatest enthusiasts recognise that if knowledge can be acquired while sleeping, then one does not thereby learn how to use it. Is the "learn while asleep" idea completely a myth, or does it have some basis in truth?

In the methods purporting to permit such learning, information is repeated while the person is sleeping through messages and texts continuously relayed through a tape recorder or a CD player. Commercial products promise phenomenal success, claiming that learning while sleeping is not only possible, but even that it is more efficient than when done while awake. The idea took hold during World War II, when it was imagined that spies would this way quickly be able to learn the dialects, accents, habits, and customs of the

---

9. This confirms other observations regarding expertise and the way in which it is reflected in the brain structures.

countries to which they were to be sent. The origin of this idea about learning is found in science fiction and imagined for the first time by Hugo Gernsback, *Ralph 124 C41+: A Romance of the Year 2660*, published in 1911. Almost twenty years later, Aldous Huxley describes children learning in their sleep in *Brave New World*. These stories of learning while sleeping in utopian (or distopian) worlds began to spread into the real world. "Theories" came forward to explain how this way of learning can work but they were vague and contradictory. One suggested that the act of learning always begins with an unconscious process and so it is more efficiently done during sleep than while awake. No serious scientific study justifies this idea.

However, some work from Russia and the former Communist Eastern Europe which sought to demonstrate successful learning during sleep deserves further consideration. Kulikov (1968) carried out his research by narrating a story written by Tolstoy for children to subjects sleeping normally. One of the twelve subjects had memories of the text upon awakening. In a second group, Kulikov first established contact with the experiment subjects while they were sleeping, through playing recorded sentences like "you are sleeping peacefully, don't wake up". After these sentences, the story was played, followed by other sentences asking them to remember the text and to continue sleeping calmly. These instructions seemed to have a real impact on the ability of the subjects to recollect the texts that had been read. The subjects to whom the texts were read during their sleep appeared to have remembered it as well as the subjects to whom the text was read while awake. Other longer studies were conducted in Russia and in Eastern Europe, always with instructions given before falling asleep (for a review, see Hoskovec, 1966, Rubin, 1968). On the basis of these findings, learning while sleeping was intensely practised in the former Soviet Union countries, above all during the 1950s and 1960s. It was claimed that languages could be learned while sleeping, not only by individuals but by entire villages thanks to nightly radio broadcasts (Bootzin, Kihlstrom and Schacter, 1990).

These studies had many faults, however, which raise doubts about the real efficiency of learning during sleep. Often, researchers indicated that the learning worked on "sensitive" subjects, without clearly defining what was meant by "sensitive", with meanings varying from a "hypnosis-sensitive subject" to a "subject persuaded by the efficiency of learning while sleeping". Another shortcoming of these experiments was the poor control of the state of sleep – were some subjects slightly awake? In general, the information to be retained was not read during deep sleep and there was no control for the state of sleep on the electroencephalogram (EEG), as in western studies. The experiments were done just after the subjects fell asleep or in the early hours, moments when EEG recordings mostly show alpha waves (Aarons, 1976). If so, it is unlikely that the subjects were deeply asleep during the reading but instead in "light sleep" and, in a way, conscious. Such factors help to explain the positive results of many earlier eastern European research studies but they do not add to confidence in their claims.

While western researchers have found no evidence to support the claims of successful learning during sleep, they have been able to identify an effect when someone is under anaesthetic (Schacter, 1996). Traditionally, patients under anaesthetic were considered to be asleep and perceiving nothing. In the 1960s, experiments were conducted whereby the surgeons behaved as if there was a serious emergency while the patient was under anaesthetic. On wakening, patients were questioned about their operation and some were very agitated. From this it was concluded that patients must have implicitly remembered something from the operation when they were asleep (Levinson, 1965). In other studies, patients recuperated better from their operation which had taken place under general anaesthetic if during it they were told that they were going to recover quickly.

However, these studies were not without fault either. For one thing, being under anaesthesia is not sleep and thus cannot be directly compared to it. Once awake, the patients could not explicitly remember what they had undergone or heard during the operation. Agitation and recovery are not sensitive enough criteria through which to evaluate memory. The nature and efficiency of the anaesthetic could cause a difference in the memorisation process (Schacter, 1996). Finally, subsequent studies did not succeed in reproducing the same results and tended to suggest that it was not possible to remember events from the operation whether the patient was conscious or not. *In sum, no study on learning while sleeping conducted in western countries, with a strict control by EEG on the state of sleep, has been able to demonstrate evidence of learning* (Bootzin, Kihlstrom and Schacter, 1990; Wood, 1992).

Therefore, we should conclude that many of the claims about learning while sleeping are myths. Although the precise role of sleep remains a mystery, recent studies have found that it plays different roles in the development of the brain and in its functioning. It is beneficial for strengthening skills such as motor learning: for example, the memorisation of specific finger-tapping sequences improves when the training is followed by periods of sleep (Kuriyama, Stickgold and Walker, 2004; Walker *et al.*, 2002). Sleep during the first half of the night favours factual memory while that during the second half favours skill memory (Gais and Born, 2004).

Until now, numerous studies found that sleep modifies the memorisation of things learned just before falling asleep (Gais and Born, 2004). Concerning the learning of facts, for long it has been found that short stories and syllables without meaning were best remembered if they slightly preceded sleep (Jenkins and Dallenbach, 1924; van Ormer, 1933). Conditioned responses – the association of two stimuli whereby a conditioning stimulus (CS) such as a ringing noise is presented simultaneously or just before an unconditioning stimulus (UCS) such as an electric shock to the finger – can also be learned while sleeping. The UCS generally provokes a strong response, like the withdrawal of the finger after the electric shock. After several tests, the subject learns to associate the two stimuli and reacts even if only the CS is produced such that the subject pulls back his/her finger at the sound of the bell, whether or not there is an electric shock. Studies show that a conditioned response can be learned while asleep and maintained when awake (Ikeda and Morotomi, 1996, Beh and Barratt, 1965). Other studies have found that a conditioned response learned while awake can be maintained when the subject is asleep (McDonald *et al.*, 1975).

In conclusion, no scientific evidence supports strong claims about learning while asleep and whether sleeping or not, one cannot rely on simple repetition for learning. To learn a foreign language, natural sciences, physics, etc., requires conscious effort. The CDs to be played while asleep promise a pathway to better learning, stopping smoking and losing weight but there is no scientific evidence to support these promises. Maybe it is not the CD that makes you stop smoking or lose weight, but motivation. Learning while asleep continues to be a myth and it is highly unlikely to see such approaches as one day recommended parts of school or university programmes.

Conclusions

The brain is fashionable. The media constantly draws in one way or another on the mysteries of this "black box". Such popularity is partly explained by the inherent interest of the subject for most people ("if you're talking about my brain, you're talking about me"), as well as by the richness of new discoveries by neuroscientific research that lend themselves to media coverage. Popular appeal can bring pitfalls. Over the past few years, there has been a growing number of misconceptions circulating about the brain, which have come to be labelled "neuromyths". They have some characteristics in common, no matter their differences in other respects.

Most neuromyths, including the main ones described in this chapter, share similar origins. They are almost always based on some element of sound science, which makes identifying and refuting them much more difficult. The results on which the neuromyths are built are, however, either misunderstood, incomplete, exaggerated, or extrapolated beyond the evidence, or indeed all of these at once. This difficulty is inherent in the very nature of scientific discourse, and in the simplifications that are all too easy to introduce when translating science into everyday language (exacerbated by the nature of media coverage). The emergence of a neuromyth may be intended or unintended. Though some are born accidentally, interests may well be served by them. Neuromyths often drive business and probably most are anything but accidental.

Potentially everyone is susceptible to neuromyths but some targets are especially important. First, there are all the educators – whether parents, teachers, or others – who are the frontline "consumers" of education and hence open to ideas being "sold". In an uncertain educational world, new ideas are readily welcomed, especially if they appear as a panacea but even if just an embryonic solution. Were education to be more confident of itself, half truths, ready-made solutions, quarter-panaceas, and myths would have less chance to proliferate. But standing at the beginning of the 21st century, educational reflection and practice are too open to infection by this fever so that neuromyths are likely to remain for the foreseeable future.

Dispelling or debunking neuromyths has been a task of this OECD/CERI work for a number of years, but it presents numerous difficulties. First, exposing neuromyths also exposes the valid insights of brain research to the fire of the neuro-sceptic who can use it to challenge any neuroscientific approach to education. It equally lays the careful user of neuroscientific evidence who rejects the myths open to attack from those whose interests are best served by continuing belief in them. It may also disappoint some of those educators who have shown a touchingly naïve faith in the promise of neuroscience.

The bridges between neuroscience and education are still too few. This analysis of myths about the workings of the brain clearly shows that greater co-operation between the two domains is needed. Any educational reform which is truly meant to be in the service of students should take into account neuroscientific studies and research, while maintaining a healthy objectivity. Similarly, brain researchers should not exclude themselves from the world of education and the broader implications of their work. They need to be ready to explain it as understandably and accessibly as possible. It will be through the exchange between the different disciplines and players (researchers, teachers, political leaders), that it will be possible to harness the burgeoning knowledge on learning to create an educational system that is both personalised to the individual and universally relevant to all.

References

Aarons, L. (1976), "Sleep-assisted Instruction", *Psychological Bulletin*, Vol. 83, pp. 1-40.

Anderson, J. (1990), *The Adaptive Character of Thought*, Erlbaum, Hillsdale, NJ.

Baron-Cohen, S. (2003), *The Essential Difference: Men, Women and the Extreme Male Brain*, Allen-Lane, London.

Beh, H.C. and P.E.H. Barratt (1965), "Discrimination and Conditioning During Sleep as Indicated by the Electroencephalogram", *Science*, 19 March, Vol. 147, pp. 1470-1471.

Bootzin, R.R., J.F. Kihlstrom and D.L. Schacter (eds.) (1990), *Sleep and Cognition*, American Psychological Association, Washington.

Bruer, J.T. (2000), *The Myth of the First Three Years, a New Understanding of Early Brain Development and Lifelong Learning*, The Free Press, New York.

DeGroot, A. (1965), "Thought and Choices in Chess", Mouton Publishers, The Hague.

Dehaene, S. (1997), *The Number Sense: How the Mind Creates Mathematics*, Allen Lane, The Penguin Press, London.

Diamond, M.C. (2001), "Successful Ageing of the Healthy Brain", article presented at the Conference of the American Society on Aging and the National Council on the Aging, 10 March, New Orleans, LA.

Gabrieli, J. (2003), "Round Table Interview", *www.brainalicious.com*.

Gais, S. and J. Born (2004), "Declarative Memory Consolidation: Mechanisms Acting During Human Sleep", *Learning and Memory*, Nov-Dec, Vol. 11, No. 6, pp. 679-685.

Gernsback, H. (2000), *Ralph 124C 41+: A Romance of the Year 2660*, Bison Books, University of Nebraska Press, Lincoln, NE.

Gopnik, A., A. Meltzoff and P. Kuhl (2005), *Comment pensent les bébés ?*, Le Pommier (traduction de Sarah Gurcel).

Guillot, A. (2005), "La bionique", *Graines de Sciences*, Vol. 7 (ouvrage collectif), Le Pommier, pp. 93-118.

Haber, R.N. and L.R. Haber (1988), "The Characteristics of Eidetic Imagery", in D. Fein and L.K. Obler (eds.), *The Exceptional Brain*, The Guilford Press, New York, pp. 218-241.

Hoskovec, J. (1966), "Hypnopaedia in the Soviet Union: A Critical Review of Recent Major Experiments", *International Journal of Clinical and Experimental Hypnosis*, Vol. 14, No. 4, pp. 308-315.

Huxley, A. (1998), *Brave New World* (Reprint edition), Perennial Classics, HarperCollins, New York.

Ikeda, K. and T. Morotomi (1996), "Classical Conditioning during Human NREM Sleep and Response Transfer to Wakefulness", *Sleep*, Vol. 19, No. 1, pp. 72-74.

Jenkins, J.G. and K.M. Dallenbach (1924), "Obliviscence during Sleep and Waking", *American Journal of Psychology*, Vol. 35, pp. 605-612.

Kim, K.H. *et al.* (1997), "Distinct Cortical Areas Associated with Native and Second Languages", *Nature*, Vol. 388, No. 6638, pp. 171-174.

Kosslyn, S.M. (1980), *Mental Imagery*, Harvard University Press, Cambridge, MA.

Kulikov, V.N. (1968), "The Question of Hypnopaedia", in F. Rubin (ed.), *Current Research in Hypnopaedia*, Elsevier, New York, pp. 132-144.

Kuriyama, K., R. Stickgold and M.P. Walker (2004), "Sleep-Dependent Learning and Motor-Skill Complexity", *Learning and Memory*, Vol. 11, No. 6, pp. 705-713.

Leask, J., R.N. Haber and R.B. Haber (1969), "Eidetic Imagery in Children: Longitudinal and Experimental Results", *Psychonomic Monograph Supplements*, Vol. 3, No. 3, pp. 25-48.

Levinson, B.W. (1965), "States of Awareness during General Anaesthesia: Preliminary Communication", *British Journal of Anaesthesia*, Vol. 37, No. 7, pp. 544-546.

Lorenz, K. (1970), *Studies in Animal and Human Behaviour*, Harvard University Press, Cambridge MA.

McDonald, D.G. *et al.* (1975), "Studies of Information Processing in Sleep", *Psychophysiology*, Vol. 12, No. 6, pp. 624-629

Neville, H.J. (2000), "Brain Mechanisms of First and Second Language Acquisition", presentation at the Brain Mechanisms and Early Learning First High Level Forum, 17 June, Sackler Institute, New York City, USA.

Neville, H.J. and J.T. Bruer (2001), "Language Processing: How Experience Affects Brain Organisation", in D.B. Bailey *et al.* (eds.), *Critical Thinking About Critical Periods*, Paul H. Brookes Publishing Co., Baltimore, pp. 151-172.

Novak, J.D. (2003), "The Promise of New Ideas and New Technology for Improving Teaching and Learning", *Cell Biology Education*, Vol. 2, Summer, American Society for Cell Biology, Bethesda, MD, pp. 122-132.

OECD (2002), "Learning Seen from a Neuroscientific Approach", *Understanding the Brain: Towards a New Learning Science*, OECD, Paris, pp. 69-77.

OECD (2004), *Learning for Tomorrow's World – First Results from PISA 2003*, OECD, Paris, pp. 95-99, *www.pisa.oecd.org*.

van Ormer, E.B. (1933), "Sleep and Retention", *Psychological Bulletin*, Vol. 30, pp. 415-439.

Ornstein, R. (1972), *The Psychology of Consciousness*, Viking, New York.

Rubin, R. (1998), *Current Research in Hypnopaedia*, MacDonald, London.

Schacter, D.L. (1996), *Searching for Memory: The Brain, the Mind and the Past*, Basic Books, New York.

Scientific American (2004), "Do We Really Only Use 10 Per Cent of Our Brains?", *Scientific American*, June.

Sperry, R.W., M.S. Gazzaniga and J.E. Bogen (1969), "Interhemispheric Relationships: The Neocortical Commissures; Syndromes of Hemisphere Disconnection", in P.J. Vincken and G.W. Bruyn (eds.), *Handbook of Clinical Neurology*, North-Holland Publishing Company, Amsterdam.

Walker, M.P. *et al.* (2002), "Practice with Sleep Makes Perfect: Sleep-Dependent Motor Skill Learning". *Neuron*, Vol. 35, No. 1, pp. 205-211.

Wood, J. *et al.* (1992), "Implicit and Explicit Memory for Verbal Information Presented during Sleep", *Psychological Science*, Vol. 3, pp. 236-239.

ISBN 978-92-64-02912-5
Understanding the Brain: The Birth of a Learning Science
© OECD 2007

PART I

Chapter 7

The Ethics and Organisation of Educational Neuroscience

Science sans conscience n'est que ruine de l'âme. (Science without conscience is but the ruin of the soul.)

François Rabelais

Human history becomes more and more a race between education and catastrophe.

Herbert George Wells

This chapter addresses the field of educational neuroscience itself. It describes how the emergence of this multi-disciplinary field has been one of the main contributions of the OECD-CERI project on "Learning Sciences and Brain Research". It highlights a variety of exemplary trans-disciplinary projects and institutions which have already been set up and are active in contributing to this new field. The research involved and its applications are also fraught with ethical challenges: these are openly laid out and some of the key choices clarified.

For educational neuroscience to become an enduring field with significant contributions to educational policy and practice, we need to address the development of this field of research itself, as a human activity creating and applying knowledge with a range of stakeholders and interests. This chapter discusses this with a dual focus. First, it considers some of the ethical questions which arise in exploring this new field, questions which are profound and central to present and future applications in education. Second, the chapter discusses organisational and methodological directions and developments. Though educational neuroscience methodology is just beginning to evolve, three key aspects of a framework for a strategic approach have already emerged: it should be trans-disciplinary, bidirectional, and international.

The ethical challenges facing educational neuroscience

> Our scientific power has outrun our spiritual power. We have guided missiles and misguided men.
>
> *Martin Luther King Jr.*

Over recent decades, enormous progress has been made in brain imaging, making the functioning of the brain more visible. This raises some fundamental ethical questions: can the human brain be examined without limitations? What are the goals of making these observations and to whom (*e.g.* to which institutions) should they be made available?

Traditionally, the ethical rules concerning biomedical research on human beings follow the Nuremberg Code of 1949 and the Declaration of Helsinki of 1964. Their main points are that:

- the voluntary and informed consent of the human subject is absolutely essential;
- considerations related to the well-being of the human subject should take precedence over the interests of science and society;
- during the course of the experiment the human subject should be at liberty to bring the experiment to an end at any time;
- during the course of the experiment the scientist in charge must be prepared to terminate the experiment at any stage.

Now, new ethical questions arise due to scientific progress relevant to neurosciences and include developments such as:

- new technologies, such as in brain imaging, permit the exploration of the human brain and its functions with excellent spatial and temporal resolutions;
- substances are able to alter the brain functions selectively, causing psychological changes;
- therapeutic studies have shown that modifications in behaviour can lead to changes in brain functions.

These developments call for new reflections about ethical rules, which should go well beyond the scientific academies: they concern the whole society and indeed each individual. To begin with, we look at progress in brain imaging and then at the development of products affecting the brain, considering the differences between medicine and stimulant products. As research of brain functioning allows us to better understand learning processes, we also need to address questions about the links between neuroscience and education.

For which purposes and for whom?

Brain imaging already allows the exploration of activated brain regions and neuronal networks during the performance of tasks or to capture the feeling of emotions under experimental conditions, but how to consider and factor in the psychological state of a patient? It is already important to think about the conditions of use of brain imaging. Should the results be limited to research or therapy? In the latter case, how to ensure that the medical information be kept confidential, as in any other medical situation. For instance, the information should not be handed over to banks, insurances, or employers who might be very interested in such information.

Brain imaging poses another problem: what to do if a pathology or a pathological risk is discovered unintentionally? A study by the Standford Center for Biomedical Ethics (USA) found that 18% of the "healthy" volunteers actually had brain anomalies (Talan, 2005). A person may be found to have a brain cyst that would never cause any problem: should the volunteer be informed about this, possibly alarming them unnecessarily, or should it be treated?

A joint report published in 2005 by INSERM (Institut National de la Santé et de la Recherche Médicale, France) advises a systematic medical screening at 36 months of age for "behaviour disorders in children and adolescents", which affect between 5% and 9% of 15-year-olds according to international estimates (INSERM, 2005). It is recommended to "locate disruptive conduct from child care centres and infant schools onwards", and for children they suggest individual therapies or "as a second resort", even the use of medication having an "anti-aggressive effect". Though this example is not specifically about brain imaging, it is directly relevant as the recommendation concerns screening and prevention. It therefore poses the same questions: should subjects be treated preventively? Who should be informed about the results (i.e., schools, institutions, educators, etc.)? The publication of the 2005 INSERM report has provoked numerous negative reactions from physicians, psychologists and psychiatrists in France. It raises immediate questions about the potential of these different techniques to open up early labelling and exclusion, and it is not fanciful to imagine a potential political use (and abuse) regarding this "medical spotting", whether of possibly disruptive young people or of others with characteristics revealed by brain imaging.

It is always difficult to foresee scientific progress. Some day, brain imaging could make it possible to determine without ambiguity the psychological state of people, and whether, for example, they are lying, scared, self-confident, etc., all of which might have profound implications if used in educational contexts. Therefore, it is important to regularly revise the rules and procedures in the light of these technologies and policy decisions should not have irremediable long-term consequences. Monitoring by the scientific community as well as informing the public must take place continually.

A global ethical issue is that brain imaging is an expensive technique whose use and benefits tend to be limited to the richest countries and populations of the world. Should we not already be thinking about ways to make it accessible to the largest possible number of people? Who will be able to take such global decisions about the equity of distribution and how will the poorer societies themselves be involved?

Ethical issues regarding the use of products affecting the brain

Products which affect the brain can be either explicitly medicinal or else stimulants and sedatives whose effects are other than therapeutic. The boundary between the two is not always clear, which leads to a number of major questions. Medicine has as its goal to heal someone, to improve their health. Stimulants and sedatives do not combat an illness, whether as cure or to prevent it, but they improve one or more functions in healthy subjects. But, it is not always easy to distinguish between the "normal" and the pathological – what is normal for one individual is not necessarily normal for another, which makes for such a fine boundary between medicine and another product. This difference can be illustrated in the decision over whether a child has a hyperactivity syndrome (ADHD: Attention Deficit Hyperactivity Disorder) – the opinions of the parents, the treating doctor, the paediatrician and the teacher can diverge. Use of sleeping pills offers another example.

Psychopharmacology is the discipline producing substances that affect the brain. In the early 20th century, a number of traditional remedies made of plants were used as psychiatric components – opium, cannabis, alcohol (Calvino, 2003). Around the 1950s, psychopharmacological products to improve mental states appeared, prominent among which were tranquilisers and antidepressants. In the second half of the 20th century, psychopharmacological research mushroomed, resulting in the production of a large number of substances. In his book "Listening to Prozac" (1997), Peter Kramer was one of the first to raise the question about using antidepressants as a supportive medication: very many people take them even though they are perfectly healthy and simply have a depressive tendency.[1] This is becoming increasingly relevant as regards the brain and learning as certain substances are also used to stimulate the brain and increase memory. Whether or not their effectiveness is proven, ethical questions are increasingly arising about the limits of use of such substances in the healthy individual. It has become necessary to regulate the use of any molecule having an effect on the brain. This means defining the different levels of responsibility held by the state, doctors, and parents and having information on commercialised substances, in which form do they appear, by whom and for whom are they produced? This is an area which touches on the boundaries of private decision-making and responsibility. Should parents have the right to give their children substances to stimulate their scholarly achievements (memory, attention), even though the dangers and even their effectiveness are not clearly defined? What are the parallels with doping in sport? These questions are increasingly relevant the better that brain functioning becomes understood.

To avoid the inappropriate use of these products, it is necessary to keep the public informed and an effort is required by the scientific community. If certain substances are found to be beneficial and safe, how should they be made available and how priced so as to

---

1. The depressive tendency is a feeling of discouragement and sadness, but is not a pathological state, in contrast to a depression.

be accessible on an equitable basis? As for brain imaging technologies, scientific and political monitoring and debate is necessary; irremediable, long-term decisions should be avoided in the absence of adequate knowledge.

Brain meets machine – the meaning of being human?

Researchers are currently trying to combine living organs with technology, thus creating "bionics" (Guillot, 2005). In this, artificial equipment (electronic or mechanical prostheses) is either integrated into a living organ or living organs are integrated into artificial ones. This research also concerns nervous components and therefore the neurosciences. For example, sensory organs, like the retina or the cochlea, may be damaged without the neuronal circuitry being affected; in these circumstances, it is possible to implant artificial sensory organs to interpret luminous, mechanical or chemical information from the environment and transform it into electrical signals to be understood by the nervous system. This research essentially aims at helping individuals with disabilities but, apart from other issues they can also cause distress. Take, for example, experiments where rats are remote-controlled by a computer programme through electrodes implanted into their brains. These experiments and developments also need to be seriously supervised, and not only by scientists but by the whole society.

In some studies, robotic structures are controlled from some distance by the brain: this way, rats have been able to command robotic arms which gave them water; monkeys have been able to move a cursor on a screen towards a target without any physical contact; even humans with the paralysis of "locked in" syndrome have been trained to use a cursor to write on a computer screen using only their brain activity. To obtain these results the brain activity had first to be registered; then a computer connected to a manipulating arm was programmed in such a way that the cerebral recording corresponding to the "desire to move" activated the movement. The advantages of developments from this research are obvious for individuals with handicaps (tetraplegics, for example), who can thus control machines from a distance. Used in a different context, for different purposes, however, it is just as obvious that this type of development needs to be closely monitored and controlled.

The risk of an excessively scientific approach to education?

As they deal with brain development and learning, neurosciences are highly relevant to teaching and learning processes, but is it dangerous to move towards an approach in learning which is "too scientific"? Neurosciences can certainly inform education, as they offer a better understanding of the brain processes involved in learning. If we understand why certain students have difficulties learning mathematics and others do not, we can adapt maths learning through change in pedagogy, for example. This way we could aspire to identify a teaching method adapted to each student for each discipline. Since the work of Howard Gardner on multiple intelligences (1983), this idea of adapted teaching has been developed, even if Gardner himself has now changed his mind on its value (Gardner, 2000). This could be extended to teachers. One day, thanks to neuroscience and brain imaging in particular, we may be able to determine what a "good" teacher is by verifying if the lessons learned have been well understood by the students by analysing their brains. Would this not run the risk of choosing only one type of teacher, corresponding to the norms of the moment, and of creating an education system which is highly scientific and highly conformist, too?

These scenarios pose the general question about the goals of education: is it about training specialised individuals or about creating and maintaining a society of citizens with a common culture? Neurosciences can deliver responses in the search of a better quality in education, open to the largest number of people possible, but they can also generate abuse. Bruer, who was one of the first to advocate education based on neurosciences ("brain-based education"), became one of the strongest critics towards this notion (Bruer, 1993).

We have seen that recent developments in brain science indeed pose numerous questions. Most of them are not specific to neurosciences: the protection of the individual as regards brain imaging is comparable to the medical secret; the control of molecules affecting the brain is like any other newly synthesised or discovered molecule; research on the brain-machine interface raises common issues and is subject to the rules governing scientific research in general. However, as neurosciences are interested in the brain, the organ that seems to be the noblest one being the source of decision and liberty, certain applications may cause unease or even opposition (the control of the living brain by a machine is a good example). In the face of these understandable fears, different monitoring bodies have been set up with agendas which align with the issues outlined in this section.

Creating a new trans-disciplinary approach to understanding learning

In the past, the trans-disciplinarity, bridging, and fusing
concepts that brought together widely divergent fields were
the exclusive privilege of genius, but in the 21st century
these tools must become more widely available.
(…) The provision of a trans-disciplinarian education that will enable
future trans-disciplinary studies is an urgent need that
we must satisfy for the benefit of future generations.

Hideaki Koizumi

Trans-disciplinarity

In ancient Greece, knowledge was not neatly contained in distinct disciplines and leading scholars moved freely among different fields. The influence of reductionism, which began with Aristoteles and has continued ever since, created disciplines with precise boundaries (as in Figure 7.1a). With this disciplinary approach, each field evolves independently as a specialised set of tools of analysis. Such specialised disciplinarity is increasingly necessary while increasingly insufficient. The enhanced precision of the disciplinary approach is needed to make significant advancements and this form of organisation enables the management of massive amounts of knowledge. As the disciplines mature, the intellectual walls between them are becoming progressively higher and thicker, but the divisions between them may well become less logical – a static disciplinary approach does not provide the means to transcend disciplinary borders when they become inappropriate for the advancement of understanding.

The current state of neuroscience and education provides a good illustration of the shortcomings of disciplinary separation. On learning, recent advances in neuroscience have produced powerful insights while educational research has accumulated a substantial knowledge base. A neuroscientific perspective on learning adds a new, important dimension to the study of learning in education, and educational knowledge could help direct neuroscience research towards more relevant areas. Because both fields are well-developed, however, they have deeply-rooted disciplinary cultures with field-specific methods and language which make it extremely difficult for experts from one field to use the knowledge from the other.

As disciplines reach maturity, a dynamic meta-structure is needed which facilitates merging and new divisions of disciplines. Under such a meta-structure, the disciplines propel the evolution of knowledge, but adapt themselves when driving forces emerge sufficient to provoke their adaptation. In the case of neuroscience and education, the more comprehensive understanding of learning is a compelling driving force as it is critical to broader goals such as sustainable economic growth, societal cohesion, and personal development.

Fusing neuroscience, education and other relevant disciplines, and creating a new trans-disciplinary field would connect work on learning across the intellectual walls dividing disciplines (Figure 7.1a). (For discussion of the terms "trans-disciplinary", "multidisciplinary", and "interdisciplinary" and how they are used here, see Koizumi, 1999.) Interdisciplinary collaboration between two established disciplines (in this case, neuroscience and education) is insufficient to catalyse disciplinary fusion. Multiple disciplines are needed to drive the emergence of a trans-disciplinary field. As bridges are built among relevant fields, a new daughter discipline with unique methods and organisation can gradually emerge. Once this new discipline has crystallised, it can enter into the dynamic meta-structure as an established discipline capable of contributing to further trans-disciplinary evolution (Figures 7.1b and c). Additionally, it can feedback to shape its parent disciplines (Figure 7.1d). Moreover, this whole process can occur with many disciplines simultaneously, creating a dynamic evolution of knowledge (Figure 7.1e).

Establishing educational neuroscience as a robust discipline is a gradual process that is already underway (see Box 7.1).[2] Multidisciplinary networks are forming to undertake trans-disciplinary work. The *Centre for Neuroscience in Education* at the University of Cambridge and *Learning Lab Denmark* (LLD) at the Danish University of Education, for example, are addressing trans-disciplinary research questions (see Boxes 7.2 and 7.3). Programmes aimed at developing trans-disciplinary educational neuroscience experts are spawning, such as those launched at Harvard University (see Box 7.4), Cambridge University, and Dartmouth College. Initially, work will consist primarily of multidisciplinary collaboration and experts can help build multiple pathways between the disciplines. As more experts are trained, the field can move toward genuine trans-disciplinary work. Many educators and neuroscientists are already paving a career path for professionals defined with new forms of expertise, establishing trans-disciplinary laboratories, schools, journals, societies, and electronic discussion forums.

Creating a common lexicon is a critical next step. At present, key terms are used differently across the fields. There is a lack of consensus about the meaning of even fundamental terms, such as "learning" (see Chapters 1 and 2). This multiplicity of definitions can lead to misinterpretations. Complementary field-specific definitions can be combined to create comprehensive terms and definitions. For example, the educational conception of learning as a social endeavour can be linked with the neuroscientific conception of learning as molecular events in the brain. Constructivist theories of learning maintain that meaning is not passively transmitted to the learner, but actively constructed by the learner. Participatory theories add the focus on how it is constructed within the

---

2. This box and the following boxes have been submitted and authored by the individual institutions in question.

Figure 7.1. **The evolution of trans-disciplinarity**

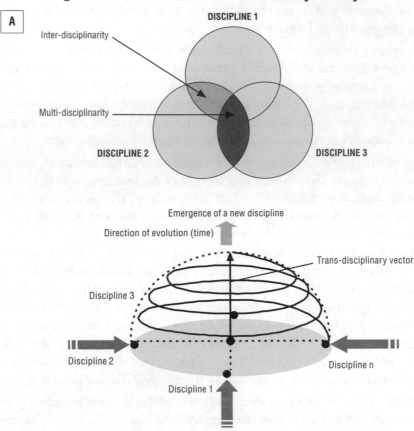

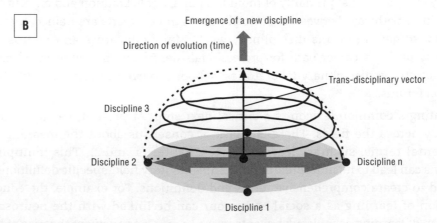

Note: This model contrasts interdisciplinary and multidisciplinary work with trans-disciplinary work. Interdisciplinary and multidisciplinary work involve overlapping two (interdisciplinary) or more (multidisciplinary) well-developed disciplines, while trans-disciplinary work involves the fusion of many disciplines and the emergences of a new, daughter discipline. Ultimately, educational neuroscience must develop as a trans-disciplinary field.

Source: Adapted from Hideaki Koizumi (1999), "A Practical Approach towards Trans-disciplinary Studies for the 21st Century", *J. Seizon and Life Sci.*, Vol. 9, pp. 5-24.

Note: As a trans-disciplinary field reaches maturity, it can enter into the dynamic meta-structure as an established discipline capable of contributing to further cross-disciplinary evolution.

Figure 7.1. **The evolution of trans-disciplinarity** (*cont.*)

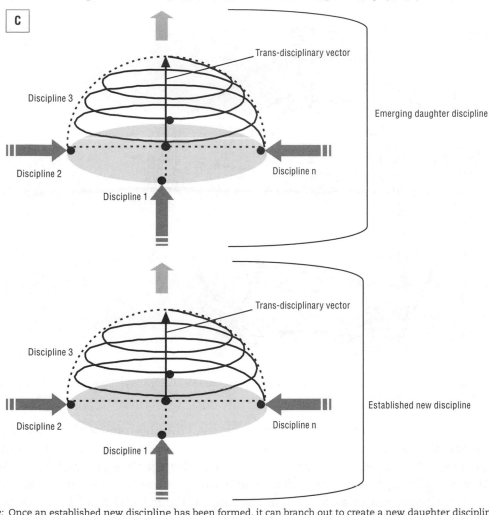

Note: Once an established new discipline has been formed, it can branch out to create a new daughter discipline.

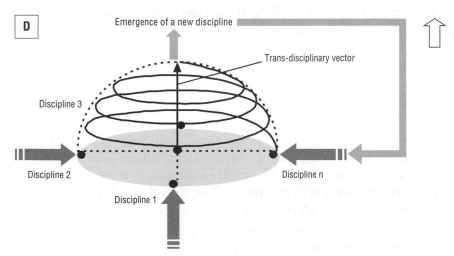

Note: In addition to contributing to continued cross-disciplinary evolution, newly established trans-disciplinary fields can also feedback to influence parent disciplines.

Figure 7.1. **The evolution of trans-disciplinarity** *(cont.)*

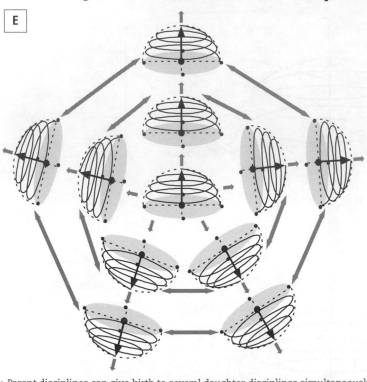

Note: Parent disciplines can give birth to several daughter disciplines simultaneously. These disciplines, in turn, can create daughter disciplines, and so forth.

constraints of a particular socio-cultural environment. With this framework of educational theory, learning is an active, socio-culturally-mediated process. From a neuroscientific perspective, learning occurs as a cascade of molecular events resulting in structural modification with significance for subsequent learning. If the two are brought together, learning can be described as a series of mediated socio-cultural adaptations of brain structure with functional consequences.

Another important step is establishing shared methodology. Education research consists of a wide variety of methods, ranging from correlative quantitative analysis to ethnography. Given the problem-driven methodological approach in education, a plurality of methods will likely emerge and persist in educational neuroscience as well. As the field develops, it will be important to make these methods explicit. It would also be beneficial to align methodological tools of measurement across the fields. For example, aligning expensive brain research with cost-effective psychological measures would enable researchers to obtain data from a larger sample size. It will be necessary to establish standards of evaluation, answering such questions as: which criteria will be used to determine if a potential research question is worth investigating? What will indicate that an intervention has been successful? What relative weight will be given to a statistically-significant laboratory result as compared with a visible effect in the classroom? These types of questions can be answered through dialogue among neuroscientists and educators across the spectrum of research and practice.

Box 7.1. **Mind, Brain, and Education (MBE)**

In this age of biology, society is looking to neuroscience, genetics, and cognitive science to inform and improve education. Several faculties at the Harvard Graduate School of Education, inspired by both societal need and burgeoning student interest, started the programme in Mind, Brain, and Education (see Mind, Brain and Education programme description below). Building usable research-based knowledge for education requires creating a reciprocal relationship between practice and research, analogous to the relation between medicine and biology. Research informs practice, and simultaneously practice informs research. The MBE programme trains people to make these connections.

Happily, outstanding young scholars flocked to the MBE programme to earn masters and doctoral degrees to help them make their own contributions to connecting both research and practice in mind, brain, and education. Independently interest in connecting biology, cognitive science, and education blossomed in many parts of the world: OECD began its programme on neuroscience and education in CERI. Japan initiated major research programmes in neuroscience and education. Conferences on Learning and the Brain for educators and scientists were organised biannually in Boston and periodically in other locations around the world. The Pontifical Academy of Sciences initiated an international conference on mind, brain, and education to celebrate its 400th anniversary (November 2003). University of Cambridge, Dartmouth College, and other universities began sister programmes to MBE. Many leading scientists and educators are working together to build the foundation for grounding education in research.

All this interest demonstrated a need for an organisation to bring people together to promote research and practice linking biology, cognitive science, and education. An international group founded the International Mind, Brain, and Education Society (IMBES) in 2004, which has organised several conferences and workshops to promote this emerging field. To create a forum for research and dialogue, IMBES has established a new journal, *Mind, Brain, and Education*, which begins publication in 2007 with Blackwell Publishing. Outstanding researchers and scholars of practice have begun submitting research reports, conceptual papers, and analyses of promising practices to the journal.

MBE, OECD, IMBES, the new journal *Mind, Brain, and Education* – all these endeavours and many more seek to create a strong base for research and practice in education. The strong interest from scientists, educators, and students bodes well for the future, but an essential need is to create an infrastructure for connecting research and practice in education. One of the primary catalysts of the research-practice link in medicine is the teaching hospital, in which researchers and practitioners work together in community hospitals to carry out research that is relevant to practice and to train young researchers and practitioners in settings of medical practice. Education needs similar institutions that help bring research into schools and practice into laboratories. MBE and IMBES, in collaboration with the Ross Institute, believe that education needs *Research Schools* to play a role analogous to teaching hospitals in building research-based education. In Research Schools, universities and schools of education will join research and practice in living community-based schools and lay a fundamental infrastructure for grounding and improving education.

Source: Kurt W. Fischer, *Mind, Brain, and Education*, Harvard Graduate School of Education.

> Box 7.2. **Centre for Neuroscience in Education:**
> **University of Cambridge, United Kingdom**
>
> Cambridge University is an international leader in basic and clinical neuroscience, with world-class expertise within the University and Addenbrookes Hospital. The opening of a Centre for Neuroscience in Education in 2005 has complemented these developments, offering a genuinely new departure on the world stage. The Centre is the first in the world to have imaging equipment within a Faculty of Education.
>
> The Centre aims are to develop research expertise in this relatively new area, to build research capacity by training researchers in applying neuroscience techniques to educational questions, to provide information about neuroscience to teachers and educators, and to communicate the potential impacts of such multidisciplinary work to the wider field. Situating the Centre within the Faculty of Education at Cambridge University has facilitated dissemination of information directly to educational researchers, trainee teachers and users of education, and also – and importantly – has enabled input from teachers and users in formulating research questions for future studies. The Centre hosted a research workshop (in 2005) on behalf of the OECD/CERI Learning and the Brain initiative, on learning to read in shallow/non-shallow orthographies (see Chapter 4).
>
> Research funding attracted by the Centre is currently £1.6 million (from ESRC [2 awards], MRC, EU Framework VI [2 awards], and The Healthcare Trust). Current projects include a large-scale longitudinal study of the brain basis of dyslexia, cross-sectional studies of typical number development and dyscalculia in children, studies of meta-cognition and executive control in very young children, and studies of children with synaesthesia, who for example experience numbers and letters with a mixture of sensory systems. The Centre is already attracting high-calibre research fellows and research students, including since inception an ESRC Research Fellow, an EU Framework VI Research Fellow, a Spanish Government Fulbright Fellow, an ESRC studentship, an NSF Visiting Research Student, a Gates Scholar, a Chilean Government Research Student, and a Taiwanese Government Research Student. There are currently 16 students and research fellows working in the Centre.
>
> The researchers at the Centre have already been consulted widely by policy makers and research users concerning the potential and impact of neuroscience for education. UK policy makers include Office for Standards in Education, Office for Science and Innovation, local government (LARCI), Her Majesty's Inspectorate for Education (Scotland), as well as specific UK initiatives such as the Rose Review of the Early Teaching of Literacy, the Government Working Group on Learning convened by the Schools Minister, and the Football Association Youth Training Scheme. Research users include voluntary and public services (*e.g.*, Cambridge Primary Head teachers, Cambridge Secondary Head teachers, Cambridge Child and Adolescent Mental Health Services, Dyslexia Scotland, National Association of Teachers of Children with Specific Learning Difficulties, National Association of Educational Psychologists, Special Schools Deputies Annual Conference, Psychology for Learning and Teaching, and the Curriculum, Evaluation and Management Centre, University of Durham).
>
> *Source:* Centre for Neuroscience in Education, University of Cambridge, United Kingdom.

Box 7.3. **Learning Lab Denmark**

Learning Lab Denmark (LLD) is part of the Danish University of Education. LLD has as its main objective to conduct interdisciplinary and practice-oriented research on learning processes in both formal and informal contexts, which can contribute to the development of teaching and learning methods. One of LLD's priority areas is neuroscience and learning. Research in this area is done by members of the research unit: *Neuroscience, Corporality and Learning*. This group has a special focus on understanding the relationship between brain, body and cognition and on learning theories that can integrate findings from evolutionary biology, neuroscience and cognitive science. Illustrative examples on current research projects include:

Tacit and implicit knowledge: The neuroscientific understanding of the mechanisms that underlie learning becomes more and more extensive. As an example, neuroscientific research suggests that we can distinguish between two modes of learning; implicit and explicit learning, which serve different evolutionary purposes. The traditional educational system, as exemplified by the public school, is almost entirely concerned with explicit learning because it produces knowledge that can be verbalised. This research project seeks to investigate the potential of implicit learning and, if possible, to develop educational guidelines that can take advantage of this learning resource.

The visual word form area: There is an ongoing debate as to whether the reading of words is based on a dedicated system in the brain not shared by other stimulus types (*e.g.*, common objects, faces); the so-called visual word form area (VWFA). Some claim that activation of the VWFA is specific to letters and letter strings whereas others argue that the area is equally – or even more – responsive to other object categories. We are currently examining this question by means of functional imaging of normal subjects engaged in processing words and pictures in tasks requiring different levels of structural processing.

Individual differences in brain maturation: With recent developments in non-invasive brain imaging techniques, it has become possible for the first time to study the dynamics of brain maturation in children during the school-age years. While much remains to be learned from such techniques, the early findings clearly demonstrate ongoing brain development. The studies also provide some evidence for differences in the progress of brain maturation from one child of the same age to the next and suggest that such differences predict the status of developing cognitive processes in individual youngsters. A plausible interpretation of these neuroimaging findings is that individual children exhibit unique patterns of brain maturation; and that a child's developing mental skills and capacities are, at least to some degree, constrained by this pattern. The proposed research programme will focus specifically on individual differences in trajectories of brain maturation, and on the relationship between these and developing academic skills.

To enhance knowledge about brain functioning and learning within pedagogical communities, we also publish popular science books in which we address how to translate neuroscience and biological insights into action. Currently, an anthology on neuroscience and pedagogy is in print as well as a book on "forest schools" in the pipeline.

Source: Learning Lab Denmark.

Box 7.4. **Harvard Graduate School of Education**

The study of learning belongs at the nexus of biology and education. Recent scientific advancements, such as brain imaging technologies, enable a new, biological perspective on learning. However, much of this work is developing only in parallel to the study of learning in education because it is difficult for experts in one field to use knowledge from another. Building bridges between these fields requires the development of hybrid professionals capable of connecting work across disciplines.

Developing such trans-disciplinary experts is precisely the goal of the *Mind, Brain and Education* programme at *Harvard's Graduate School of Education*. Now in its fifth year, the *Mind, Brain and Education* programme trains students to synthesise research on learning across relevant fields. The curriculum draws on a wide variety of disciplines, including neuroscience, genetics, cognitive psychology, and education, and all students are enrolled in a core course designed to facilitate synthesis. This course, entitled *Cognitive Development, Education, and the Brain*, is team-taught by experts in relevant fields who encourage students to approach learning through an integrated lens, using various disciplinary perspectives together to analyse educational issues. Graduates are poised to make significant contributions to educational research and practice as Kurt Fischer, programme director and professor, emphasises the importance of a reciprocal integration of research with practice. Harvard's *Mind, Brain and Education* programme provides a valuable model for other institutions aspiring to develop trans-disciplinary programmes.

Source: Christina Hinton, *Mind, Brain and Education* graduate, Harvard Graduate School of Education.

Reciprocal inputs from both sides – bi-directional progress

Neuroscience alone cannot provide the knowledge necessary to design effective approaches to education and so educational neuroscience will not consist of inserting brain-based techniques into classrooms. Rather, a reciprocal relationship must be established between educational practice and research on learning which is analogous to the relationship between medicine and biology (Figure 7.2). This reciprocal relationship will sustain the continuous, bi-directional flow of information necessary to support brain-informed, evidence-based educational practice. Researchers and practitioners can then work together to identify educationally-relevant research goals and discuss potential implications of research results. Once brain-informed approaches are implemented, practitioners should systematically examine their effectiveness and provide classroom results as feedback to refine the directions of research.

Figure 7.2. **Bi-directional exchange between research and practice**

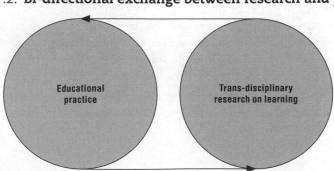

Note: A bi-directional flow of information between trans-disciplinary research on learning and educational practice. Research findings shape educational practice, and practical results, in turn, refine research goals.

Practitioners need some knowledge of the brain both to interpret neuroscience findings and to communicate classroom results to neuroscientists (see Box 7.5). Structures are therefore needed to educate practitioners about the brain, including those in teacher education and professional development programmes, and there needs to be initiatives communicating with civil society more broadly. The interdisciplinary neuroscience and education programmes at Harvard University, Cambridge University, and Dartmouth College provide examples of intensive teacher education programmes designed to create trans-disciplinary expertise. It is also possible to integrate information about the brain into conventional initial teacher education and short-term professional development programmes. One possible organisational strategy would be to follow a trans-disciplinary curricular sequence – from molecular to cellular to brain systems to individual body systems to social systems. (Such a sequence would emphasise connections between brain and society, making clear the iterative process whereby society shapes brain structure and therefore influences behaviour, which, in turn, shapes society.) Initiatives might best include components to generate motivation and build positive attitudes with the aim of ensuring that knowledge is infused with practice and programmes could also help practitioners design curricula on the brain for students to bolster meta-cognitive awareness of the learning process.

Because the field of neuroscience is expanding so rapidly, any educational programme on neuroscience should include training designed to facilitate continuing to learn about the brain after the programme ends. This training should indicate where to find accurate information about the brain. It could also involve cautioning practitioners about prevalent neuromyths and teaching them how to interpret neuroscience in the media with a critical lens. Initiatives aimed at communicating with civil society more broadly can make use of the Internet (see Box 7.6). Available web-based educational neuroscience tools developed by CERI already include a useable knowledge database and a monitored discussion forum (*www.ceri-forums.org/forums*).

As theoretical frameworks informed by neuroscience are developed and as practices based on those models are implemented, practitioners will need to track the progress of these practices because classroom results supply invaluable data that can be used to fine-tune models. If an intervention works in one context and not another, for example, this result provides important insights into the nature of the intervention-problem interaction. It helps to frame new research questions, such as: what are the critical ingredients in the intervention that promote success? How can this model be adapted to other problems without these elements? Consider, for example, an intervention that helps only certain children with Attention Deficit Hyperactivity Disorder (ADHD) improve attention. Identifying the conditions under which the intervention is successful guides researchers toward a more sophisticated and differentiated understanding of ADHD. Practitioners can collect data in many different ways, including using journals informally to document observations, engaging in semi-formalised discussions with colleagues to analyse classroom experiences, and publishing essays that reflect on their practice.

It is difficult for practitioners to stay up-to-date with laboratory results and for researchers to stay informed about classroom results within the constraints of traditional structures. While trans-disciplinary conferences, journals and associations provide bridges between the two fields, perhaps the most ideal solution is to integrate laboratories and schools as much as possible. The *Transfer Centre for Neuroscience and Learning* in Ulm is among the first institutions that integrate neuroscience research and educational practice (see Box 7.7). Establishing research schools with educational practice intimately connected to brain research is a promising way to stabilise trans-disciplinary work.

Box 7.5. **Educators' views of the role of neuroscience in education**

Practitioners play a vital role in developing brain-informed didactics as they are in a unique position to evaluate its success *in vivo*. Understanding educators' expectations, preconceptions, needs and attitudes is important for successful cross-disciplinary collaboration. A pioneering study conducted by Susan J. Pickering and Paul Howard-Jones at the *University of Bristol* provides important insights into educators' views on connecting neuroscience and education.

198 participants from the UK and abroad completed a questionnaire concerning their thoughts, beliefs, and knowledge of the link between neuroscience and education. Additionally, a small sample of participants took part in a semi-structured interview, which was based on the questionnaire but allowed participants to develop the discussion in directions of interest to them. Participants were recruited from the 2005 *Learning and Brain Europe* conference held in Manchester, UK, the 2005 *Education and Brain Research* conference held in Cambridge, UK, and the *Brain and Learning* homepage of the OECD/CERI.

The questionnaire included short-answer, free-response and *Likert Scale* questions designed to obtain information about:

- Educators' views on the importance of an understanding of the workings of the brain in a range of educational activities for children and adults.

- The sources of information educators used to obtain information about neuroscience and education.

- Educational ideas that educators had come across involving the brain.

- Whether participants' institutions had used initiatives based on ideas about the brain, and if such ideas were useful.

- The importance of a number of issues in the application of neuroscience to education, such as: communication between interested parties, relevance, accessibility of information and ethics.

The data collected provide a rich representation of practitioners' views on connecting neuroscience and education. Some key findings include the following:

- Educators believe that information about the brain is highly relevant to a wide range of educational activities, including the design and delivery of educational programmes for children and adults, for those with and without learning problems, and in the understanding of the role of nutrition in educational performance.

- Educators rely on a variety of sources of information on the brain, including academic journals, conferences, professional journals, books, in-service training. Differences in source preference may reflect issues of access, previous experience, and preconceived notions about the role of neuroscience in education.

- A vast array of ideas on the brain and education are floating around educational communities. Ideas that have been used by educational institutions range from scientifically-based approaches, such as the theory of multiple intelligences, to neuromyths, such as left-brain/right-brain learning styles. Nonetheless, respondents were generally positive about the usefulness of most of the ideas that they have come across.

- Practitioners are more interested in results that are directly relevant to classroom practice than theoretical developments that are not yet applicable to practice. However, practitioners do not seem satisfied with being "told what works", but are seeking to understand how and why certain brain-informed practices are useful.

- Participants reported that they were short of time, had limited access to resources such as academic journals, and felt vulnerable to individuals claiming to be experts on brain-based practices. They identified the following solutions: creating hybrid professionals who can facilitate cross-disciplinary communication, including neuroscience in initial teacher training, and launching initiatives aimed at helping teachers develop critical skills for questioning new brain-based practices.

This study shows an encouraging enthusiasm among educators, and a desire to understand how the brain learns and use this knowledge to strengthen their practice.

Source: Paul Howard-Jones and Susan J. Pickering, University of Bristol.

Box 7.6. **Technology and a world wide education perspective**

Recent discoveries related to brain mechanisms in attention, numeracy and literacy have led a number of researchers under the leadership of the OECD's Centre for Educational Research and Innovation (CERI) to develop a website. This *www.teach-the-brain.org* website is designed to start the process of developing educational experience based on brain research that can be delivered with the aid of the web.

So far two programmes have been developed for web delivery, one in attention training and another on literacy. These programmes exist so far only in downloadable form. The user needs to access the website and then download the programmes for use with his/her own computer. The goal has been to implement fully interactive programmes that could at the same time deliver services and collect relevant data on their use. In that way the programmes can foster the research that can lead to updates and improvements.

Although there is research showing that the use of these programmes can lead to improvement and can change aspects of brain function, there is no reason to suppose that they are optimal. Rather, it is hoped to use the site to inform various audiences about the nature of the research and the links between the research and the programmes that have been developed. In addition many commercial programmes of a similar nature are being sold to the public. Our intention has been to make available links to these programmes, when possible, and when they have been shown to have substantial support. A further aim is to foster discussion of the programmes that have not been tested or have failed to live up to their commercial expectations.

While many of the implications from brain research have been developed for very early education either before the start of formal schooling or in the early years of school there have also been findings that might relate to the development of later skills. For example, some research has tried to trace the development of the visual word form system from four years through ten years of age. The findings suggest that the system starts to develop rather late and changes from a system mainly operating on familiar words, which operates on the rules that govern English orthography. This kind of study has relevance to the methodology that might be best to acquire genuine expertise in the skill of reading. Other skills such as the perception of visual objects also seem to show that they proceed by gaining access to systems that may have earlier been used for other stimuli. Future research may help in the development of high levels of expertise for complex concepts. Brain imaging may help by allowing us to determine how a given form of practice influences a particular part of the network underlying the skill.

A reasonable goal for the web-based system would be to synthesise reading, attention and numeracy exercises into the development of improved comprehension of scientific documentation, including the weighing of evidence, the appreciation of graphs and charts and mathematic formulations, and the understanding of scientific communication. It is widely believed that such knowledge is important in the development of a global workforce appropriate for the 21st century. It would be a worthy goal of a curriculum to aid children in all countries to obtain this kind of knowledge in the most efficient way possible.

Source: Michael I. Posner, University of Oregon, USA.

Box 7.7. **Transfer Centre for Neuroscience and Learning, Ulm, Germany**

To a large extent, neuroscience research is about learning. The more we know about how the brain accomplishes learning, the better we should be able to use this knowledge in order to improve learning in any social setting, from preschool to school, occupational learning, universities and continuing lifelong education. While the importance of neuroscience for education is widely realised, basic research findings cannot be applied readily to the classroom. Intermediate steps are necessary. In 2004 the "Transfer Centre for Neuroscience and Learning" (ZNL) was founded to implement these steps and to carry out research that links the neuroscience knowledge to the way learning and teaching is done.

In the ZNL a multidisciplinary team consisting of psychologists, educational scientists as well as specialists in the areas of medicine, biochemistry, and linguistics works in the transfer of neuroscience knowledge to practitioners. As there is no established field of "transfer science", we approach this problem on many levels and with respect to many aspects of teaching and learning. In particular, transfer-research topics include:

Dyslexia: In which way do dyslexic children and adults differ from normal people? Are there early markers? What is the most efficient treatment? We develop the Computer Assisted Speech Assessment and Remediation (CASPAR) project: an Internet platform to identify at risk pre-school children and offer computer games to alleviate their symptoms.

Physical exercise and learning: Physical exercise aids brain functioning. What type of intervention (physical exercise) has the largest impact on attention and learning? In elementary school? In vocational school? We evaluate school programmes dedicated to add physical exercise to a student's day.

Emotion and learning: Our previous work has highlighted the importance of emotions for learning. We proceed in this work and carry out further fMRI studies on emotion regulation. In addition, we use ambulatory assessment of emotional heart rate to understand where and when school children are emotionally engaged.

Learning and memory: Multimodal teaching and learning is seen by many teachers as their gold standard. Using neurophysiologic techniques, we investigate the scientific underpinnings of multimodal learning to understand the working mechanisms. Furthermore, we evaluate the impact of multimodal teaching in classroom.

Memory consolidation: Memory consolidation is a topic of high relevance for institutional learning. We investigate its neurophysiologic underpinnings using fMRI. What are the effects of different activities on memory consolidation within school settings?

Nutrition and learning: The brain as the "hardware" for learning and thinking is influenced by nutrition. Empirical data suggest that many school students skip breakfast and have an unhealthy diet. We investigate the effects of breakfast as well as of Omega 3 fatty acids on attention and learning.

Source: Zentrum für Neurowissenschaft und Lernen (ZNL), Ulm, Germany.

Moving beyond national boundaries through international initiatives

The examples of emerging trans-disciplinary and bi-directional initiatives presented thus far are mainly national in scope and this is an essential first step towards the development of educational neuroscience. Given the momentum of these emerging initiatives, however, it is becoming necessary to consider moving beyond these to develop international networks. Their advantages are three-fold. First, international networks permit national initiatives to better learn from what others are doing, particularly as each initiative is operated by a different disciplinary research mix, within different environments, and often with a differing focus. Such varied experience, if shared systematically, would offer a rich information base to which each initiative can have access. Second, international networks would permit individual research efforts to go beyond "learning from others" and provide opportunities for cross-fertilisation, hence facilitating the emergence of new ideas and models. Finally, such a network would help to stimulate debate on ethical issues which ultimately need to be dealt with nationally but reflection on which would benefit from bringing together diverse perspectives.

Educational neuroscience research cannot generate a universal, prescriptive pedagogical approach, but it can inform the construction of educational methods within each cultural context. Research priorities may vary across countries depending on educational goals, and results may be interpreted differently through different cultural lenses. Not all research indeed is transferable across cultures. Consider the case of dyslexia, which is manifest in different ways according to the orthographic structure of each language (see Chapter 4). Conducting research in different countries is necessary to assess the transferability of results. Furthermore, examining cross-cultural differences provides powerful insights into gene-environment interactions. Therefore, cross-country longitudinal and cohort studies provide a valuable method of investigation, and international research collaboration should be encouraged.

As the trans-disciplinary field of educational neuroscience emerges, an international co-ordinator will be important. The OECD's Centre for Educational Research and Innovation (CERI) has served this function to date, but other bodies should now assume this responsibility. Governments of member countries can take an active role in co-ordinating research initiatives in their country, as the Japanese and Dutch governments have done (see Boxes 7.8 and 7.9 respectively). Developing cohesive research networks among multiple disciplines within countries can facilitate cross-country collaboration. International societies, such as the newly formed *International Mind, Brain and Education Society* (IMBES), can co-ordinate initiatives in different countries (see Box 7.1).

Box 7.8. **JST-RISTEX, Japan Science and Technology-Research Institute of Science and Technology for Society, Japan**

Founded on the understanding that brain science based cohort studies are critical for understanding mechanisms of development and ageing in the brain, the Research Institute of Science and Technology for Society (RISTEX) has been expanding brain science programmes initiated in 2001 to include cohort studies from 2004 onwards. The following briefly describes seven ongoing cohort studies:

1. *Tokyo Twin Cohort Project (ToTCoP):* The aim of this project is to uncover the genetic and epigenetic (similar to environmental) factors that affect development: temperament, motor skills, cognitive-linguistic abilities and other behavioural traits during the early stages of human development. This project, which is based on a five-year longitudinal study of twins (with about 1 000 pairs) between infancy and childhood, is expected to elucidate the relative contribution of, as well as the interaction between environmental and genetic factors.

2. *Cohort study of autism spectrum disorders (ASDs):* The aim of this cohort study is to explore the social origins of typical and atypical development among children. This is carried out by investigating the pathogenesis and variability in manifestations based on data on behavioural development and corresponding neural networks. This project is expected to contribute to the early detection and intervention for children with autism spectrum disorders (ASDs), and to an understanding of the variations in social development that would help to provide a solution to current problems in the school environment.

Box 7.8. **JST-RISTEX, Japan Science and Technology-Research Institute of Science and Technology for Society, Japan** (cont.)

3. *Cohort studies of higher brain functions of normal elders and children with learning disabilities:* The broad goal is to address one of the difficulties that is emerging in an increasingly ageing society with low fertility rates. This is done by analysing issues including anti-ageing methods to maintain and improve brain functions in elderly persons, and intervention methods to develop healthy brain functions in children with learning disabilities.

4. *Cohort studies on language acquisition, cerebral specialisation and language education:* The aim of these studies is three-fold. The first is to investigate the mechanisms of first and second language acquisition in relation to cerebral specialisation and functional plasticity in the brain. The second is to identify "sensitive period(s)" for second language learning. The third is to propose a cognitive neuroscience-based guideline for second language learning and education, especially for English, including optimal ages and conditions. To this end, four-year cohort studies have been developed for three distinct populations: *a)* native English speakers who have learned Japanese; *b)* native Japanese speakers who have learned a foreign language in Japan; and *c)* native speakers of Japanese from ages 2 to 5.

5. *Development of a new biomedical tool for student mental health:* This study develops a new biomedical tool in order to easily and objectively assess stress response. This is done by adopting high-throughoutput analysis of gene expression by micro-array which has the potential advantage of studying complex stress response.

6. *Cohort study on the motivation for learning and learning efficiency using functional neuroimaging:* This cohort study explores brain mechanisms relevant to the motivation for learning among children and adults. This is performed by analysing the neural basis of motivation using functional MRI and by simultaneously measuring the extent of fatigue experienced by the subjects during tasks on motivational learning. This study also looks at genetic and environmental aspects among children with learning disabilities, to assess whether their problem lies in the mechanisms of motivation. The main goal of this study is to develop and/or propose methods for high-efficiency learning by maintaining a high level of motivation and lowering fatigue during the learning process.

7. *The Japan Children's Study (JCS):* This study elucidates the developmental mechanisms behind "sociability" or "social abilities", and identifies factors which make a nurturing environment suitable or unsuitable for babies and children. Cohort studies will be used to test the findings on social development from previous laboratory studies at the population level. Several types of preliminary cohorts will be launched at the same time: studies of infant cohorts starting from four months old, and a study of preschool cohorts starting from five months old for example. The results obtained from these preliminary cohort studies carried out with about 500 children over several years, will bring about invaluable information for further large-scale cohort studies.

These cohort studies based on the concept of "Brain-Science and Education" are likely to have three major implications:

- Acquire and present scientific evidence for policy-making, especially on education.
- Provide assessment of the potential effects of new technologies on babies, children, and adolescents.
- Provide assessment of the applicability of hypotheses drawn from animal and genetic case studies to humans.

Finally, drawing on an analogy from the concept of environmental assessment which emerged in the 1980s, society became aware of the necessity to evaluate the environmental impact of science and technology. Drastic changes in our society may be causing human sociological and psychological problems as well. An environmental evaluation from a metaphysical aspect may also be essential if we are to leave a legacy of a sustainable society to future generations.

Source: Hideaki Koizumi, JST-RISTEX.

Box 7.9. **Educational Neuroscience Initiatives in the Netherlands**

At the end of 2002, the Dutch Science Council (NWO) in consultation with the Dutch Ministry of Education, Culture and Science set up the Brain and Learning committee. The committee was to undertake initiatives to stimulate an active exchange between brain scientists, cognitive scientists and educational scientists and the practice of education. The initiative has led to two major activities which have been very influential.

The first activity was a so-called "Week of Brain and Learning" which was organised in February 2004 with the title "Learning to Know the Brain". The core activity was an invitational conference for 45 opinion leaders. There was also an international scientific symposium and a symposium for educational practitioners and the lay public. The purpose of the invitational conference was to identify difficulties, obstacles and concrete targets for innovation of education based upon insights from neuroscience and cognitive science. Participants from science, education, and societal institutions met in workshops and plenary discussions. They reached consensus about major routes to follow and an "agenda for the future" and were unanimous with respect to the statement: "Yes: the time is ripe for an active exchange between the various disciplines and domains."

The second activity was the production of a book entitled "Learning to know the brain" (May 2005). This describes the consensus which was reached in the invitational conference on the major topics "Individual differences", "Learning in adolescence", "Mathematics", "Motivational processes", "Learning processes" and "Adult learning". In addition, recommendations were done for development of the theme in the form of twenty propositions (to be downloaded from *www.jellejolles.nl*).

Both the book and the conference have had a major impact. By autumn 2006, concrete progress can be seen at three levels. With respect to the level of scientists and science institutions (NWO), a multidisciplinary and multidimensional "National Initiative" on Brain and Cognition has been formed across scientific domains. The ambition is to become a so-called "National Research Initiative" (NRI) with a total budget of € 290 million. The domain of brain, learning and education – "The Learning Mind" – is one of the three core themes of the NRI next to "The Healthy Mind" and "The Working Mind".

With respect to the level of the ministry of education and institutions involved in educational development, the ministry has organised a working conference on the theme in June 2006. Based upon the consensus which was reached, various institutions and organisations which have "development and innovation of education" as their duty have become active on the field and now search for the best routes to follow.

With respect to the level of practitioners and school organisations, the topic appears "to be alive": workshops, lectures, congresses are now organised over the country for various organisations for teachers and for school institutions. A start has been made with execution of evidence-based educational interventions in the school setting in collaboration between University and School.

In summary: the subject "Brain, learning and education" is recognised as a very important one in the Netherlands. Optimism exists with respect to the possibility to mobilise the financial resources needed, and "to bridge gaps" between science and the educational field. It is regarded essential that the representatives of the various domains learn to listen to the language of the others. The progress in the Netherlands is directly related to the OECD/CERI initiative on "Learning sciences and the brain".

Source: Jellemer Jolles, University of Maastricht, Netherlands.

Cautions and limitations

While neuroscience can provide valuable insights into learning, it is important to recognise its limitations. Educators should be cautious when transferring results from controlled laboratory settings to the complex classroom. As the field of educational neuroscience develops, neuroscientists will likely modify research tasks so that they are more representative of complex educational settings. In addition, research-driven policies should be implemented through the seeding of educational trials in which the effectiveness of these policies is systematically examined. This reciprocal integration of research and practice would seek to ensure the validity of research-based practices.

The educational implications of neuroscience results are conditional on the values and goals of each learning community. For example, though neuroscience suggests that learning a foreign language at the primary stage is likely to be more efficient and effective than in secondary school (see Chapters 2 and 4,) this does not necessarily imply that all schools should teach foreign language in primary school. If the relative value placed on learning foreign language is lower than that placed on learning other age-sensitive skills in certain learning communities, the latter would take priority. Accommodating how the brain functions is only one of many factors that must be taken into account when constructing educational programmes and teaching. Neuroscience is a tool with specific strengths and weaknesses which is extremely useful for tackling certain questions but relatively ineffective for others. Neuroscience can, for example, address the question as to when foreign language can be learned most easily but it is less useful for answering which foreign languages should be taught.

In the development of a trans-disciplinary approach to educational policy design, it is important to be clear about the purpose of developing neuroscientific knowledge. It cannot generate a universal, prescriptive pedagogical approach but it can inform the construction of pedagogy and educational programmes within each context. Educational policy informed by neurobiology cannot simply be imposed on schools – the educational implications of any line of research must engage in a synergistic interaction with each educational community such that policy appropriate for each culture is developed. Therefore, neuroscientific knowledge needs to be widely accessible to those engaged in designing educational policy so that they could use this information to construct policy that is appropriate for a given school culture. They would need systematically to examine the effectiveness of such policy once implemented.

References

Bruer, J. (1993), "Schools for Thought: A Science of Learning in the Classroom", MIT Press.

Calvino, B. (2003), "Les médicaments du cerveau", *Graines de Sciences*, Vol. 5 (ouvrage collectif), Le Pommier, pp. 111-134.

Gardner, H. (1983), *Frames of Mind: Theory of multiple intelligences*, Basic Books.

Gardner, H. (2000), *Intelligence Reframed: Multiple Intelligences for the 21st Century*, Basic Books.

Guillot, A. (2005), "La bionique", *Graines de Sciences*, Vol. 7 (ouvrage collectif), Le Pommier, pp. 93-118.

INSERM expertise collective (2005), "Troubles de conduites chez l'enfant et l'adolescent", Éditions INSERM.

Koizumi, H. (1999), "A Practical Approach to Transdisciplinary Studies for the 21st Century – The Centennial of the Discovery of the Radium by the Curies", J. Seizon and Life Sci., Vol. 9, pp. 19-20.

Kramer, P.D. (1997), *Listening to Prozac: The Landmark Book about Antidepressants and the Remaking of the Self*, Penguin Books, New York.

OECD (2002), "Learning Seen from a Neuroscientific Approach", *Understanding the Brain: Towards a New Learning Science*, OECD, Paris, pp. 69-77.

OECD (2004), *Learning for Tomorrow's World – First Results from PISA 2003*, OECD, Paris, pp. 95-99, *www.pisa.oecd.org*.

Talan, J. (2005), "Rethinking What May Look Like a Normal Brain", *www.newsday.com*, 1 February.

ISBN 978-92-64-02912-5
Understanding the Brain: The Birth of a Learning Science
© OECD 2007

PART I

Conclusions and Future Prospects

It is no good to try to stop knowledge from going forward. Ignorance is never better than knowledge.

Enrico Fermi

This chapter concludes Part I of this report by bringing together the key messages and potential policy implications, which show how neuroscientific research is already contributing to education and learning policy and practice. The themes include discussion of lifelong learning; ageing; holistic approaches to education; the nature of adolescence; ages for particular forms of learning and the curriculum; addressing the "3 Ds" (dyslexia, dyscalculia, and dementia); and assessment and selection issues in which neuroscience might increasingly be involved. The chapter also points to areas needing further educational neuroscientific research that have emerged from the different chapters of the report.

After seven years of a pioneering activity on learning sciences, it would be tempting on the one hand to exaggerate the claims that can be made, but also easy to hide behind the plea that further research is needed before we can reach any conclusions. On the latter, it is certainly true that more research is needed, and some key lines for further research are suggested below. On the former, this concluding chapter largely abstains from specific recommendations. The field is still too young, and the connections between neuroscience and education too complex, for this to be justified. There are few instances where neuroscientific findings, however rich intellectually and promising for the future, can be used categorically to justify specific recommendations for policy or practice. Indeed, one of the messages from this activity, made already in the 2002 *Understanding the Brain – Towards a New Learning Science* report, is that we should beware of simplistic or reductionist approaches, which may grab headlines or offer lucrative opportunities but which are a distortion of the knowledge base.

This chapter brings together the main themes and conclusions from the preceding analysis. It is possible to put forward some broad propositions or challenges which can open up and refresh the debate on the future shape and character of our education systems. If we witness the birth of a science of learning, new ideas and evidence will rapidly arise and transform the current landscape. We do not need to wait for that research; part of CERI's mission has always been to help OECD countries think through their future agendas. The conclusions are at quite a high level of generality, precisely in order to give the necessary impetus to carry the discussion across the very broad terrain mapped out in the preceding chapters.

Key messages and conclusions

> The most important scientific revolutions all include, as their only common feature, the dethronement of human arrogance from one pedestal after another of previous convictions about our centrality in the cosmos.
>
> *Stephen Jay Gould*

Educational neuroscience is generating valuable new knowledge to inform educational policy and practice

The sweep of this volume – from the learning that takes place in the earliest years of infancy through to that of the elderly, from knowledge related to specific subject areas through to that concerned with emotions and motivation, from the remedial to the more general understanding of learning – shows how wide-ranging is the contribution that neuroscience can make to educational policy and practice. It has shown that the contribution of neuroscience to education takes different forms.

On many questions, neuroscience *builds on the conclusions of existing knowledge* from other sources, such as psychological study, classroom observation or achievement surveys. Examples discussed in this volume – such as the role of diet to improve educational performance, the turbulence of puberty, or that confidence and motivation can be critical to educational success – are not new. But the neuroscientific contribution is important even for results already known because:

- it is opening up understanding of *"causation" not just "correlation"*; and moving important questions from the realm of the intuitive or ideological into that of evidence;

- by revealing the mechanisms through which effects are produced, it can *help identify effective interventions and solutions.*

On other questions, neuroscience is *generating new knowledge, opening up new avenues.* Without understanding the brain, for instance, it would not be possible to know the different patterns of brain activities associated with expert performers compared with novices (as a means to understanding comprehension and mastery), or how learning can be an effective response to the decline of ageing, or why certain learning difficulties are apparent in particular students even when they seem to be coping well with other educational demands.

To these two key contributions can be added a third – that of *dispelling neuromyths.* Such distortions, discussed in detail in Chapter 6, risk to distract serious educational practice with the faddish, off-the-shelf solutions of the airport lounge bookshop.

Another key set of distinct contributions by neuroscience to education is:

- Research which is *deepening the knowledge base of what constitutes learning* as a central aspect of human and social life, and in ways which cut across the different institutional arrangements called "education".

- Neuroscience is developing the means *for revealing hitherto hidden characteristics in individuals*, which may be used for remedial purposes – to overcome reading problems or dyscalculia for instance. Eventually, they may also be used to select or improve performance or exclude, raising a raft of thorny ethical issues as discussed in Chapter 7.

- It is, along with other disciplines, able *to inform how best to design and arrange different educational practices*, especially as regards the match between findings on how best learning takes place and when, on the one hand, and how education is conventionally organised, on the other. It is another question whether that knowledge is being sufficiently acted upon at present.

Brain research provides important neurological evidence to support the broad aim of lifelong learning and confirms the wider benefits of learning, especially for ageing populations

One of the most powerful set of findings concerned with learning concerns the brain's remarkable properties of "plasticity" – to adapt, to grow in relation to experienced needs and practice, and to prune itself when parts become unnecessary – which continues throughout the lifespan, including far further into old age than had previously been imagined. The demands made on the individual and on his or her learning are key to the plasticity – the more you learn, the more you can learn. Far from supporting ageist notions that education is best concentrated on the young – the powerful learning capacity of young people notwithstanding – neuroscience has shown that *learning is a lifelong activity and that the more that it continues the more effective it is.*

As the demands for having an evidence-base on which to ground policy and practice grow, so has it become even more important to broaden the understanding of the "wider benefits" of education beyond the economic criteria which so often dominate policy cost-benefit analyses. There is growing evidence to show for instance that educational participation can have powerful benefits in terms of health or civic participation (see also CERI work on the "Social Outcomes of Learning"). This report has underpinned the arguments about the wider benefits of learning: the enormous and costly problems represented by senile dementia in ever-ageing populations can be addressed through the learning interventions being identified through neuroscience.

Combinations of improved diagnostics, opportunities to exercise, appropriate and validated pharmacological treatment, and good educational intervention can do much to maintain positive well-being and to prevent deterioration.

We need holistic approaches based on the interdependence of body and mind, the emotional and the cognitive

With such a strong focus on cognitive performance – in countries and internationally – there is the risk of developing a narrow understanding of what education is for. Far from the focus on the brain reinforcing an exclusively cognitive, performance-driven bias, it actually suggests the need for holistic approaches which recognise the close interdependence of physical and intellectual well-being, and the close interplay of the emotional and cognitive, the analytical and the creative arts.

The ways in which the benefits of good diet, exercise, and sleep impact on learning are increasingly understood through their effects in the brain. For older people, cognitive engagement (such as playing chess or doing crossword puzzles), regular physical exercise, and an active social life promote learning and can delay degeneration of the ageing brain (see Chapter 2).

The analysis of this report shows not only how emotions play a key part in the functioning of the brain, but the processes whereby the emotions affect all the others. Especially important for educational purposes is the analysis of fear and stress, which shows how they, for instance, reduce analytical capacity, and *vice versa* how positive emotions open doors within the brain.

This is just as relevant for the adult student confronted by an uncomfortable return to education as it does for the young person confronted by the unfamiliar demands of secondary or higher education. It has an equity dimension, for fear of failure, lack of confidence, and such problems as "maths anxiety" (Chapters 3 and 5) are likely to be found in significantly greater measure among those from less privileged backgrounds.

We need to understand better what adolescence is (high horsepower, poor steering)

This report is particularly revealing about the nature of adolescence in terms of the stage of brain development in the teenage years and particularly in terms of emotional maturation.

The insights provided by neuroscience on adolescence and the changes which take place during the teenage years are especially important as this is the period when so much takes place in an individual's educational career. The secondary phase of education is conventionally covered by this phase, with key decisions to be made with long-lasting consequences regarding personal, educational, and career options. At this time, young people are in the midst of adolescence, with well-developed cognitive capacity (high horsepower) but emotional immaturity (poor steering).

Clearly, this cannot imply that important choices should simply be delayed until adulthood. It does suggest, with the additional powerful weight of neurological evidence, that the options taken should not take the form of definitively closing doors. There needs to be stronger differentiation of further learning opportunities (formal and informal) and greater recognition of the trajectories of adolescent maturation.

Neuroscience also has developed the key concept "emotional regulation". Managing emotions is one of the key skills of being an effective learner. Emotional regulation affects complex factors such as the ability to focus attention, solve problems, and support relationships. Given the "poor steering" of adolescence and the value of fostering emotional maturity in young people at this key stage, it may well be fruitful to consider how this might be introduced into the curriculum and to develop programmes to do this.

We need to consider timing and periodicity when dealing with curriculum issues

The work of psychologists like Piaget has long influenced our understanding of learning linked to individual development. Educational neuroscience is now permitting the qualification of the Piagetian models (including demonstration of the capacities already possessed by young infants), while broadening the understanding of timing and optimal learning through the study of "sensitive" periods.

The message emerging from this report is a nuanced one: there are no "critical periods" when learning *must* take place, and indeed the neuroscientific understanding of lifetime "plasticity" shows that people are always open to new learning. On the other hand, it has given precision to the notion of "sensitive periods" – the ages when the individual is particularly primed to engage in specific learning activities.

The example of language learning has featured prominently in this report, and is a key subject in an increasingly global world. In general, the earlier foreign language instruction begins, the more efficient and effective it can be. Such learning shows distinct patterns of brain activity in infants compared with school-age children compared with adults: at older ages more areas of the brain are activated and learning is less efficient. Even so, adults are perfectly capable of learning a new language.

This report has also dispelled myths about the dangers of multilingual learning interfering with native language competence; indeed, children learning another language reinforce the competences in their mother tongue.

These are important questions for education. These findings deepen the basis on which to pose questions about when in the lifespan certain types of learning should best be undertaken, grounded on evidence rather than tradition. They support the importance of laying a very strong foundation for lifelong learning, hence further emphasise the key role of early childhood education and basic schooling, not as ends in themselves but as giving the best possible start.

At the same time, the report (Chapter 6) has warned against *over-emphasising* the determining importance of the ages birth to three years on later learning.

Neuroscience can make a key contribution to major learning challenges

The contribution that neuroscience is already making to the diagnosis and identification of effective interventions is most clear in what might be termed the "3-Ds": dyslexia, dyscalculia, and dementia.

Dyslexia: Until recently, the causes of dyslexia were unknown, but now it is understood that it results primarily from atypical features of the auditory cortex (and maybe, sometimes, of the visual cortex). Only recently has it been possible to identify these features at a very young age. Early interventions are usually more successful than later interventions, but both are possible.

Dyscalculia is now understood to have comparable causes as dyslexia, though early identification and hence interventions are well less developed.

Dementia: The very significant findings about learning and dementia have been mentioned above, and education is being identified as an effective, desirable source of "prevention" to among other things delay the onset of Alzheimer's symptoms and reduce their gravity.

On the more general understanding of literacy (Chapter 4), the dual importance of phonological and direct semantic processing in the brain during reading in English suggests that a balanced approach to literacy instruction may be most effective for non-shallow alphabetic languages. As for shallow orthographies, neuroscience seems to confirm the appropriateness of "syllabic methods" to learn reading, and there is interesting potential to be explored in comparisons between alphabetic and non-alphabetic languages on reading acquisition.

On numeracy (Chapter 5), since humans are born with a biological inclination to understand the world numerically, formal mathematics instruction should build upon existing informal numerical understandings. Because number and space are tightly linked in the brain, instructional methods that link number with space are powerful teaching tools.

More personalised assessment to improve learning, not to select and exclude

The potential of brain imaging could have very far-reaching consequences for education, as well as raising critical ethical issues. Knowledge about how the brain functions, and about how competence and mastery are reflected in brain structures and processes, can be applied at a *system-wide level*, interrogating conventional educational arrangements and practices to ask whether we organise them for optimal learning. Many conventional forms of assessment, where success can be boosted by cramming, have been shown to be "brain-unfriendly" with low retained comprehension.

But beyond these general findings, the results of neuroscience may eventually also be applied *on individual learners* to find out such matters as whether they really comprehend certain material, or about their levels of motivation or anxiety. Used properly, this individual focus may add fundamentally powerful diagnostic tools to the process of formative assessment (OECD, 2005) and personalised learning.

This relates to the pursuit in a number of countries of greater "personalisation" of curricula and educational practices (OECD, 2006). Neuroimaging potentially offers a powerful additional mechanism on which to base personalisation. At the same time, studies of the brain show that individual characteristics are far from fixed – there is constant interaction between genetic function and experience and plasticity, such that the notion of what constitutes an individual's capacities should be treated with considerable caution.

But, on the other side, such individual applications of neuroimaging may also lead to even more powerful devices for selection and exclusion than are currently available. A biological CV would be open to profound risks, while being potentially attractive to such users as universities or employers. It would be an abuse of the valuable tools of neuroimaging if they were deployed in the negative ways of rejecting students or candidates on the grounds that they do not show sufficient learning capacity or potential (especially when the plasticity of the brain shows how open to development is the capacity to learn). The excessively "scientific" conception of education described in Chapter 7 – used as the basis for selecting students and teachers alike – would be anathema to many.

Key areas for further educational neuroscientific research

> If we value the pursuit of knowledge, we must be free to follow wherever that search may lead us.
>
> *Adlai E. Stevenson Jr.*

The below research areas do not pretend to be exhaustive of interesting fields of educational neuroscientific enquiry; instead, they have emerged from the analysis of this report as priority areas. Some of them represent the need to deepen knowledge where at present our understanding is sketchy.

It is also about setting an educational agenda for neuroscience as well as the medical agenda which has naturally tended to dominate it up to now. It is for the neuroscientific community to realise how valuable its contributions are to better understanding the key human activity of learning for educational purposes as it applies for all – gifted through disabled, young through old – and not only for those requiring remediation.

- Better understanding of optimal timing for what forms of learning, especially in relation to adolescents and older adults where the review shows that the knowledge base is not yet well developed (Chapter 2). This includes "sensitive periods" – when the capacity to learn is greatest – in specific areas such as language learning.

- Understanding the interaction between increasing knowledge and declining executive function and memory. More research into the ageing process, and not only among the elderly but also adults in their middle years – both in terms of capacity to learn and in terms of the role of learning to delay the deleterious effects of ageing.

- Much is needed on emotions in the brain. Further investigation using psychological and neuroimaging studies is needed of the neurobiological mechanisms which underlie the impact of stress on learning and memory, and the factors that could reduce or regulate it. A specific question for further investigation is how the adolescent's emotional brain interacts with different kinds of classroom environments.

- Better understanding of how laboratory conditions influence findings, and the applicability and transferability of results in different settings other than those in which they were generated. The key role of appropriate learning materials and specific environments needs to be analysed so as to move away from the crude formulations which ask whether environment does or does not make a difference.

- Confirmatory studies showing how beneficial nutrition can impact positively on brain development and more studies directly related to the educational domain. The same applies to physical exercise, sleep, music and creative expression.

- Much more is needed on what types of learning requires the interaction of others and on the role of cultural differences. This should be further broken down in terms of student demographic (especially gender) and socio-cultural differences, but it is also a minefield for misinterpretation. Neuroscience should certainly not be brought into the service of racist or sexist stereotypes.

- Research can help lead to the better understanding of multi-dimensional pathways to competence, for instance in reading. The need to expand the focus to real-world educational situations and applications, *e.g.* whole sentences rather than single words or characters.

- It would be very useful to further build up the differentiated map of mathematics in the brain, which builds on the insights gained already on the apparently paradoxical combination of dissociable skills and brain functions, on the one hand, and interconnectivity, on the other. Identification of approaches to overcome "mathematics anxiety" would be very useful.

- Understanding different brain activity – neural networks, role of cognitive function and memory – among "experts" as compared with average learners as compared with those with genuine problems. This will inform both the identification of successful learning and of effective, targeted teaching methods.

Birth of a learning science

Recent advances in neuroscience have produced powerful insights while educational research has accumulated a substantial knowledge base. A neuroscientific perspective adds a new, important dimension to the study of learning in education, and educational knowledge could help direct neuroscience research towards more relevant areas. Because both fields are well-developed, however, they have deeply-rooted disciplinary cultures with field-specific methods and language which make it extremely difficult for experts from one field to use the knowledge from the other. A new trans-disciplinarity is needed which brings the different communities and perspectives together. This needs it to be a reciprocal relationship, analogous to the relationship between medicine and biology, to sustain the continuous, bi-directional flow of information necessary to support brain-informed, evidence-based educational practice. Researchers and practitioners can work together to identify educationally-relevant research goals and discuss potential implications of research results. Once brain-informed educational practices are implemented, practitioners should systematically examine their effectiveness and provide classroom results as feedback to refine research directions. Establishing research schools with educational practice intimately connected to brain research is a promising way to stabilise trans-disciplinary work.

Educational neuroscience can help to drive the creation of a real learning science. It might even serve as a model of trans-disciplinarity for other fields to emulate. We hope that this publication will help give birth to this real learning science, as well as a model for continued trans-disciplinary fusion.

References

OECD (2002), *Understanding the Brain – Towards a New Learning Science*, OECD, Paris.

OECD (2005), *Formative Assessment – Improving Learning in Secondary Classrooms*, OECD, Paris.

OECD (2006), *Personalising Education*, OECD, Paris.

PART II

Collaborative Articles

ISBN 978-92-64-02912-5
Understanding the Brain: The Birth of a Learning Science
© OECD 2007

PART II

Article A

The Brain, Development and Learning in Early Childhood

by
Collette Tayler, School of Early Childhood, Queensland University of Technology, Australia
Nuria Sebastian-Galles, Faculty of Psychology, University of Barcelona, Spain
Bharti, National Council for Educational Research and Training, India

A.1. Introduction

The emergence of new and non-invasive brain imaging and scanning technologies has informed an unprecedented expansion in brain science, particularly in the area of developmental neurobiology. We have known for decades that brain growth and development is programmed from conception by information contained in our genes. Yet we are only beginning to observe and understand, at the cellular level of the brain, how stimuli from the external environment affect and control the use of that genetic information. Only in recent times has the brain come to the forefront of educational research and ideologies, particularly in regard to development and learning in early childhood. Decades of educational research involving young children give insights about early learning from different standpoints and some results complement ideas emerging through studies in neurology while others have no apparent connection at this time (Ansari, 2005; Slavin, 2002; Bruer, 1997).

This paper summarises what emerging neuroscience is revealing about the development of the brain's architecture and functions in early life, discussing the relevance of this new knowledge for promoting and supporting learning and development in early childhood. We outline ideas and findings about the importance of the early childhood phase of learning (birth-to 96 months) and the constitution of learning environments for young children. We conclude with some ideas for future research in the area of early childhood development and learning that may integrate the concerns, interests and skills of neuroscientists and educators and advance scientific research in education.

A.2. What do we know about brain development in neonates, infants and young children?

A.2.1. The onset and process of brain development

From the point of conception, genetic information controls the formation and reproduction of cells in the growing foetus. Brain development is a result to billions of neurons, the cellular building blocks of the brain, and trillions of synapses, the connections that receive and send electrochemical signals (Shore, 1997). In the human foetus, neurons overproduce in massive amounts during the first two trimesters and peak when the foetus is at seven months gestation. As neurons expand, the brain grows in volume and weight, with the vast excess of neurons produced in the first two trimesters then "pruned back" in the neonate through a process of natural selection at the cellular level. The selective pruning that takes place allows for fine-tuning of both the structure and function of the human brain. The neurons that exist in the infant at birth thus ensure a neuronal network capable of facilitating learning and adaptation from that time onwards. The number of neurons existing at birth remains stable although the number of synaptic connections increases at a remarkable rate after birth (Goswami, 2004).

The neuronal circuitry formation in the hippocampus shortly after birth enables the infant to begin attending to and storing sensory information. By two to three months of age, neuronal circuitry has begun forming at an increasing rate in the parietal, occipital and temporal lobes. These lobes receive no sensory information directly from the external environment. Their function is to integrate information transmitted from primary sensory and motor areas, and facilitate increasingly sophisticated and co-ordinated movements by the infant as s/he interacts with the external environment. The formation of permanent neuronal circuitry in the frontal lobes, which perform the most complex processing of information, typically begins at six months of age, when the infant begins to plan and execute goal-directed behaviour. Much of the volume of brain development occurs in the early years of life.

Normal human brain development follows similar universal sequences. At birth, the neuronal circuitry that established in utero continues to form and undergo myelination in the brainstem, thalamus, primary sensory and motor areas, and parts of the cerebellum. This enables the newborn to breathe, cry, wake, sleep, recognise his/her mother's smell and voice, suck, swallow, excrete, and perform basic motor movements of the legs, arms and hands – all of which are vital functions for initial survival. On-going neurological development in infancy and early childhood ensures the refinement and learning of these functions, increasingly enabling the child to attend, respond, relate, move freely, speak and represent ideas using different symbolic forms (see Box A.1).

Box A.1. **Emotions and memory (learning)**

When we are angry or distressed, our capacity to learn diminishes. While adults may be better able to control feelings, children have greater difficulties in doing so. Why?

One crucial structure in the brain, involved in the regulation of emotions is called the amygdala. The name comes from Latin – *almold* – because it is a small almond located in the middle of the brain. The amygdala forms, together with the hypothalamus, the most important "drug release" centres of the brain. It receives inputs from different perceptual areas, as well as secondary brain centres and it is directly responsible of increasing heart rate and blood pressure. Together with the hypothalamus, it controls a wide variety of hormones.

Hormones are important, not only in determining different aspects of sexual life, or in boosting or reducing growth, but also they are responsible of modifying how information is transported along the nervous system. The release of some particular hormones increases our capacity to transmit information, and therefore to learn. In addition, the release of other hormones will reduce this capacity. So when a person is angry or distressed it is more difficult to learn. This can particularly affect young children as they are yet to reach full capacity in control of their emotions.

Another component in the amygdala-hypothalamus system that makes the difference between childhood and adulthood emotional reactions is the frontal lobe. This is a part of our brain that develops relatively late, in fact, it is the last part of the brain to mature. Being responsible of the most "rational" part of our cognition – planning, reasoning – the frontal lobe "tempers" the functioning of the amygdala-hypothalamus system. Indeed, as adults, we are able (expected) to control our emotions and be rational, even in unpleasant circumstances (see control of emotions in Chapter 3).

Yet there are large individual differences among developing brains. The infant's brain may be construed as an information-processing unit, with the brain's neuronal network conceptualised as circuitry with infinite wiring possibilities. The hard-wiring outcomes that result for any child are specific to the individual and dependent on a combination of genetic and environmental endowments. At birth, each neuron in the cerebral cortex has approximately 2 500 synapses, but by the age of 3, a child's neurons have around 15 000 *synaptic connections*, which is said to be about twice that of the average adult brain (Gopnik, Meltzoff and Kuhl, 1999). In the year following birth – the period of infancy – the size of the brain increases some 2.5 times, from about 400-1 000 g (Reid and Belsky, 2002). This period represents the most intensive growth in the brain with most of the increase in brain weight being due to the growth of "support systems" such as myelination. It is clear that the periods of infancy, and early childhood to age 3-4 years account for remarkable development.

Nelson (2000) records the path of development of vision, language and higher cognitive functions, confirming the speed of development in the first 2-3 years of life. In this period, only the synapses that are stabilised or consolidated through usage will be maintained (Changeux and Dehaene, 1989). The process of selective pruning of synapses is driven by an interaction between genetics and the environment (Reid and Belsky, 2002). Establishing the biology of brain development as probabilistic, this interactive developmental process is described by Rutter:

> "… there is a genetic programming of the general pattern and course, but extensive opportunities to correct the process of development in accord with both environmental input and the workings of the brain, in terms of cell-cell interactions" (Rutter, 2002, p. 11).

Those experiences to which the individual is repeatedly exposed will result in the myelination of the relevant synapses. These "hard-wired" synapses become responsible for recording and coding the life experiences and learning encounters of the infant and young child.

A.2.2. *The part played by experience*

Experience is said to begin when sensory neurons are activated, with activation in turn, connecting cells in deeper layers of the system. Clark (2005) describes this process: "For example, if eating an apple activates a particular set of cells then these in turn will activate the cell in the second layer which has become 'hardwired' through learning to detect the taste of apple. The cells of the second layer are not genetically ordained to detect apple or lemon, but do so as a result of experience" (p. 681). Hence, it is likely that thought about any sensory experience, or on any topic, is the result of complex patterns and distributed processing across many neural components in the brain.

The process of children's neural growth and development – drawing on genetic capacity and the environment – is affected from well before birth. For this reason, programmes that support maternal well-being during pregnancy (including adequate nutrition, exercise and rest), contribute to the healthy development of the neonate's brain. In turn, an infant's nutrition, healthy unpolluted physical environments and positive psycho-social environments are important elements in the healthy development of the brain from birth.

A.2.3. *Timing and sequencing – important factors in brain development*

Synaptic growth and myelination constitute major brain activities in infancy and early childhood, the timing varying for different areas of the brain. This process is fundamental to cognitive, physical-motor, linguistic, social, cultural and emotional development. According to the age and level of development of the child, sensory information received via touch, taste, sound, sight, and smell, stimulates the brain's neurons and synapses to form an increasingly sophisticated set of neural pathways which convey, process, integrate, and store the information for present and future reference. To this end, infants and young children encounter the world through a variety of sensory experiences vital to the developing brain. The brain changes as a function of experience with repeated experience strengthening neuronal networks.

The interactive process between genetics and the sensory environment occurs before birth to some extent, through exposure to sensory information in utero. For example, playing music during the third trimester of pregnancy influences the activity and behaviour of the foetus (Kisilisky *et al.*, 2004). However, as Thompson and Nelson (2001) report, the interaction between genetics and environment that determines the timing and course of synaptic production is, by no means clear. How neuronal and synaptic overproduction and pruning processes occur, in which sequences, and at which periods of growth, is yet to be understood despite the rapid growth in knowledge about the developing brain through the application of new technologies. As yet there are limited research methods that can account for multiple and complex stimuli in the environment concurrently with dynamic and complex activity within the brain.

Ethical procedures that protect participants in the study of animal and human development place reasonable limits on the methods and techniques applied. Basic research normally occurs first with animals, particularly rats, with the subsequent transfer of findings to humans being highly qualified and subject to long periods of further research. Human autopsy specimens, too, play a part, providing "estimates of age-related differences in synaptic density (although) with sometimes only a handful of samples in any particular age" (Thompson and Nelson, 2001, p. 9). Such results estimate synaptic density as static figures and cannot indicate whether the synapses that are counted owe their existence to a genetic programme or to experience.

A.2.4. Plasticity – a key feature of the infant brain

At birth, the brain's neuronal circuitry is extremely "plastic" in regard to processing and storing sensory information – circuits are readily formed, broken, weakened and strengthened. This suggests the importance of the period of infancy and early childhood to brain development and the growth of physical, cognitive, linguistic, social and emotional capacity. Plasticity, however, does not imply that any part of a child's brain can learn anything. Gazzaniga (1998) argues that because the brain responds differently according to different kinds of stimulation and experience, the brain networks of an individual are personal and unique. Yet this does not construe plasticity as a brain having "re-wired itself". Crick (1994, p. 10) encapsulates this idea: *"The brain at birth, we now know, is not a tabula rasa but an elaborate structure with many of its parts already in place. Experience then tunes this rough and ready apparatus until it can do a precision job"*.

Experimental and pathological studies of brain growth and development in animals, as well as observations of human brain activity at the neuronal level, indicate that the plasticity of neuronal circuitry is genetically scheduled to vary at certain ages and stages of life (Dyckman and McDowell, 2005; Kolb and Wishaw, 1998). After birth, brain plasticity supports the *formation* and *development* of the neuronal circuitry in unique ways according to individual genetics and stimuli. The growth and development of neural circuitry is susceptible to specific "epigenetic" processes (gene expression, affected by environmental stimuli). These processes are influenced by the person's age and stage of life and account for the description of "critical" and "sensitive" periods of brain development.

A.2.5. *Critical or sensitive periods in neural development?*

Educational research has already confirmed the importance of early experiences and environments on children's learning and development (Sylva *et al.*, 2004; Thorpe *et al.*, 2004). Neuroscientific research on critical periods and the consequences of early sensory deprivation confirms that in some species, structural and functional development of certain aspects of the brain may require certain experience at particular times. For example, birdsong learning is restricted to particular periods, different for different species and dependent on certain conditions (Brainard and Doupe, 2002). Failure in establishing specific brain circuitry for some functions, such as the visual cortex, at a critical period may produce irreversible loss of visually driven activity in the cortex (Fagiolini and Hensch, 2000). For different species the duration of critical periods seems to vary in relation to life expectancy.

However, the concept of critical periods in human neurological development is no longer supported as it has become clear it is never to late to learn (Blakemore and Frith, 2005). In humans, so called "critical" periods are more likely to be seen as "sensitive" and to extend over years (such as the period of early childhood). "One of the major concepts currently being investigated in neuroscience is that such critical periods represent heightened epochs of brain plasticity, where particular sensory experiences during these periods produce permanent, large-scale changes in neuronal circuits" (Ito, 2004, p. 431). What is apparent is that these more broadly defined "sensitive periods" in human development follow the same chronology for all human beings and it is important to identify sensory problems such as hearing and vision impairment in children as early as possible because of likely lasting effects.

> "The findings suggest that early sensory deprivation can have lasting consequences, possibly very subtle ones, undetectable in everyday life. They also suggest that even after sensory deprivation, recovery and learning can still occur. Such late learning may be different from the type of learning that occurs naturally during sensitive periods" (Blakemore and Frith, 2005, p. 461).

The first two years are reported as highly sensitive for the creation of the brain's pathways for attention, perception, memory, motor control, modulation of emotion, and for the capacity to form relationships and language (Davies, 2002). The acquisition of language is the most well documented process related to specific periods of children's development, with the capacity to acquire language seeming to decline beyond the period of middle childhood. The specific environment experienced during a sensitive phase in infancy, for example, affects the infant's ability to make later phonological discriminations. The declining ability of humans to learn new languages or produce new speech sounds after adolescence correlates with the circuitry limitations found after certain sensitive periods have elapsed.

Throughout life, but particularly during sensitive periods, repeated stimulation, formation and reformation of neuronal circuits enable neurons and synapses to become progressively specialised in the type of information they process, integrate, and store. Myelination preserves the functioning circuitry and improves the speed and efficiency of information transmission. Once myelinated, a neuronal circuit or synapse is "fixed" for life, unless brain damage or cellular degeneration occurs. At the same time as circuits are being fixed through myelination, unused synapses are being eliminated or "pruned", to ensure maximum efficiency in neurotransmitter deployment. The availability of synapses to facilitate the formation of new permanent neuronal circuits consequently decreases over

Dáta

Údar

Teideal

Leabhar ☐   **Iris** ☐   (Cuir tic sa bhosca cuí, led' thoil)

Imleabhar   Bliain

Láthair:

Stóras an tSéipéil
(An Mháirt & Déardaoin amháin)

Stór Bhóthar na nDuganna
(An Mháirt & Déardaoin amháin)

Uimhir Sheilfmhairc

Iarrtha ag   **Uimhir Aitheantais**

ial language development

ıg, s/he has already acquired a fairly sophisticated
dy at birth, infants are able to notice the differences
ese and Dutch), as other mammals, like cotton-top
ifferent brain imaging techniques have revealed that
ths of age) show greater activation of left hemisphere
presented with normal speech, but no superior left
l with speech played backwards (adults also do not
kwards speech, although the pattern of activation for
nfants and adults). Human infants are also able to
sting in any language in the world, even if parents
t (like Japanese very young infants, who can perceive
rents are unable to perceive). Thus, at birth, humans
ions" with the speech signal; these computations are
at they allow newborns to learn equally well any
ıg months, infants will start tuning their language
es of the language of the environment. Indeed, from
eir capacity to perceive some foreign contrasts (like
ity to perceive /r-l/) and at the same time, they will
categories. It can be said that infants are becoming
mediocre universal listeners). There is some evidence
initial perceptual capacities to the properties of the
o later successful language acquisition: that is, the
se the perception of foreign contrasts, the faster they

ire the sound system, infants are also extracting
hat will eventually lead to the discovery of words and
d morphological properties of their language. One of
t their disposal to this end is that of keeping track of
or instance, the average probability that a particular
is higher within words than across words. Different
oung as 8 months are able to use this type of cue to
gnal. It is important to notice that unlike written
eliable markers indicating word boundaries; the
development clearly show the complexities of the
d that engineers have failed to build machines able to
the same efficiency that human infants show by

s implications for learning and development, and
rning opportunities encountered in early childhood,
nd sensitive periods can occur. The degree to which
ess may be more limited, is not clearly determined.

plasticity

The two concepts – *sensitive periods* and *brain plasticity* – at a fundamental level seem to
represent different and opposing conceptualisations of development (Hannon, 2003). On
one hand, plasticity suggests that learning and development can take place at any age. It is

never too late for intervention and learning. On the other hand, sensitive periods imply a crucial role for timing, early intervention and learning. With the exception of studies of language development, the evidence confirming sensitive periods appears to draw mainly from experiments with animals (see, for example, Mitchell, 1989). Based on the research studies to date, there is support for both these constructs. Further studies are needed to reveal the particular conditions in which learning is facilitated by sensitive periods and learning is achieved at different times, primarily as a result of brain plasticity.

A.2.7. Learning in and beyond early childhood

Infants and very young children implicitly learn language and become mobile. A mix of genetic endowment and environmental experience is conditional to success but the relationships among attention, implicit and explicit learning remain unclear. It is apparent that the brain notices things that the mind does not, yet as Goswami (2005) points out, "children spend much of their day in classrooms and their brains do not automatically notice how to read or how to do sums. These skills must be directly taught" (p. 468). Current knowledge of implicit learning, or learning without attention, is mainly from studies of the motor system, although unconscious learning of an artificial grammar is also noted by Blakemore and Frith (2005). What does developmental programming contribute to learning in and beyond early childhood? How does the acquisition of implicit information regarding values, attitudes and beliefs, and growth of social cognition and emotional regulation, affect functional skill development? The reciprocal relationship between implicit and explicit learning seems of vital importance to advance the science of education. Interdisciplinary studies, particularly focused in the early years of life, are necessary to advance current understandings.

There is some evidence that learning after early childhood may be mediated in different ways. Rutter (2002) points out that after the early childhood period the extent of brain plasticity, and the extent of variance in plasticity across different parts of the brain or brain systems, is as yet unknown. For example learning a second language beyond the infancy phase involves different parts of the brain from those employed in acquiring the first language (Kim *et al.*, 1997) and learning from new or unusual individual experiences continues to have neurological effects throughout life. "This learning is different from developmental programming, but also involves experiental effects on the brain (yet) remarkably little is known about structure-function links" (Rutter, 2002, p. 13).

A.3. How important are the early years of development and learning?

Learning and development at this stage of life takes a particularly large role, relative to performance. Marking out early childhood as an important and sensitive period to influence development and learning has intrinsic appeal, particularly to early childhood educators and paediatricians. We know that conditions in early life affect the differentiation and function of billions of neurons in the brain, and that early experience sets up pathways among different (distributed) centres in the brain. In the absence of appropriate environmental stimuli, brain development and functioning is likely to be altered, under-achieved, and/or delayed in a child, with concomitant impact on child learning and development. Fundamentally, there are significant implications for practices related to child-rearing, adult-child engagement and children's early education and care programmes, which is why early childhood is placed high on the political agenda of many countries. Reviews of early childhood education and care policies and services have taken place in 20 OECD countries, as

Box A.3. **Mirror neurons**

Children easily assimilate and imitate what they see and hear. Parents and educators show concern for children's safety and the risks including exposure to negative role-models. Many parents express concern about the extreme violence showed in media. Governments try to prevent children from accessing to explicit webpages. Indeed, school education is meant to provide children not only with knowledge, but with positive role-models. Why are models so important?

Humans are particularly gifted at imitating others. We feel sad or even cry when seeing someone suffering, we smile and laugh when we hear other people laughing (even if we hear someone unknown laughing next to us in a bus... some laughs are particularly "sticky"). A precursor of all these behaviors is already present at birth. If we pull our tongue to a newborn, s/he will do the same!

The question of knowing the precise neuronal mechanism underlying imitation has been very elusive and until quite recently it was a mistery. In 1996 an Italian neuroscientist made an extraordinary discovery allowing us to start understanding the basic mechanisms of imitation. Giacomo Rizzolatti and his colleagues at the University of Parma, Italy have discovered the existence in the brain of monkeys of "mirror neurons". What Rizzolatti observed was the existence of a particular type of neurons that fired when monkeys performed a very specific task with their hand: for instance, taking a peanut and putting it into the mouth. But, interestingly mirror neurons also fired when the monkey saw another monkey doing the very same task. The specificity was very high, for instance, if they saw the experimenter doing the same hand movement, but without any peanut to be taken, the corresponding mirror neuron did not fire.

The discovery of these neurons (located in the brain of the monkeys in an area analogous to human's Broca area, one of the main centres of language processing) has boosted a lot of research. Currently neuroscientists are working with the hypothesis that we are able to understand other people's actions (and perhaps feelings) because when we see them performing these actions (and having particular feelings) our mirror neurons could be firing, making us feel as if we were actually making these actions (or having these feelings). Could there be an abnormal functioning of mirror neurons at the basis of some personality pathologies? This is an exciting field of current research.

reported in *Starting Strong II* (OECD, 2006). Economists have made the case for investment early in children's lives for the maximum benefit to society, measured with a focus on human, social and identity capital (Lynch, 2004; Heckman and Lochner, 1999). Cunha *et al.* (2005) acknowledge that skill and motivation in one life phase begets skill and motivation in subsequent phases. Hence, investment in foundational stages is seen to increase productivity, with complementary or later stages being weakened by poor, earlier conditions or investments. In sum, the early nurturance of very young children is considered to be of major importance because of the extraordinary growth and development that occurs at that time. This knowledge is one of the mainstays underpinning arguments for the importance of the early years and for universal, high-quality early childhood and family support services.

A.3.1. *The case for early intervention and education programmes*

Findings in developmental neuroscience have highlighted unique developmental opportunities and vulnerabilities in infancy and early childhood. As a result, there are implications for, and actions taken by, directors of early childhood intervention and education programmes. Some applications are contentious because of the over-extension of findings from animal studies that have no justifiable application to humans. Existing interventions, many noted by Hannon (2003), relate to:

● Prenatal development – There is a case for targeted anti-natal interventions, especially for mothers in circumstances of disadvantage where children are at risk of being born less ready to thrive. Interventions that minimise or eliminate the negative effects of nutrient deficiency, neurotoxins and infectious diseases, such as rubella, are clearly beneficial (Shonkoff and Phillips, 2000).

● Synaptogenesis and synaptic loss – Some speculate that if there were high levels of stimulation and enrichment in the period 0-3 years, more synapses would be retained and this could be an overall benefit to children. A plethora of products (electrical, mechanical, nutritional) and programmes are marketed under claims of increasing the intelligence of babies and toddlers. There is no evidence for such reasoning (Bruer, 1999a and b; Goswami, 2004, 2005). On the other hand, deprivation of infants and young children from basic contact with human and material environments, such as in the case of severe abuse and neglect, is clearly damaging (Rutter and O'Connor, 2004).

● Social cognition and emotional regulation – The capacity of young children in preschool years to monitor the reliability of the information they receive has often been underestimated as children extend trust with appropriate selectivity (Koenig and Harris, 2005). Increasingly in research, very young children are being found to be reliable informants of their experience and understanding. Although we are only beginning to contemplate the complexities of social cognition and social interaction (Davis, 2004), answers to new questions about the origins and scope of very young children's trust will provide a stronger scientific base for early intervention programmes.

● Sensitive periods – Although observations from the original studies of Weisel and Hubel – on visual deprivation in kittens – have no implications for early childhood intervention, there may be time-windows in early childhood where certain kinds of development need to take place. Some later catching up through changed circumstances may also be possible. The existence of sensitive periods for language acquisition and development has been documented in studies from psychology. As yet, neuroscience evidence on sensitive periods in human development is limited.

● Environmental complexity – The studies by Greenhouth et al. (1972, 1987) report environmental complexity as having an influence on the brains of rats – those reared in complex environments were superior when learning maze tasks. Although advocates of early childhood intervention programmes argue for "enriched" environments, direct evidence is unavailable. The extent to which well-intentioned carers and educators surround infants and young children with multiple, bright and "busy" objects, toys and gadgets in the cause of enhancing learning has to be questioned. Whereas extreme environmental and sensory deprivation can clearly be harmful, enriched environments do not necessarily improve brain development (Blakemore and Frith, 2005).

● Neural plasticity – Findings in this area have relevance in showing that early development and intervention programmes in early childhood need neither to take full responsibility for securing long-term outcomes in learning nor to claim full credit for outcomes later in life. Brains go on changing and abilities never developed or thought to be lost may, to some extent, recover after the early years.

A.3.2. Children's learning dominates early childhood

It goes without saying that learning in the early years is critical for all humans. Educational practices are moving towards gathering a stronger empirical evidence base of key periods and interventions that facilitate development and learning, with neuroscientific research advancing this evidence base. Blakemore and Frith (2005) argue that the brain has evolved "to educate and to be educated" (p. 459). What adults have taken for granted in the past, particularly regarding the development and learning of infants and toddlers, is now a point of new learning and realisation.

The findings about synaptic development indicate that early social and emotional experiences are the "seeds of human intelligence" (Hancock and Wingert, 1997, p. 36). However, the seduction of seeking to have babies, toddlers and young children generate new knowledge about the world by, for example, "googling" in the virtual world, needs to be tempered with empirical evidence of the merit for effective and sustained development and learning.

In the early childhood phase, studies of children's learning are based upon a variety of complementary and contrasting theories of development and learning. Four overall findings are reported from micro-genetic studies of children's learning, notwithstanding the diverse theoretical orientations, task contents, and varying ages of children (Siegler, 2000). The first, that *change is gradual*, is especially so if an adopted approach, such as finger counting for addition, is easy and effective for the child, albeit time-consuming. The second finding – that *discoveries follow success as well as failure* – confirms that children generate novel strategies to solve problems as they arise, especially in the absence of external pressure. Typically, this is evident in periods of free play. The third, that early *variability is related to later learning*, and fourth, that *discoveries are constrained by conceptual understanding* – have implications for early childhood pedagogy. Experience brings changes in relative reliance on a child's existing strategies, although diverse and contradictory strategies and ways of thinking can co-exist over prolonged periods of time. New ways of thinking, more frequent use of more effective thinking and increasingly effective execution of alternative approaches, are hallmarks of young children's learning over time. During this period skilled early childhood pedagogues can facilitate children's conceptual understanding. Given high levels of neural network production in infants and very young children, and long-term recall now known to emerge well before the verbal ability to describe past experiences (Bauer, 2002), there is strong evidence and argument to take seriously interactions with infants and toddlers, both in the care of families at home and during periods of non-parental care such as in early childhood settings.

A.3.3. Negative contexts for learning

Because of the heightened importance of the early years for learning, impoverished early experiences can be debilitating and need to change. Rutter (2002, p. 9) confirms the consistency of findings indicating psychopathological risks for children, namely:

- persistent discord and conflict, particularly scapegoating or other forms of focused negativity directed towards an individual child;
- a lack of individualised personal care-giving involving continuity over time (*e.g.* in institutional upbringing);
- a lack of reciprocal conversation and play; and
- negative social ethos or social group that fosters maladaptive behaviour of any kind.

In all these instances, it is the *erosion of relationships* which manifests negative consequences for young children. The consequences of poor and damaging relationships impact negatively on children's lives far into the future. Davies reports on the brains of Rumanian orphans who have grown up with profound social and emotional problems: "Nearly all of these children show distinct functional eccentricities in various areas connected with emotion" (Davies, 2002, p. 425). Yet positively, Rutter and O'Connor (2004) report that the heterogeneity of outcome indicates that effects of early biological programming and neural damage stemming from institutional deprivation are not deterministic. How does carer-child relationship mediate the effects of the child's early material environment? Given knowledge to date, it is clearly important that those surrounding young children – families, parents and non-parental caregivers, service providers, pedagogues, community members – enact nurturing behaviour and build reciprocal relationships that enhance the human potential of children.

A.3.4. Early childhood education and care – important but no "magic bullet"

It is clear that growth in the brain does not stop after early life and plasticity does not vanish. To a considerable extent, there has been uncritical acceptance or assertion about the early childhood experiences and their later impact on behaviour. The degree to which crime, anti-social behaviour and other societal problems may be eliminated by "prevention" programmes in early childhood is as yet, not the subject of empirical investigation. The claims stem from a relatively small number of longitudinal studies, based on US populations, attesting an array of benefits which accrue from high quality early childhood programmes (Lynch, 2004). Yet it is also clear that brief interventions in the preschool years are unlikely to achieve life-changing and life-lasting benefit for all children (Karoly *et al.*, 1998, 2001). Yet early engagement with children is clearly important given the fluidity of development at this period. There is ample evidence that gains from early intervention programmes that address cognitive and other areas, particularly centre-based programmes, are maintained for long periods for many children (Barnett, 1995; Brooks-Gunn, 1995; Karoly *et al.*, 1998). The effects of early childhood programmes continue into primary school, although the rate at which effects are sustained may depend on the quality of learning environment encountered by children at school. The early years of a child's life are pivotal to his/her later social, academic and physiological development.

A.4. What do we know about learning environments to facilitate early childhood development?

A.4.1. The subtleties of play and learning in this period

Brain development and the cognitive achievements of infants and young children are disguised in the apparently innocuous guise of a child's play. Developmental psychology research makes it clear that a child in the first three years of life already has an enormously complex and inter-linked knowledge base about the world (NSCDC, 2005, 2004a, 2004b), a knowledge-base normally acquired in highly informal ways. Rushton and Larkin (2001) explain that the child's environment is central to the manner of learning with studies indicating, for example, that children involved in learner-centred environments have better receptive verbal skills (Dunn, Slomkowski and Beardsall, 1994) and are more confident in their cognitive abilities. Play-based programmes for young children are strongly supported in the early education literature, although such programmes vary widely in application and in the assumptions made about the place of play in learning.

Observing play environments can enhance understandings about young children's intuitive thinking and their engagement with others. The relative balance of child-initiated strategies (including play) and adult-led experiences may also affect child learning outcomes. In general, early childhood educators prefer to err on the side of child-initiated learning events and to use strategies that rely on spontaneous teaching moments. Five principles underscore the importance of play to animal and human brain development and learning in early childhood (Frost, 1998):

- All healthy young mammals play – Animal infants initiate games and frivolity, mediated by adult caregivers; human infants – whose period of motor immaturity after birth is longer – depend more on parents and others to give structure and direction to initial play, this being a scaffold for development.

- The range and complexity of play quickly increases in correspondence with neural development.

- Early games and frivolity in animals (escape, stalking, pouncing) and humans (movement, language, negotiation) equip them for skills needed later.

- Play is essential for healthy development as it seems to facilitate the linkages of language, emotion, movement, socialisation and cognition. "It is playful activity, not direct instruction, seclusion, deprivation or abuse that makes a positive difference in brain development and human functioning" (Frost, 1998, p. 8).

- Play deprivation may result in aberrant behaviour. Yet, Smith and Pellegrini (2004) question the voracity of play, particularly socio-dramatic play in human learning, pointing to "play ethos" devotion in early childhood development programmes in modern western societies that is not fully verified. Their review points to:

"... the importance of adults (often parents) in their attitudes towards children's play. Whether a child engages in play – perhaps pretend play especially – varies greatly by culture and buy the extent to which adults discourage such play (by imposing work or even caregiving demands on even very young children); tolerate it for a while (as a way of reducing direct child demands on them for caregiving; or actively encourage it (as a way of developing cognitive and social skills – a form of 'parent investment')" (p. 296).

Environmental conditions are known to effect the development of brain circuitry and learning, prompting a need for careful attention to the type and diversity of environments in which children grow (Eming-Young, 2002, 2000). Much of the literature in early childhood education concludes that participation in well-run, play-based early childhood programmes has measurable positive impact on intellectual performance, social achievements, self-esteem and task orientation upon entry to school. Attention to the quality of interactions is most noted (Katz, 2003).

A.4.2. Curriculum and pedagogical focus and young children's development

Different educator understandings of the child can result in substantial differences in Early Childhood Education and Care (ECEC) curriculum and pedagogical processes, and the outcomes for children. Adults engaging with young children may view a child as: a unique personality in the present; a citizen with individual rights and agency to choose his/her participation in events and activities; an individual of the future who must be prepared for pre-determined challenges ahead; or, because of age, a person who is incapable or incompetent and needing specific direction; or justifying being ignored by adults. Both parent and educator assumptions about the child and parent and educator knowledge of what children know or should be able to do at particular stages (infancy, toddlerhood, pre-kindergarten, early school) shape the kind of learning environment, that children encounter at home and in early childhood centres. Clearly, informal learning is a central part of the lives of infants, toddlers and preschoolers. Programmes that enable unscripted opportunities for play and social engagement, that surround children in language-rich environments, and support inquiry are generally considered important by early childhood educators.

Findings such as the development of the pre-frontal cortex in infants depending on loving and pleasurable relationships with significant carers; the release of high levels of cortisol in situations of prolonged stress; the higher metabolic rate of the child's brain and its greater capacity for synaptic connectivity in early childhood (NSCDC, 2005, 2004a, 2004b) suggest important elements of a healthy early childhood environment – secure attachment and primary care, varied sensory experience, responsive care-giving. When basic needs of children are met (health and safety, nutrition, nurture and care) optimum conditions are in place for the growth of healthy dispositions, the development of confidence, critical skills, problem-solving and co-operation (Ramey and Ramey, 2000).

From an educational perspective, the ability to pay attention changes significantly across the preschool years and consequently enables changes to be made to the preschool curriculum and pedagogical focus as children get older and move towards primary schooling. Growth from toddler to preschooler normally results in children paying more sustained attention, although retaining attention for things that are not salient to them is difficult. This is important in informing how children will learn effectively in education settings at different stages of development. Programmes that seek to build upon the interests of children and exploit the spontaneous teaching moments frequently arise in play-based settings and gradually (over years) move to become more formal, accordingly matching children's development.

A.4.3. Learning environments that support language development

Early childhood is a seminal period of language acquisition and development. For children growing, learning and engaging in contemporary society, literacy development requires the growth of proficiencies in varied communication and multi-text environments, in multi-lingual settings. Learning the nuances of the culture and environment and the capacity to communicate through multi-symbol systems (verbal, written, dramatic, artistic…) are part of the normal learning and development of young children. Burchinal *et al.* (2000, 2002) cite evidence that the quality of infant care and the training of staff in care centres are linked to higher measures of cognitive and language development. Direct experiences in different communicative events, with peers and adults help children's language and communicative competencies – the relative balance of stimulation, consistency, and encouragement being important to children's attainments. Children's curiosity (and the questions they ask) provides a key to learning about their understandings and thinking. The process of doing, interacting and talking to clarify thinking, is evident in different kinds of informal early learning environment.

> "Growth-promoting relationships are based on the child continuous give-and-take ('action and interaction') with a human partner who provides what nothing else in the world can offer – experiences that are individualised to the child's unique personality style; that build on his or her own interests, capabilities and initiative; that shape the child's self-awareness; and that stimulate the growth of his or her heart and mind" (NSCDC, 2004a, p. 1).

Sensitivity to context should inoculate early educators against the use and transmission of pre-determined, "format-programmes" for young children – those prepared for a "typical" child, for children of a specific chronological age, or popular programmes taken from a different location and context. Such programmes, applied without sensitivity to context, are normally of poor quality because they lack consideration of local children and circumstances. Context sensitivity enables educators to recognise and employ diverse ways of learning, seeing, creating and representing ideas with children.

A.4.4. Strategies used by educators to support learning in early childhood

Contemporary early childhood educator practices developed mainly across the 20th century, influenced by studies of child development, educational and psychological theory and research, the interests and directions of government policy around children's services, and through ritual and routine. Particular programmes and practices have not normally developed from a strong scientific research base or careful evaluation of competing theories. The strategies typically used by early educators in ECEC learning environments include:

- *Listening to children* – seen as fundamental to finding out about the learning strategies children bring to a problem or activity, the strengths they have, and the points where more knowledgeable others can help.

- *Listening to and co-ordinating with parents and family members* – to establish children's strengths and interests, to learn of their dispositions and find out how the child's development and learning is mediated at home.

- *Establishing common knowledge* – exploiting situations when children are together in groups, such as in kindergarten, provides common experience and common ground for expanding children's thinking. Encouraging children to recall experiences that relate to a current task is seen to build learning continuity and establish new concepts and understandings.

- *Using positive modelling* – educator modelling of meta-cognitive strategies that regulate task achievement (*e.g.* What is my problem? What is my plan? How am I going to proceed? What worked? How do I know?) is seen to impact positively on young children's learning behaviours. Asking children to predict and build theories to explain events in which they show interest is thought to keep children interested and active (What do you think? Why do you think that?).

- *"Re-cognising"* (Meade and Cubey, 1995). Reflecting in words to young children what they are doing in action is seen to help clarify processes and ideas. Scaffolding a child's "hands on" experiences is a key mediation role of educators in early learning environments.

- *Giving specific instruction* in certain skill areas is thought to be important. For example, for older preschoolers emergent literacy research suggests that children need increasing phonemic awareness and grapho-phonic knowledge for successful reading. Games (rhymes, sorting, odd-one-out, etc.) may be used to practise and build repetition. Teaching routines that ensure personal safety and hygiene are considered important.

- *Spending time in observation* – educators may stand aside from, or join in, children's tasks according to the learning events taking place. Generally, it is seen as important for children to have control of their learning in a supported way.

- *Celebrating diversity* – educators may act to endorse and expand children's knowledge of diverse language and dialects, and become thoughtful about approaching topics in a variety of ways (musical, story-based, play, discovery, pictorial, artistic, logical deductive) because of realisations that children learn in diverse ways and can show understanding by using different symbolic media.

● *"Focusing" through recall and restatement* – early educators' questions, explanations and the linking together of different events are seen to help children focus and progress their understanding.

● *Ensuring children experience different speaking and listening situations* in order to broaden communication experiences. This is seen by some early educators as an important role for adults when they engage with young children.

Current understandings of brain growth provide an appreciation of how biology and the environment – the learning context – are inextricably linked. Early educators view the young child as an active learner and consider each child's developmental level and individual characteristics in the context of the child's family and community (Gilkerson, 2001).

A.5. What challenges exist in synthesising research in neuroscience and early education?

There are differences in purpose and focus in the neuroscientific and educational research. Neuroscientists are only beginning to learn which experiences wire the brain in which ways. The extent to which research with animals may have application for humans means such research is a starting point for serious speculation on the operation of the human brain and the learning mechanisms of young children. New technologies allow the observation of the human brain during active processing and cognition but drawing implications of neuroscientific studies for education and child development is far from an exact science (Frost, 1998, p. 12).

Trans-disciplinary design of questions and methods to study young children and the brain offers enhanced opportunity for understanding the key elements and mechanisms of human development and learning. However, there are problems when linking education to neuroscience. First, the notion of "sensitive" periods in brain development may be well-established, but the extent to which this is transformed in popular press and policy advocacy as the reason to support enhanced early childhood education is problematic. Bruer (1999a and b) cites this as an instance where neuroscientists speculated about the implications of their work for education and where educators uncritically embraced that speculation. Second, competing theories within neuroscience and within education require careful research designs that address the theories scientifically in order to move the field forward. Third, much of the research on human subjects that studies the place of experience in shaping brain development has focused on the effects of abuse and neglect in childhood. Empirical evidence on the role of experience in shaping brain development in children without depriving experiences remains virtually non-existent. Work with normally developing young children can help to illuminate the processes of brain development, something of interest in its own right, as well as for its potential to promote health and well-being (Reid and Belksy, 2002, p. 584).

Some argue that it is premature to apply cognitive neuroscience findings directly to teaching. Bruer (1999a and b) maintains that early childhood education is best served by the *application* of cognitive teaching practices rather than neurological findings. Further, neuroscientists are said not to have enough information about relationships between neural functioning and instructional practice to assist educators (Winters, 2001, p. 4). Undeniably, it is necessary to exert a level of prudence when approaching neuroscientific findings in an educational context.

As early childhood educators take a perspective that encompasses all aspects of development – integrating knowledge of the brain with ideas about the developing child – understandings of development and the child should increase (Gilkerson and Kopel, 2004). Practices currently advocated in early childhood programmes, and reported in this paper (e.g. multi-sensory experience and play, scaffolding thinking, child-initiated activity, social relationships), give direction primarily on the basis of tradition, anecdotal experience and educational research. Much of the advocated practice appears not to be in conflict with the current neuroscientific findings. The key in advancing educational practices lies in the integration of theories of learning and neuroscientific study that gathers evidence on efficient and enriching programmes for children.

Nuria Sebastian-Galles and Collette Tayler

A.6. Practitioner's response

The neuroscience observations and their explanations of educational phenomenon is like looking at education through mysticism, this was the first thought that came to my mind after browsing the paper. It is like looking at education from the eyes of a seer or sage, who is aware of each and every happening inside the organs, tissues, and cells or even further down the line, when an action is taken or an idea is thought. The question is – like wise men, can the research findings in neuroscience influence the learning in desired manner? If yes then how and where to begin? As far as "where" is concerned, nothing is more apt than early childhood education and care (ECEC) platform.

The early life experience has a greater impact on later life. Seeing this in the new light of neuronal circuitry leads to the idea that the early life environment should have as many multi-sensory experiences as possible. This may mean that the child rearing practices around the world needs to redirect their efforts a little towards providing different touch, sound, taste, visual and smell stimuli as early as six months or even sooner. India has a rich tradition of playing harmonious music or ancient Vedic wisdom to the expectant mother, with a belief that doing so eases the little ones journey through gestation, positively influences the life after birth, and also helps in keeping the mother calm and happy. Many new borne are found to react to the before birth environmental sounds.

The sooner an infant/child learns to control the emotions the better it is for learning. The controlling of emotions can be facilitated through practising Indian system for wholistic development of mind and body known as yoga and meditation. Still more scientific research evidence regarding the effect of yoga and medication on the development of frontal lobe is needed.

"Individual's repeated exposure to experience will result in the myelination of the relevant synapses." This seems to provide justification for exposing infants and toddlers to structured meaningful learning experiences through the prescribed play school/ECEC centre activities. Though the opportunities for many experiences are provided, yet which one will attract which infant and to what extent can't be foretold with surety. This brings us to individual differences in very young (say one year or sooner) children.

The part played by the early experience advocates very strongly towards strengthening, improving and establishing effective ECEC system all over the world.

"According to the age level (…) neuronal networks." This appears to suggests two things: one, the more diverse the sensory experiences are the more developed the brain will be i.e. the more synapses will be formed and myelinated. And secondly the more an experience is repeated the better it is for neuronal network strength.

The *timing* and *sequence* appear to provide new explanations to the basic ECEC concepts like school readiness, reading readiness, etc., based on the findings in the neuroscience.

Plasticity – The infant and toddlers adjust readily to new situations/experiences/ideas as compared to adults. This might be due to the freshly forming synapses, along with easy availability of neurons in infants and toddlers, whereas in adults the difficulty may be due to the not so plastic brain, commonly talked as having fixed ideas or notions.

Critical or sensitive periods in neuronal development – have strong significance for children with different abilities, disabilities, gifts and talents. This brings to forefront a whole new set of research questions, say for *e.g.* can giftedness be introduced? Can disabilities be prevented? Is talent a result of finding specific stimuli at certain critical or sensitive periods in neural development? Can the belief in genetic genius be simply explained on the basis of right stimuli at the right time?

The broadly defined and chronologically similar sensitive periods for all human beings may imply that each individual child has the inherent potential to become genius. The significance of first two years of life may result in rethinking and re-deciding about the entry age to preschool.

Language acquisition is one of the major developmental landmarks. Research in neuroscience is now confirming what has been proved earlier by researches in other fields *i.e.* infants and toddlers, ability to learn any language with equal ease.

"Practice leads to perfection, appears to be true at the level of neurons and synapses." The success of adult education programme is clear evidence for learning beyond sensitive periods. The unavailability of ready neurons and myelinated synapses may account for comparatively large energy and efforts requirement at this age.

If myelinated synapses or neuronal circuits indicate learning then can we say that the loss of memory or temporary leaning is due to pruning?

Learning from new or unusual individual experience continue to have neurological effects throughout life, beautifully supports the learning by doing approach to education.

The *mirror neurons* seems to suggest the possibility for inculcation of empathy through carefully structured experiences, in turn leading to morally better community and society. At this point of time we need to be careful about the negative impact say delinquent tendencies. The field of mirror neurons is very exciting and appears to hold many promises. Can we say that professionals like counsellors who seem to understand others' feeling may apparently have learnt to do so as a result of their training but in reality they might be fine tuning their mirror neurons?

The neuroscientific discovery of the effects of early life experiences on later life, gives a fresh claim to the cause of ECEC programme. Infants or toddlers reared in environment full of gentle story reading often acquire a hobby of reading in later life, whereas absence of such does not necessarily result in dislike for books.

Much needed emphasis is justly placed world wide on the prenatal, child and mother care programmes, involving free vaccination and awareness campaigns towards proper care, health hygiene and nutrition of the expectant mothers. Once the baby is born the focus shifts towards nursing the baby for at least six months on breast milk. In the light of the findings of neuroscience the period between 6 months and 3 years (the play school age) needs to be properly taken care of.

Studies related to nuclear and joint families may throw some light on the effects of *environmental complexity*.

Neuronal plasticity – Truly ECEC may not be allowed to take neither the full responsibility nor the credit for outcomes in later life, due to the dynamic nature of the brain, still the influence of rich early life experience cannot be fully ignored. An effective ECEC programme may save the much efforts and energy in later life, just like taking precautions may save occurrence of disease or disaster at a later time.

Children's learning dominates early childhood – so far whatever suggestions educational research was making, is now backed by neuroscientific research findings. It is like gaining new important witness.

Seeds of human intelligence till date are looked upon as abstract concepts measured through intelligent quotient, emotional quotient, and spiritual quotient. This now needs to be further viewed in terms of synaptic connections at a much deeper but tangible level.

The initial finding that the neuronal circuitry in the early stages can be influenced may help in discovering preventive disability measures or may lead towards new interventions approaches.

ECEC pedagogy aims towards attaining self-dependency and laying strong foundation for later life. This is thought to be the time for nurturing healthy eating, sleeping and hygiene related habits. In the light of micro–genetic findings of children's learning, the pedagogic practices in ECEC should provide more and more opportunities for multi-sensory experiences and discovery learning. The focus needs to be shifted slightly from attaining basic self-dependency to facilitating conceptual understanding, regarding morality.

If we need to take seriously *interactions with infants and toddlers*, then much needs to be invested in the effective skill training of the carers, along with campaigns of mass awareness towards improving the social status of carers. More and more ways of attracting highly educated and literate persons towards the profession need to be thought upon.

Erosion of relationship – brings to forefront the concern for children affected by natural disasters like tsunami, abused in prison, forced into child labour, victimised by war and accidents.

ECEC no magic bullet but a significant step towards strong and healthy foundation.

Subtleties of play and learning in this period – learning should be fun and earlier the fun start the better it is.

The overall idea of building upon the interests of children and exploiting the spontaneously arising teaching moments should be extended to at least primary school, with very less formality or structure in school's daily activities.

Now that the neuroscientific research is also supporting the fact that young ones are more apt in learning new languages as compared to adults, the ECEC should aim at providing experiences for one language other than mother tongue. The ample efforts towards providing opportunities for self-expression through art and drama should be made.

Why not to appoint bilingual or multilingual or multitalented persons as carers in ECEC centres?

Much effort is required in the area of developing training package for ECEC educators, with a focus to develop multitalents, interests and skills necessary for the job in care centres.

Sensitivity to context – the parent and community leaders also need to be inoculated against the use and transmission of pre-determined, format programmes for young children. Currently the ECEC seems to be governed by the rules of the market, i.e. whatever is demanded by the client, the supplier provides, in this case the client (parents) needs to be made aware of the later consequences of their current desires.

Strategies used by the early educators – the strategies used by the early educators aptly listed in the paper, can be regrouped as following, with celebrating diversity finding its place in more than one group:

● Probing the phenomenological world – through listening, observation and co-ordinating with parents and family members.

● Curriculum – is defined or planned when we aim for establishment of common knowledge, and celebrate diversity.

● Class room instruction – through positive modeling, re-cognising, giving specific instruction, focusing through recall and restatement, celebrating diversity.

● Language development – is facilitated by ensuring that the children experience different speaking and listening situations and also by celebrating diversity.

All these strategies may help in building and myelination of the synapses. In a way whatever is being done in ECEC so far can be beautifully explained neuroscientifically.

Sensitive periods – clearly need to be given its due importance through further research both in ECEC and neuroscience. This makes a very strong case for strengthening and funding the ECEC all over the world.

The existence of competing theories within both the neuroscience and education definitely requires careful research design leading to a general consensus. This in turn may bring to surface fresh thinking, ideas and approaches to make ECEC more effective.

Apart from studying the place of experience in shaping the brain development with respect to normally growing children, similar studies need to be taken to prevent learning difficulties and disabilities.

In my view it is not premature to apply the cognitive neuroscience findings to teaching. The neuroscience findings so far are very encouraging and seem to be in direct agreement with educational thinking. The gathering of more information about the relationship between neural functioning and instructional practice to assist educators, seems to lead to the aboriginal issue of who came first the chicken or the egg?

Bharti

References

Ansari, D. (2005), "Commentaries. Paving the Way towards Meaningful Interactions between Neuroscience and Education", Blackwell Publishing, pp. 466-467, *www.dartmouth.edu/~numcog/pdf/Blakemore%20and%20Frith%20Commentary.pdf?sid=587019*.

Barnett, W.S. (1995), "Long-term Outcomes of Early Childhood Programs", *Future of Children*, Vol. 5(3), pp. 25-50.

Bauer, P.J. (2002), "Long-term Recall Memory: Behavioral and Neuro-developmental Changes in the First Two Years of Life", *Current Directions in Psychological Science*, Vol. 11(4), pp. 137-141.

Blakemore, S.J. and U. Frith (2005), "The Learning Brain. Lessons for Education: A précis", *Developmental Science*, Vol. 8(6), pp. 459-471.

Brainard, M.S. and A.J. Doupe (2002), "What Songbirds Teach us about Learning", *Nature*, Vol. 417, 16 May, pp. 351-358.

Brandt, R. (1999), "Educators Need to Know about the Human Brain", *Phi Delta Kappan*, November, pp. 235-238.

Brooks-Gunn, J. (1995), "Strategies for Altering the Outcomes of Poor Children and their Families", in P.L. Chase-Lansdale and J. Brooks-Gunn (eds.), *Escape from Poverty: What Makes the Difference for Children?*, Cambridge University Press, New York, pp. 87-117.

Brooks-Gunn, J. (2003), "Do you Believe in Magic? What we Can Expect from Early Childhood Intervention Programs", *Social Policy Report*, Vol. 17(1), pp. 3-14.

Bruer, J.T. (1997), "Education and the Brain: A Bridge too Far", *Educational Researcher*, Vol. 26(8), pp. 4-16.

Bruer, J.T. (1999a), "In Search of Brain-Based Education", *Phi Delta Kappan*, May, pp. 649-657.

Bruer, J.T. (1999b), *The Myth of the First Three Years*, Free Press, New York.

Burchinal, M.R., J.E. Roberts, R.Jr. Riggins, S.A. Zeisel, E. Neebe and D. Bryant (2000), "Relating Quality of Center-based Child Care to Early Cognitive and Language Development Longitudinally", *Child Development*, Vol. 7(2), pp. 339-357.

Burchinal, M.R., D. Cryer, R.M. Clifford and C. Howes (2002), "Caregiver Training and Classroom Quality in Child Care Centers", *Applied Developmental Science*, Vol. 6(1), pp. 2-11.

Changeux, J.P. and S. Dehaene (1989), *Cognition*, Vol. 33, pp. 63-109.

Clark, J. (2005), "Explaining Learning: From Analysis to Paralysis to Hippocampus", *Educational Philosophy and Theory*, Vol. 37(5), pp. 667-687.

Crick, F. (1994), *The Astonishing Hypothesis: The Scientific Search for the Soul*, Scribner, New York.

Cunha, F., J. Heckman, L. Lochner and D.V. Masterov (2005), "Interpreting the Evidence of Life-Cycle Skill Formation", IZA Discussion Paper Series, No. 1575, Institute for the Study of Labour, Bonn, Germany, July.

Davies, M. (2002), "A Few Thoughts about the Mind, the Brain, and a Child in Early Deprivation", *Journal of Analytical Psychology*, Vol. 47, pp. 421-435.

Davis, A. (2004), "The Credentials of Brain-based Learning", *Journal of Philosophy of Education*, Vol. 38(1), pp. 21-35.

Dunn, J., C. Slomkowski and L. Beardsall (1994), "Sibling Relationships from the Preschool Period through Middle Childhood and Early Adolescence", *Developmental Psychology*, Vol. 30, pp. 315-324.

Dyckman, K.A. and J.E. McDowell (2005), "Behavioral Plasticity of Antesaccade Performance Following Daily Practice", *Experimental Brain Research*, Vol. 162, pp. 63-69.

Eming-Young, M. (2000), *From Early Child Development to Human Development*, World Bank, Washington DC.

Eming-Young, M. (2002), *Early Childhood Development: A Stepping-stone to Success in School and Life-long Learning*, Human Development Network Education Group.

Fagiolini, M. and T.K. Hensch (2000), "Inhibitory Threshold for Cortical-period Activation in Primary Visual Cortex", *Nature*, Vol. 404, pp. 183-186, March.

Frost, J.L. (1998), "Neuroscience, Play and Child Development", Paper presented by the IPA/USA Triennial National Conference, ERIC Document 427 845, PS 027 328.

Gazzaniga, M. (1998), *The Mind's Past*, University of California Press, Berkeley.

Gilkerson, L. (2001), "Integrating and Understanding of Brain Development into Early Childhood Education", *Infant Mental Health Journal*, Vol. 22(1-2), pp. 174-187.

Gilkerson, L. and C.C. Kopel (2004), "Relationship-based Systems Change: Illinois' Model for Promoting the Social-emotional Development in Part C Early Intervention", Occasional Paper No. 5, Erikson Institute, Herr Research Centre.

Gopnik, A., A. Meltzoff and P. Kuhl (1999), *The Scientist in the Crib. What Early Learning Tells us about the Mind*, Harper Collins, New York.

Goswami, U. (2004), "Neuroscience and Education", *British Journal of Educational Psychology*, Vol. 74, pp. 1-14.

Goswami, U. (2005), "The Brain in the Classroom? The State of the Art. Commentaries", Blackwell Publishing, pp. 467-469, *www.blackwell-synergy.com/doi/pdf/10.1111/j.1467-7687.2005.00436.x.*

Greenhough, W.T., J.E. Black and C.S. Wallace (1987), "Experience and Brain Development", *Child Development*, Vol. 58(3), pp. 539-559.

Greenhough, W.T., T.C. Maddon and T.B. Fleischmann (1972), "Effects of Isolation, Daily Handling and Enriched Rearing on Maze-learning", *Psychenomic Science*, Vol. 27, pp. 279-280.

Hancock, L. and P. Wingert (1997), "The New Preschool" (Special Issue), *Newsweek*, 129, 3637, Spring-Summer.

Hannon, P. (2003), "Developmental Neuroscience: Implications for Early Childhood Intervention and Education", *Current Paediatrics*, Vol. 13, pp. 58-63.

Heckman, J.J. and L. Lochner (1999), "Rethinking Education and Training Policy: Understanding the Sources of Skill Formation in a Modern Economy", Mimeograph, October.

Ito, M. (2004), "'Nurturing the Brain' as an Emerging Research Field Involving Child Neurology", *Brain and Development*, Vol. 26, pp. 429-433.

Karoly, L.A., P.W. Greenwood, S.S. Everingham, J. Hoube, M.R. Kilburn, M. Rydell, M. Saunders and J. Chieas (1998), *Investing in our Children: What we Know and Don't Know about the Costs and Benefits of Early Childhood Interventions*, RAND, New York.

Karoly, L., R. Kilburn, J. Bigelow, J. Caulkins and J. Cannon (2001), *Assessing the Costs and Benefits of Early Childhood Intervention Programs: Overview and Application to the Starting Early Starting Smart Program*, RAND Publication MR1336, New York.

Katz, L. (2003), "State of the Art in Early Childhood Education 2003", ERIC Document, No. 475 599.

Kim, K.H.S., N.R. Relkin, K.M. Lee and J. Hirsch (1997), "Distinct Cortical Areas Associated with Native and Second Languages", *Nature*, Vol. 388, pp. 171-174.

Kisilisky, B.S., S.M.J. Hains, A.Y. Jacquet, C. Granier-Deferre and J.P. Lecanuet (2004), "Maturation of Fetal Responses to Music", *Developmental Science*, Vol. 7(5), pp. 550-559.

Koenig, M.A. and P. Harris (2005), "The Role of Social Cognition in Early Trust", *Trends in Cognitive Sciences*, Vol. 9(10), pp. 457-459.

Kolb, B. and I.Q. Wishaw (1998), "Brain Plasticity and Behaviour", *Annual Review of Psychology*, Vol. 49, pp. 43-64.

Kuhl, P.K. (2004), "Early Language Acquisition: Cracking the Speech Code", *Nature Reviews Neuroscience*, Vol. 5, pp. 831-843.

Lally, J.R. (1998), "Brain Research, Infant Learning and Child Care Curriculum", *Child Care Information Exchange*, Vol. 5, pp. 46-48.

LeDoux, J. (2003), *Synaptic Self: How our Brains Become Who We Are*, Viking Penguin, New York.

Lindsay, G. (1998), "Brain Research and Implications for Early Childhood Education", *Childhood Education*, Vol. 75(2), pp. 97-101.

Lynch, R. (2004), "Exceptional Returns. Economic, Fiscal, and Social Benefits of Investment in Early Childhood Development", Economic Policy Institute, Washington DC.

Meade, A. and P. Cubey (1995), *Thinking Children: Learning about Schemas*, NZCER, W.C.E. and Victoria University, Wellington.

Mitchell, D.E. (1989), "Normal and Abnormal Visual Development in Kittens: Insights into the Mechanisms that Underlie Visual Perceptual Development in Humans", *Canadian Journal of Psychology*, Vol. 43(2), pp. 141-164.

Nelson, C.A. *et al.* (2000), "The Neurobiological Bases of Early Intervention", in J.P. Shonkoff and S.J. Meisels (eds.), *Handbook of Early Childhood Intervention*, second edition, Cambridge University Press, Cambridge, Mass.

NSCDC (National Scientific Council on the Developing Child) (2004a), "Young Children Develop in an Environment of Relationships", Working Paper 1, Summer, NSCDC, *www.developingchild.net*.

NSCDC (National Scientific Council on the Developing Child) (2004b), "Children's Emotional Development is Built into the Architecture of their Brains", Working Paper 2, NSCDC, *www.developingchild.net*.

NSCDC (National Scientific Council on the Developing Child) (2005), "Excessive Stress Disrupts the Architecture of the Brain", Working Paper 3, Summer, NSCDC, *www.developingchild.net*.

OECD (2006), *Starting Strong II: Early Childhood Education and Care*, OECD, Paris.

Posner, M.J. (2004), "Neural Systems and Individual Differences", *Teachers College Record*, Vol. 106(1), pp. 24-30.

Ramey, S.L. and C.T. Ramey (2000), "Early Childhood Experiences and Developmental Competence", in J. Wolfagel and S. Danzigner (eds.), *Securing the Future. Investing in Children from Birth to College*, Russell Sage Foundation, New York, pp. 122-150.

Reid, V. and J. Belsky (2002), "Neuroscience: Environmental Influence on Child Development", *Current Paediatrics*, Vol. 12, pp. 581-585.

Rushton, S. and E. Larkin (2001), "Shaping the Learning Environment. Connecting Developmentally Appropriate Practice to Brain Research", *Early Childhood Education Journal*, Vol. 29(1), pp. 25 33.

Rutter, M. (2002), "Nature, Nurture and Development: From Evangelism through Science towards Policy and Practice", *Child Development*, Vol. 73(1), pp. 1-21.

Rutter, M., T. O'Connor and the English Romanian Adoptees Study Team (2004), "Are there Biological Programming Effects for Psychological Development? Findings from a Study of Romanian Adoptees", *Developmental Psychology*, Vol. 40(1), pp. 81-94.

Shonkoff, J.P. and D.A. Phillips (2000), *From Neurons to Neighbourhoods: The Science of Early Child Development*, National Academy Press, Washington DC.

Shore, R. (1997), *Rethinking the Brain. New Insights into Early Development*, Families and Work Institute, New York.

Siegler, R.S. (2000), "The Re-birth of Children's Learning", *Child Development*, Vol. 71(1), pp. 26-35.

Slavin, R.E. (2002), "Evidence-based Education Policies: Transforming Educational Practices and Research", *Educational Researcher*, Vol. 31(7), pp. 15-21.

Smith, P.K. and A.D. Pellegrini (2004), "Play in Great Apes and Humans", in A.D. Pellegrini and P.K. Smith (eds.), *The Nature of Play: Great Apes and Humans*, pp. 285-298.

Sylva, K., E. Melhuish, P. Sammons, I. Siraj-Blatchford and B. Taggart (2004), "The Effective Provision of Preschool Education (EPPE) Project. Final Report", Department for Education and Skills, London, December.

Thompson, R.A. and C.A. Nelson (2001), "Developmental Science and the Media. Early Brain Development", *American Psychologist*, Vol. 56(1), pp. 5-15.

Thorpe, K., C. Tayler, R. Bridgstock, S. Grieshaber, P. Skoien, S. Danby and A. Petriwskyj (2004), "Preparing for School. Report of the Queensland School Trials 2003/4", Department of Education and the Arts, Queensland Government, Australia.

Werker, J. and H.H. Yeung (2005), "Infant Speech Perception Bootstraps Word Learning", *Trends in Cognitive Sciences*, Vol. 9, pp. 520-527.

Winters, C.A. (2001), "Brain-based Teaching: Fad or Promising Teaching Method", ERIC Document 455 218, SP 040 143.

ISBN 978-92-64-02912-5
Understanding the Brain: The Birth of a Learning Science
© OECD 2007

PART II

Article B

The Brain and Learning in Adolescence

by
Karen Evans, School of Lifelong Learning and International Development,
University of London, UK
Christian Gerlach, Learning Lab Denmark, Denmark
Sandrine Kelner, High School Teacher, Nancy, France

B.1. Introduction

The brain consists of a vast amount of cells, or neurons, which constitute the basic operative unit in the brain. During the period of the highest prenatal brain development (10-26 weeks after conception), it is estimated that the brain grows at a rate of 250 000 neurons per minute. At birth the brain contains the majority of the cells it will ever have, with estimates ranging from 15-32 billions. This span does not only reflect that cell counting is imprecise, but also that the number of cells varies considerably from person to person. After birth, new neurons are only produced in limited numbers. By far most conspicuous changes in the brain following birth occur in the connections between neurons; new ones are formed and old ones are either strengthened or eliminated. And there is plenty of room for change given that any particular neuron is often connected with several thousands other neurons. For a long time it was assumed that such changes primarily happened in childhood, because the brain is already 90% of the adult size by the age of 6. Today this belief has clearly changed. It is now evident that the brain undergoes significant changes throughout life. In this writing we will focus on the neural changes that occur during adolescence; a period that spans, roughly, from 12 to 18 years of age. It will be examined how these neural changes relate to the significant behavioural changes that also occur during adolescence and which include altered affective regulation, risk-taking behaviour, decision-making abilities and development of independence. The ultimate objective will be to consider what implications these developmental changes have for learning, teaching and education. Before we embark on this we will provide some background knowledge on brain development at both the microscopic (neuron) and macroscopic (brain system) level.

B.2. Understanding brain development – what are we looking at?

B.2.1. Brain development at the microscopic level

In order to understand the changes that the brain undergoes through time it is necessary to know a little about how the brain is composed. The smallest functional unit in the brain is the neuron. It consists of three main parts: **a body**, which contains the machinery for maintaining the cell, an **axon** and several **dendrites** that extend away from the cell body. The function of dendrites is to receive input from other neurons while the function of the axon is to provide input to other neurons. A neuron communicates with another neuron by releasing a chemical substance (a **neurotransmitter**) from its axon, which may have several endings (**terminals**). This neurotransmitter then passes a small cleft before it gets attached to receptors on the surface of a dendrite belonging to a neighbouring neuron. The locations where axons and dendrites are separated by these small clefts where neurotransmitters can pass are called **synapses**.

The communication between neurons is modulated by several factors. The most conspicuous is that neurons may increase their number of synapses (**synaptogenesis**). However, the number of synapses may also decrease, which is termed "**pruning**". At a more subtle level, the "strength" of communication between two neurons may also be modulated by the amount of neurotransmitters released from the axon terminals, how quickly the neurotransmitter is removed from the synaptic cleft, or by how many receptors the receiving neuron has on its surface. These latter changes are often referred to as **strengthening** or **weakening** of the existing synaptic connections.

The changes that neurons undergo are affected by the experience of the individual. Apparently, this happens in a Darwinian manner (survival of the fittest), so that connections that are not used are lost or weakened, whereas connections that are used frequently are strengthened, providing more efficient and robust ways for communication. Thus, learning is achieved either through the growth of new synapses or through the strengthening or weakening (even elimination) of existing ones. In fact, there is good evidence for both mechanisms, with the former being more prominent during childhood and adolescence and the latter being more prominent during adulthood. Metaphorically speaking then, the brain is a sculpture carved by experience.

Besides the synaptic changes the neurons also undergo another change. To understand this change we must take a look at what happens when neurons communicate. As mentioned, communication comes about through the release of a neurotransmitter from the axon of a neuron. However, the axon must get a signal as to when it has to release the neurotransmitter. This comes in the shape of an electric impulse that travels from the body of the neuron down through the axon. This happens in the following way: neuron A releases a neurotransmitter in the synaptic cleft between neuron A and B. Some of this neurotransmitter will pass the synaptic cleft and attach itself to neuron B's receptors. This causes pumps in neuron B's membrane to open so that chemical substances outside the cell now pass into the cell, whereas others leave the cell. If the influence on neuron B is strong enough – that is, if a sufficient number of pumps are activated – the electric voltage of the cell will change in such a manner that a serial reaction will occur down the axon whereby the electric impulse can travel from the body of the cell down through the axon. The axon acts, in a manner of speaking, as a kind of wire and just like a wire the axon can pass the current (the electric impulse) more optimally (faster) if it is insulated. At birth, most axons are not insulated, but they will become so with time. This happens when a sheath of fat is wrapped around them. This type of development is referred to as **myelination** (myelin is the fatty substance that constitutes the insulation). When the axon is myelinated, the electric impulse can "jump" down the axon in the gaps between the fatty sheaths rather than laboriously crawl its way down. This means that myelinated axons can transmit information up to 100 times faster than unmyelinated axons. As is the case for the changes of the synapses (synaptogenesis/pruning/strengthening/weakening), it appears that the process of myelination can be experience dependent (Stevens and Fields, 2000).

B.2.2. Brain development at the macroscopic level

The neurons in the brain are not connected at random. A common principle is that neurons which serve the same or similar functions are located close to each other in assemblies. These assemblies again are connected with other assemblies, causing a given brain area to be connected directly or indirectly with numerous other areas in complicated circuits. Not surprisingly, this does not mean that all assemblies serve the same function. On the contrary, a lot of brain areas are highly specialised, subserving very specific functions. As an example, some groups in the visual cortex code for colour, while other and separate groups code for motion, shape, etc. Whenever we see a given object it follows that this is a product of many specialised areas that each contribute with a given aspect of our perception. When many areas need to co-operate to provide a given function, we refer them as **cognitive networks**. Some functions appear to be in place at birth. This is probably the case for the operation that segments speech into words (Simos and Molfese, 1997). (Often there is no space between the pronunciation of individual words so it is quite a task to figure out where one ends and a new one begins). Other functions must be "built". The ability to read demands a complex network that involves many different areas in the brain. This network is not in place at birth but must be formed by connecting and co-ordinating the activity of many specialised areas. Partly because of this, the ability to read requires a lot of instruction, whereas comprehension of spoken language seems to occur spontaneously, that is, without formal instruction.

B.2.3. Brain development can be examined at multiple levels

Knowledge of brain development at the microscopic level comes from the study of neurons in animals. Because there is not much difference between a neuron in an animal and in a human, this knowledge is valid for humans also. Studies at this level are less informative when we are interested in the relationship between cognitive functions and neural changes. This is so because neurons behave in the same manner, and their development is guided by the same general principles, regardless of whether they are located in the front or the back of the brain, even if the functions they subserve are very different. If we want to know something about cognitive functions, larger regions – and preferably the whole brain – must be examined at the same time. Moreover, if we want to understand human development, it is most natural to study human brains rather than monkey brains (even when they are quite similar). Until approximately two decades ago this meant that studies of human brain development was limited to *postmortem* examinations. Even though one can gain a lot of knowledge about the brain by cutting it into thin slices, this method suffers from the limitation that such brains have often belonged to sick or old people. This complicates the study of normal brain development. Having said this, it must be admitted that substantial parts of human brain development were already mapped in 1901, based on this method. The work was conducted by Paul Flechsig in the late 18th century. By examination of the degree of myelination in different brain areas at different ages, Flechsig worked out maps showing the progressive maturation of different brain areas. A lot of useful information can be gained from these maps. As an example it can be seen that the areas that subserve consolidation of knowledge in long-term memory (the mesial parts of the temporal lobes) mature before frontal areas that mediate working memory (the part of short-term memory where information is manipulated actively). Even though Flechsig's findings have been confirmed in newer studies, his maps leave a lot to be desired. For one thing, it would be useful to know at which particular time in ontogenesis (the developmental history of the individual) the different areas mature and when this development is completed. It would also be interesting to know whether children and adults – due to developmental differences – use different parts of the brain to solve a given task. Within the last 20 years, information about such aspects has become obtainable due to the development of imaging techniques (brain scans). These techniques allow brain development to be examined in healthy (and living) individuals. Given that these techniques are somewhat different and, hence, reveal different aspects of brain development, they will be described below.

B.2.4. Imaging techniques

Imaging techniques can be divided into two to types: the **functional** ones and the **structural** ones. The structural imaging techniques provide pictures of the brain's anatomy by showing the distribution of grey matter (cell bodies) and white matter (axons). The structural scanning technique most used today is Magnetic Resonance Imaging (MRI) (for details on these methods see Box B.1).

Box B.1. **The principle behind MRI**

The human body contains quite a lot of water, but the concentration of water is different in different types of tissue. In the brain we can discriminate between two types of tissue; grey matter and white matter. Grey matter consists of cell bodies, while white matter consists of axons (the axons are white because they are covered by fatty sheaths/myelin). The principles behind MRI are somewhat complicated but, by and large, they are based on the following: in water there is a great incidence of hydrogen and in the hydrogen atoms so called protons are located. Because protons are magnetic, they behave like small compass needles and because they are normally moving they point in all directions. If they are placed in a strong magnetic field, which is what the MRI scanner creates, they will align in the direction of the strong magnetic field pointing now in the same direction. Even though the protons are now kept at their places they still spin around their own axis with a given frequency. If a radio wave with the same frequency (as their spin) is now directed at the protons, the protons will tilt slightly, absorbing the energy in the wave. When the radio wave is terminated, the protons will reorient themselves, thereby sending back a radio wave with the same frequency as the one they were affected by. These radio waves can be detected by a system of antennas. Because the intensity of the radio signal will depend upon the concentration of protons, and because this concentrations differs in grey and white matter, intensity differences can be reconstructed as images showing the areas of grey and white matter.

Box B.2. **The principle behind PET and fMRI**

When neurons are operating they are in need of increased amounts of energy. This energy comes from sugar, which is transported to the neurons via the blood. By measuring the metabolism or the cerebral blood flow, one gets an indirect measure of the brain's level of activity. This can be done by the use of Positron Emission Tomography (PET), where one injects a radioactive substance that either binds itself to sugar or diffuses in the blood. The most hard working areas in the brain will also be the ones with the highest metabolism, the highest cerebral blood flow and, accordingly, also the highest concentration of the radioactive substance. By counting the number of radioactive decays and representing these figures as images, we get images of how activation varies in different areas of the brain at a given time. There is one major drawback of the PET technique; it requires the injection of radioactive substances (albeit in small doses) which one would like to do without. Fortunately this is possible today as it has turned out that one can also use MRI to measure cerebral blood flow. This is possible because oxygenated and deoxygenated blood have different degrees of magnetisation, which again will yield different signals when the blood is affected by radio waves. Given that neurons must utilise oxygen in order to convert sugar to energy, this means that the radio signal from a given area will be different depending on whether it is working or not. By measuring the ratio between oxygenated and deoxygenated blood, one gets an index of the degree of work activation performed by a given area. The technique used to measure activation by means of MRI is called fMRI (functional MRI).

In comparison with the structural imaging techniques, functional imaging techniques provide pictures of the brain's activity at a given time. With functional imaging one can identify which areas are more active in given conditions and hence work out which areas subserve particular functions (for details on this method see Box B.2).

B.2.5. It is difficult to tell to what extent development is caused by nature or/and by nurture

The invention of imaging techniques has caused a major step ahead in our understanding of brain development because it allows brain development to be studied in healthy and living people. We are now in a position to examine when different areas of the brain develop (mature) and also whether children use different areas in the brain than adults. Before we take a look at the findings acquired with these techniques, it is important to mention some of the problems that face the interpretation of these findings.

Until now we have talked about brain development as if it was **one** process. It is clear, however, that brain development is driven by several factors. On the one hand, it is obvious that some aspects of brain development are genetically given and therefore not affected by individual experience in a direct sense. As an example, it is not accidental that the visual cortex is located in the same parts of the brain and wired in the same manner in all individuals.[1] This type of development is a genetically caused process of biological **maturation**. On the other hand, it is also the case that some aspects of brain architecture are actually affected by experience and therefore differ from one individual to another. Not even monozygotic twins have identical brains (White, Andreasen and Nopoulos, 2002). Brain development then is clearly a product of both maturation (nature) and experience (nurture). This means that it is very hard to establish which changes in the brain are caused by biological maturation and which are not.

Another problem concerns the relationship between development and age. If we only examine brain development as a function of chronological age we will end up with imprecise measures. This is due to the fact that there will be variation in development owing to individual differences. In other words, even if area "x" matures before area "y" in all individuals, the exact chronological time at which area "x" undergoes maturation will vary from individual to individual. In one person it may happen around the age of 6 while in another it may happen at the age of 8. If we only base our studies on the mean development observed in a group with a particular age, say 12, our estimates will be noisy because individual differences may overshadow real differences. This problem is present in so-called **cross-sectional studies** where different age groups are examined and then compared. There is only one way to overcome this problem and it is to study the same group of individuals at different ages. In this case we get rid of some of variability due to developmental differences between individuals at a given age. Studies based on such measures are termed **longitudinal studies**. These types of studies are the exception rather than the rule because imaging techniques have only been available in the last 10-15 years.

Even though longitudinal studies yield more precise measures than cross-sectional studies, they do not escape the fundamental problem concerning the relationship between maturational-driven and experience-driven development. It remains difficult to establish

---

1. As an example, our visual cortex will develop normally even though each of us will receive very different visual impressions. Only in extraordinary cases, as in total blindness caused by eye deficits, will this genetically specified development be disturbed.

whether a given change observed in the same individual between the age of 6 and 8 is caused by maturation or experience (or both). After all, quite a few things happen experience-wise in two years; in school, for example.

With these problems of interpretation in the back of our minds, we will take a look at what is known about ontogenetic brain development.

B.3. The brain is a sculpture carved by experience

B.3.1. Brain activity seen over time

A lot of information concerning brain development comes from studies of brain metabolism measured with PET (Chugani and Phelps, 1986; Chugani, Phelps and Mazziotta, 1987; Chugani, 1998). These studies provide a picture of the brain's synaptic activity. They show that metabolism in newborns (< 1 month) is highest in the brainstem, in parts of the cerebellum and the thalamus, as well as in the primary sensory and motor areas. In terms of functionality, this means that newborns are able to regulate basal functions such as respiration, arousal, etc. (brainstem); register touch, visual impressions, etc. (thalamus and primary sensory areas); and can perform elementary actions (primary motor areas and the cerebellum). Besides this, relatively high metabolism is also found in some of the areas that support memory and attention (hippocampus, cingulate gyri and the basal ganglia). By the second to third month, metabolism increases in the secondary and tertiary regions of the parietal, temporal and occipital lobes; that is, in regions that do not receive sense impressions directly but instead process information from primary areas. In terms of behaviour, this indicates that the child becomes better able to integrate information from different modalities and to co-ordinate movements. This increase in metabolism, however, is not uniform in all secondary and tertiary regions. Characteristically, metabolism in the frontal parts of the brain only starts to increase when the child is about six months old. It is the frontal lobes that perform the most complex information processing and these areas are normally associated with executive functions; that is, the ability to plan ahead and perform complex goal directed actions.

The pattern of development described above does not differ significantly from that described by Flechsig in 1901, although it does give a better impression of the chronology. Both the PET studies and the study by Flechsig demonstrate how ontogenetic development in general mirrors phylogenetic development. Nevertheless, the PET studies do reveal a surprising pattern. Even though the rate of metabolism in newborns is about 30% lower than in adults, it rises steeply from then on and over the following four years. So steep is the rise that the rate of metabolism by the fourth year is twice that found in adults. From the age of 4 to the age of 9-10, the rate of metabolism is stable. It then decreases to adult levels, which happens by around 16-18 years of age. Because metabolism primarily reflects synaptic activity, this pattern indicates that there are far more synapses available by the age of 4 than are actually needed and that these superfluous synapses are eliminated in time. A natural part of brain development thus appears to consist of the elimination of superfluous connections between neurons. Only synapses that are in use survive and this depends on the individual's experience (Rauschecker and Marler, 1987). From a genetic point of view this arrangement is rather neat. Instead of specifying all connections in the brain, only the most essential are given from the start. The rest is a question of environmental influence. In this way it is assured that the brain's functionality corresponds to the needs of the organism.

B.3.2. Brain structure seen over time

The studies referred to above are based on PET studies of brain activity. Even though there is a tight connection between metabolism and the number of synapses in operation, the PET technique can only indirectly reveal how the brain changes structurally over time. To obtain direct information about structural changes we must turn to the MRI technique that can identify changes in grey and white matter. At present there are several such studies reported (see Durston et al., 2001; Paus et al., 2001; Casey et al., 2005; Paus, 2005 for reviews).

In general there is a close correspondence between findings obtained with MRI and PET studies. In one of the most extensive studies (Giedd et al., 1999; Giedd, 2004), which is based on a longitudinal examination of over 161 persons, it was found that the grey matter volume (GMV) exhibited a characteristic ∩-shaped pattern; that is, GMV increases in childhood, peaks in adolescence and decreases from this point and onwards. This development, however, is different in different regions in the brain. In the parietal lobes the GMV peaks around the age of 11 (average for girls = 10.2 and average for boys = 11.8). In the frontal lobes the GMV peaks around the age of 12-and-a-half (average for girls = 11.0 and average for boys = 12.1) and in the temporal lobes around the age of 16-and-a-half (average for girls = 16.7 and average for boys = 16.5). (Curiously, no decrease in GMV was detected in the occipital lobes which mediate visual functions.) It should be noted that differences are found within these regions (frontal, temporal and parietal) because GMV peaks earlier in primary than in secondary and tertiary areas. As an example, the dorsolateral part of the frontal lobe, which is associated with executive functioning, is one of the last areas to mature. Another region, which also peaks at an advanced age, is located in the temporal lobes (more specifically in the lateral parts and especially in the left hemisphere) (Sowell et al., 2003). This area is believed to play a special role in the storage of semantic knowledge, that is, knowledge about what things are and how they work (a bit like a lexicon). The finding that GMV peaks at a relatively advanced age in this area, in fact around the late twenties, is consistent with the fact that we acquire semantic knowledge throughout life but especially during childhood, adolescence and early adulthood.

As a general principle it thus seems as if GMV reduces between adolescence and early adulthood. This is not what is found when changes in white matter volume (WMV) are examined; that is, the parts of the neurons that send information onwards and which connect the different areas of the brain. WMV seems to increase at least until the age of 40 (Sowell et al., 2003) and it does not seem as if there is significant change in WMV across different brain regions (Giedd, 2004).

If we hold together the changes we observe in GMV and WMV over time and relate these changes to functional aspects we can summarise by saying that: with time we lose plasticity/learning potential (GMV reduces) but increase functionality (superfluous synapses are eliminated concurrently with optimisation of communication pathways – the myelination of axons yielding increased WMV).

B.3.3. Relationships between brain and behaviour are often indirectly derived

It is a given that MRI has increased our knowledge regarding brain development. Many of the studies, however, are characterised by a specific limitation. As a general rule they examine brain development without studying cognitive development at the same time. This means that we have to make qualified guesses as to which behavioural changes these physiological changes may cause. While it is certainly possible to obtain some knowledge regarding relationships between brain and behaviour in this way, these derived relationships remain indirect. Direct relationships can only be revealed if we examine development in brain and behaviour in the same people at the same time. While such studies are possible (Tyler, Marslen-Wilson and Stamatakis, 2005) they are still sparse.

B.3.4. Children and adults do not use the brain in the same way

In the studies referred to above the focus has been on structural changes in the brain as a function of age. Another way to study the relationship between cognitive functions, brain development, and age is by means of **activation studies** performed with functional imaging techniques (PET and fMRI). These types of studies reveal brain activity (as opposed to structural changes) associated with different tasks (e.g. reading, object recognition, etc.). The advantage of these studies is that observed differences between groups (e.g. adolescents vs. adults) are directly related to the task being solved by the subjects. This is not necessarily so if one has only found an area in which brain development seems to mirror the development of some cognitive function. Such a correlation, say between the area A and the cognitive function B, could in principle be caused by changes in the cognitive function C if function C develops in the same way as B over time. In activation studies this can be controlled directly because the researcher can decide which cognitive functions are to be called upon – it will depend on the task chosen.

The number of activations studies examining age differences far exceeds the number of corresponding structural studies. A review of these studies is therefore beyond the scope of this paper. In general it can be concluded that these studies indicate that children/ adolescents activate more areas of the brain than adults do and that these activations are more diffuse (Casey et al., 2005). This is consistent with: i) the finding that the overcapacity of synapses in children and adolescents is eliminated in time; and ii) that there is continuing growth of white matter that ties regions more effectively. Both aspects should yield more focal activation as a function of age.

B.3.5. The adolescent brain and changes in adolescent behaviour

As shown above, the whole brain is developing during adolescence. Nevertheless, some areas appear to undergo more radical development during adolescence than during other periods. This is especially the case for the dorsolateral parts of the prefrontal cortex. Usually, these parts of the brain are associated with cognitive processes such as working memory (where on-line information is actively manipulated), the allocation of attention, response inhibition, and temporal structuring of new or complex goal directed actions (Fuster, 2002). (Note that these operations are intimately related in everyday behaviour.) Not surprisingly perhaps, it has been known for some time that abilities that require such operations continue to develop in adolescence (Keating and Bobbitt, 1978). More recently it has been shown these developmental changes do indeed correlate with physiological changes (both structurally and functionally) (Casey, Giedd and Thomas, 2000). The developmental changes discussed here are very likely to underlie four of the characteristics that usually tell children and adolescents apart, namely the ability to: i) reason hypothetically; ii) think about thinking (meta-cognition); iii) plan ahead; and iv) think beyond conventional limits (Cole and Cole, 2001). Consequently, by mid-adolescence, adolescents' decision-making abilities are brought to adult levels. Interestingly, this does not necessarily mean that adolescents make decisions in the same manner as adults. Indeed, adolescents are known to engage in more risk-taking behaviour than adults, which again may be associated with increased sensation and novelty seeking behaviour (Spear, 2000). Although these behavioural differences are probably related to biological factors, it is not clear how they are related to brain development because they can be observed in other species as well. Altered risk-taking behaviour is usually associated with structures mediating emotional regulation (the ventromedial prefrontal cortex and amygdala) (Bechara et al., 1997; Krawczyk, 2002) and developmental changes in the brain during adolescence are not known to be associated with these areas. This, however, could be caused by (at least) two factors. Firstly, as compared with the experimental decision-making problems usually investigated in developmental studies, decision-making in real life takes place in unstructured situations characterised by a lack of explicit rules and a variety of potential solutions. Accordingly, decision-making abilities in mid-adolescents might have been overestimated in developmental studies and might in fact first reach adult levels in late adolescence (that is, the observed discrepancy between decision-making abilities and risk-taking behaviour may be more apparent than real). If so, this would be in line with studies of brain development which indicate that development of the dorsolateral prefrontal continues in late adolescence. Secondly, the possibility exists that reduced risk-taking behaviour (including sensation and novelty seeking behaviour) might be related to maturation of the ventromedial prefrontal cortex, which, as mentioned, is associated with emotional regulation. It is not impossible that such maturation could go unnoticed because this structure is not easily imaged with fMRI (Devlin et al., 2000). Whatever the biological causes of risk-taking, sensation-seeking and novelty-seeking behaviour are, this behaviour facilitates the adolescent emigration from the natal group by providing the impetus to explore novel and broader areas away from the home (Spear, 2000).

B.3.6. Summary and general implications

Despite the problems associated with interpreting whether biology (maturation) or experience determine brain development, a general pattern evolves. The most striking part of brain development is probably that it continues into adulthood and that it is a dynamic process. To the extent that it is meaningful to term some processes as strictly biological, it would be the overproduction of synapses that seems to be present from birth until adolescence. Experience, however, also plays a major role in as far as it appears that it is experience that determines which synapses are to be kept, which are to be eliminated and which communication pathways (axons) are to be made more effective. These "tuning" processes operate well into adulthood. This, however, does not mean that we cannot learn after the age of 30. We can; synapses can still be strengthened and weakened and new synapses can still be produced (Draganski et al., 2004) although probably in fewer numbers. What it does mean is that we lose the plasticity, the learning potential that the overproduction of synapses represents. We become slower learners and less flexible. On the other hand, comfort is to be found for those of us who have passed the twenties: while learning potential is lost we get better at doing what we have already learned.

The finding that significant brain development continues well into adulthood indicates that the time frame in which experience and hence education determine the abilities we end up having is rather outstretched. This does not mean that we are born as blank slates just waiting to be written upon. By nature's hand we are "built" to learn some things more quickly and easily than others and we are born with some knowledge/ expectations in advance.[2] It does question, however, the notion that our development is

---

2. It is beyond the scope of this paper to address this issue. Interested readers are referred to Pinker (2002) and Premack and Premack (2003).

determined from very early on. If this is so, then why all these biological changes well into adulthood? It also means that we must re-evaluate the idea of **critical periods**, according to which a limited time frame exists in which we **must** learn certain things if we are ever to learn them. At present there is little evidence supporting this viewpoint within neuroscience.[3] If one speaks of periods at all, one usually speaks of **sensitive periods**, that is, periods in which it is optimal to learn something, but which can be surpassed without devastating effects. A common example of a sensitive period is that it is easier to acquire a language early in life than later on. It is doubtful, nevertheless, whether this actually reflects a sensitive period characterised by a special readiness in terms of brain development and language acquisition. Some evidence suggests that greater difficulties in acquiring a second language merely reflect that we already master one in advance (the mother tongue). It is more likely that the acquisition of our native language will already have shaped those areas of our brains – *e.g.* through the elimination of synapses – that are also necessary for the acquisition of a second language (Johnson and Munakata, 2005). On this account, brain plasticity is lost **as a consequence** of learning and not because there are certain periods in which learning is optimal. Having said this, it must be noted that we cannot learn every thing with the same ease at any point in time. As we have seen, there are differences between brain regions in terms of maturation. Likewise, there are individual differences in brain maturation. Accordingly, we must adapt education to brain development; we simply cannot expect a person to master something the brain is not yet ready for.

B.4. Theories of learning in adolescence and the life course

We started this analysis with a review of what is known about the brain and what that might tell us about adolescent behaviour. An opposite approach is to review what is known, observed and understood about adolescent learning, and consider how "brain research" contributes to that understanding. Where neuroscientific research reaches the limits already indicated in the preceding sections, we can begin to identify where a greater synthesis might be needed between the insights of brain research, cognitive science, social research and evolutionary psychology if our knowledge is to advance further.

Adolescence has generally been held to be a social construction, "invented" when the changes at puberty became the starting points for longer and longer periods of transitions into socially accepted roles and expectations of adulthood. Life stages called "postadolescence" or "pre-adulthood" were subsequently "invented" as transitions into adult roles and responsibilities extended up to the mid-twenties in many western societies. Adolescence came to be viewed as a kind of socially constructed "psycho-social moratorium", a time necessary for "preparing and rehearsing to be adults" in an increasingly complex society (Cockram and Beloff, 1978). This period was held to extend well beyond physical "maturity" for social reasons. The first major contribution of the new research into the brain reported in earlier sections of this paper is to show that there is a physical as well as social basis to lengthy periods of "transition", with the brain continuing to develop and change in characteristic ways up to and into the third decade of life and, subsequently, well into adult life.

---

3. At least if one does not include extraordinary situations. It is likely, for example, that a person who is blindfolded at birth and for the following four years will never learn to see properly once the blindfold is removed. Usually, however, it is not such cases that critical periods refer to.

The understanding of adolescent development in the 20th century was dominated for a long time by life stage theories that emphasised a natural developmental stage of "storm and stress" that has to be worked through in adolescence. Life stage theories also focused on age related normative tasks that bring stability and identity resolution if certain adolescent "growth tasks" are completed satisfactorily for the individual and society. The key growth tasks include, for example, construction of a social, sexual, physical, philosophical and vocational self (see also Havighurst and Kohlberg for stages of moral development). For Erikson, the key feature of the adolescence in the "seven stages of man" was identity formation *versus* identity confusion, a view still widely influential. Peer groups become increasingly important in adolescence as the search for identity and meaning for others develops (Coleman, 1961; Bandura, 1997).

Social and cultural theories brought different dimensions to the analysis of adolescent experiences, focusing on social structures and power relations in society, the social processes (including those of home family and schooling) that shape the life chances, and the relative vulnerability and powerlessness of youth in relation to these. Metaphors of "trajectories" in adolescence, which became popular in the 1980s and 1990s, are often associated with theories that emphasise the longitudinal effects on adolescents of the social and cultural reproduction of roles and life-chances. These, and related cultural theories, often emphasised reactions of alienation and resistance in adolescence that are played out in the context of schooling as well as home family and community.

Metaphors of "navigations" in adolescence were an invention of the turn of the 21st century (Evans and Furlong, 1998). These were associated with greater recognition of the complexities of life course development and the growth of exposure of young people to "secondary stimuli" outside the child's immediate world of home, family, school and neighbourhood. The processes of transition from childhood, where one's environment is controlled by others, to adulthood, where environments and influences are more subject to the person's own volition, become more complex with, for example, modern media and communications. Coleman's (1970) "focal theory" started to counter simple "stage theories" by showing that the "storm and stress" adolescent stage, if it exists, is not a time of generalised turmoil, but that young people are progressively engaged with different issues and challenges at different times. Sensitive periods may exist for different tasks and issues to be focused on, but these are very individualised, proceed quite differently for different individuals, and are not irrevocably lost if not fully worked through at the "sensitive" time. Without fully resolving each life issue, the young person moves on to focus on the next. Stress is linked to the issue in hand at the present time, with residual stress left behind by previous engagements to which the young person may return. The emphasis in theories linked to the "navigation" metaphor is also on how young people negotiate structures of risk and opportunity, how these provide sources of stability or instability in the later life course, on biographical construction and on "individualisation" (Baethge, 1989; Evans and Heinz, 1994). Self efficacy (Bandura, 1997) and personal agency (Evans, 2002) are important dimensions of the learning biographies of adolescents and adults.

These theories have tended not to consider learning processes directly, although they embody many assumptions and implications for learning. Learning theories themselves follow certain traditions and illuminate certain aspects of learning. Psychological traditions that viewed learning as individual, rational, abstract, detached from perception and action and governed by general principles have been challenged by socio-cultural and "situated cognition" theories that argue cognition is essentially social, embodied, located, context dependent and specific to environments. The significance of the psychodynamic dimensions of the emotions in learning has increasingly been recognised, and cognitive science has brought into the debate new understandings of "multiple intelligences" in individual learning and human action (Gardner, 1984; Bruer, 1993).

Research into the brain can neither confirm nor refute these theories. We can however identify where findings of the kind reported in the previous sections are consistent or variable with theories based on and informed by the methods of social science. In some cases, consideration of the findings of brain research can tell us which theories it may now be more or less fruitful to pursue.

Theories of adolescent development are best linked to questions of learning in adolescence through an integrative theoretical perspective that recognises that:

1. Learning is a natural process for human beings.

2. Learning is much more than content acquisition or the development of cognitive skills.

Learning may be defined as the process of expansion of a person's capacities. Learning always involves the interaction of cognitive and emotional processes, and learning always occurs in social contexts through interaction between learners and their environments.

As a recent social science research synthesis has shown, whether what people have learned at a particular time is retained in the long term depends on their engagement and how important it is to them. "Whether we attend to any particular stimuli in the environment at all depends on the brain's assessment of their importance to us. Central to the unconscious selection of what to attend to and what to learn is the self, which itself is learned and developed through our interactions with others" (Hallam, 2005).

The changes in capacity brought about through learning will often be observable through behaviour, but not always. Changes in capacity in adolescence involve the interplay of dispositions and orientations, maturing mental functions, identity development and beliefs about the self, motivations, strategies for achieving autonomy and control. All are strongly affected by environmental influences.

Several social science theories emphasise the dispositional aspects of this learning, showing how people have generalised dispositions or orientations towards the world that shape their learning. Some mental predispositions appear to be built in from birth (Pinker, 2002, Premack and Premack, 2003), and are developed and shaped by subsequent experiences. From a sociological standpoint, socially constructed dispositions have been shown by researchers such as Hodkinson and Bloomer (2002) to have particular significance in adolescent learning, reflecting the internalisation of cultures and the development of the young person's gender, sexual and racialised identities, as well as emergent individual interests, desires and wishes. For example, this research has shown how the career aspirations and dispositions to learning of late adolescent learners changed in ways that suggested interrelationships with experiences, at the same time as underlying continuity with earlier identity formation. Dispositions appear to be highly significant in the shaping of the young person's "learning career", including whether they engage or withdraw from organised learning, yet cultures and identities are often given insufficient attention in the organisation of secondary and postsecondary schooling.

Social science has also shown that much learning is subconscious and that tacit skills and knowledge underpin behaviour, actions and performances of people in educational and everyday environments. Conscious learning interacts with tacit learning all the time.

Recognition (by the learner) of the tacit dimensions of learning is necessary before this can be controlled and purposefully applied (Evans *et al.*, 2004; Illeris, 2004). It is likely that this capacity increases with age, with metacognitive abilities (thinking about thinking) emerging in adolescence and continuing to increase into adult life. The increasing significance of prior learning and the development of metacognition in late adolescence appear to be consistent with the neuroscientific findings concerning the nature and timing of the development of different regions of the brain and the overall process of "sculpting" outlined earlier.

The process of maturation in learning has long been recognised through the work of Piaget (1967) and his successors. Subsequent research has shown higher mental functions gradually become differentiated in adolescence, with the ability to think logically and deductively maturing and greater cognitive capacity developing. There is evidence that the emotional and cognitive processes are closely intertwined (Damasio, 1994), although it is held that adolescents becoming capable of greater "maturity" in behaviour is a manifestation of insulation increasing in the brain, as well as the effects of increased experience. Here again, nature and nurture are difficult to separate. There is also consistency between these accounts of the emergence of higher level cognitive functions and the findings concerning the pattern of increase and reduction in WMV and GMV in adolescence and beyond.

We know now that the brain development of the individual reflects his or her learning experiences and the activities engaged in, that extent of change depends on time spent in learning and that significant, permanent and specific changes in brain functioning occur when considerable time is spent in learning and practice of particular skills and ways of doing things. For example, comparison of brain activation in 13- to 15-year-old musically trained learners with other learners of the same age led Altenmuller *et al.* (1997) to conclude that the brain substrates of processing reflect the "learning biography" of each individual; not only what we have learned but how we have learned it.

While learning biographies are highly individual and unique, some patterns are indicated. Mental "schemata" have been held by psychologists to be most flexible in children and to become cemented in adulthood (Illeris, 2004). This links with earlier notions of "fluid" intelligence in childhood and "crystallised" intelligence in mature adults, with the extent of experiential learning and ability to form judgements and decisions a key factor in differentiating adolescents from adults (Davies, 1971). In adolescence, schemata are becoming less flexible and identity defences start to come more into play – in ways that can limit learning – as people become adults. These accounts of "fluid" and "crystallised" intelligence also appear consistent with neuroscientific findings about GMV, the number of areas of activity shown in the brain, and their relative diffuseness or focus at different ages.

Further insights into learning as the development of capacities can be gained from the literature on expertise. The processes of acclimatisation, competence, proficiency (Alexander, 2003) all need increasing levels of personal motivation, as well as the move from surface to deep learning if the higher levels are to be attained (Entwhistle, 1984; Hallam, 2005)

Another aspect of that progression to "proficiency" has been identified by Csikzentmihalyi (1996), who observed the experience of "flow" in learning in adolescents as occurring with intense engagement in a chosen activity. Irrespective of the type of activity that is chosen and carried out, this "state of flow" manifests itself in intellectual and emotional excitement, and an adaptation in which body chemistry is modified and the person moves into what has been termed a kind of "mental overdrive" in seeking competence and mastery. This leads Csikzentmihalyi to propose that enjoying mastery and competence is evolutionarily adaptive. This is consistent with other research (see for example, Bandura, 1989) which has shown that people learn best when trying to do things that are challenging and of deep interest to them, again reflecting the close interplay of the emotional in cognition and the development of capacity.

Interest has to be internalised and become part of identity to be sustained over time. Since adolescence is a critical time for identity formation, the "identification" phase with particular domains of learning tends to occur in the adolescent years (Hallam, 2005). All of the above are parts of the drive towards control of learning in adolescence, the single biggest differentiator between childhood and adult learning. Adolescent development theories portray adolescents as striving to exercise control over themselves and their environment; this goes hand in hand with identity formation, or the development of the young person's sense of who they are and how they are seen (and want to be seen) by others.

Erikson (1968) proposed that identity formation reaches its most critical stage in adolescence, as this is the time of life in which the young persons struggle with identity confusion to find "who they are". This is highly significant for learning, since identity is an important driver for motivation and learning, and identity defence is often found in resistance to or withdrawal from learning. As the fundamental differences between learning in childhood and in adulthood stem from the degree of control the learner has in the situation, as well as differences in prior learning and capacity, the drive towards autonomy in adolescence is of crucial importance in learning and in the way the adolescent learner co-constructs his or her learning in interaction with the social environment. This process again appears to be consistent with the late adolescent emergence of more focused problem-solving and decision-making abilities. The continuing development of these capacities into adulthood is consistent with the evidence cited earlier concerning later maturing of the relevant region of the brain (the prefrontal cortex).

The drive towards autonomy and independence, allied with identity formation process, is commonly held also to involve experimentation and risk-taking, activities which may themselves expand capacities and constitute significant learning. It has been suggested that adolescence may be a "sensitive" period for these types of learning, providing for the essential "rupturing" of parental ties that allows the mature personality to emerge. As we have seen from the earlier discussion, there is little evidence from brain research for this, but there does seem to be a link to the drive to move outwards from the natal group that may be better explained by recourse to evolutionary psychology. It is also reasonable to suppose that while adolescence may be a sensitive stage for this kind of learning, failure to complete these particular growth tasks during the teenage years is unlikely to result in irrevocable problems that are impossible to resolve later, given the degree of plasticity that is retained into and through adult life.

B.4.1. Implications for teaching and learning in the adolescent years

So far it has been argued that important changes in capacity occur as people move from childhood through adolescence to adulthood, and that these are strongly influenced by interactions in the wider social environment. The social scientific evidence is consistent with the evidence from neuroscience concerning the way in which the brain is uniquely sculpted according not only to what is learned, but also how it is learned. The question of whether adolescence can be considered a critical period for the development of autonomy and control, driving the adolescent away from dependence on the family of origin, has also been considered.

Some, such as the team engaged with the 21st Century Learning initiative in the United Kingdom, argue that these findings are already pointing the way to major shifts in the way society should organise teaching and learning in the adolescent years; the organisation of schooling, they argue, must go with the "grain of the brain" if it is to be fully effective.

Changing the metaphor, Sylvester is cited as saying (1995) that Edelman's model of the brain as a rich, layered, messy, unplanned jungle ecosystem suggests that a jungle-like brain might thrive best in a jungle-like classroom that includes "many sensory, cultural and problem layers that are closely tied to the real world environment in which we live – the environment that best stimulates the neuro-networks that are genetically tuned to it".

A key consideration is the fundamental relationship between motivation and learning. It is here that many of the challenges arise in the way in which teaching and learning is currently organised in the secondary phase of education in most advanced economies. We have seen that motivation to learn is closely linked to identity and the goals people have for themselves. The value attached to learning particular things depends on how they relate to identity and goals at particular times. There is a difficulty when the curriculum is pre-determined and the relationship between what is on offer and the adolescent's personal goals is poor. The more closely the goals of teachers, learners and educational systems are matched, the more effective the learning will be, argues Hallam (2005). Furthermore, the more closely this learning is linked to the multiple stimuli of the "real world environment", the more it will engage and stimulate the learner.

What changes would have to be made to the organisation of learning in adolescence to go with the "grain of the brain", according to the studies available to us?

1. Reconceptualise learning as the lifelong development of human capacities, in which the changes in the brain of adolescence and late adolescence are at least as important for learning and potential as those of early childhood. A system of education operating on the basis of progressive selection according to "ability" at particular ages does not fit with the evidence we now have about how human capacities develop. A lifelong system of education is required.

2. Mastery approaches to learning are needed, where learners aim to improve on their own performance and/or that of their work group, without reference to the relative progress of other individual learners. This also links to the fact that there can be considerable variations in brain maturation across individuals in any given class or group organised by age. Principles of cognitive apprenticeship can be adopted more fully. Learners are "scaffolded" up to the next level through support from teachers, who are themselves supported in providing the necessary informal support and encouragement.

3. Attention should be paid to the beliefs young people have about the brain and ability. Many young people believe intelligence and ability to be fixed and unchanging. Dweck and Leggett (1988) have shown that young people who have these "entity" beliefs about intelligence are more likely to have performance goals involving the achievement of positive comparisons when measured against others, while those who have "incremental" views about intelligence tend to have learning goals linked to their own progressive mastery of content or tasks. Those young people who have "entity" theories of fixed ability/intelligence tended to minimise the amount of time they put into school work, believing ability is innate, avoiding activities they think they will fail in, putting in little effort where they are weak and also taking "being good at something" as an indication that they do not need to work hard at it (Hallam, 2005).

4. Labelling people as particular types of learners or as having particular learning styles is likely to limit rather than enhance learning. Coffield's (2004) work on learning styles has shown significant flaws in the underpinnings and evidence for many of the models currently in use, and highlights the dangers in their uncritical application.

5. Prior learning should be built on progressively, to facilitate the achievement of competence and proficiency. Current schooling arrangements in many education systems do not do this effectively, but instead introduce disjunctions at the adolescent stages that may significantly impede learning progress.

6. Adolescents need to continue to develop a range of perspectives gained through experience and different types of learning, to enhance their problem-solving capacities and decision-making abilities. This extends into late adolescence and early adulthood. Breadth needs to go hand in hand with opportunities for the intensely focused activities that allow and encourage "flow". The extension of support for these types of learning should extend well into the third decade of life for all young people, not only those who have entered higher education, but also those who have entered work.

7. Out of school/extra-curricular activities should be given much greater significance. Many of the experiences of "flow" and exhilaration in learning are currently experienced outside formal learning contexts rather than within them and young people who "fail" within school can, and often do, develop their identities and gain peer approval through unconventional activities that provide this experience and enjoyment of learning. How "extra-curricular" and optional activities are valued and supported by teachers is of crucial importance in adolescence. Their potential should be built on for all learners, and not confined to those who have the parental resources and access. Learning will also be enhanced when the prior learning and experience that comes through these optional activities are recognised by the learners and their teachers and built on in other learning environments, enabling connections to be consolidated and expanded.

8. Learners should be given choice in what and how they learn, wherever possible and where this is consistent with the need for breadth of experience and perspective, during the adolescent years.

Additionally, Hallam, drawing on a combination of neuroscientific findings and the literature on "expertise" has also suggested that those engaged in teaching will enhance motivation and learning by:

- Avoiding overloading the curriculum.
- Setting tasks that are challenging but not too difficult.
- Providing for different levels of prior knowledge and experience.
- Avoiding making attributions related to "ability".
- Encouraging students to take responsibility for their own learning, whilst recognising that this is something to be learned in itself and not something that can be demanded by children.

B.5. Challenges and future directions: towards a new synthesis?

This final section aims to:

1. Summarise significant gaps and missing dimensions in current knowledge; what is known; what is not known and what it is important to find out next.

2. Consider whether a new synthesis can be found between brain research, cognitive science, other social sciences and evolutionary psychology, in understanding how far aspects of learning in adolescence can be regarded as key elements in a critical life stage.

We have already shown that, despite difficulties in disentangling the effects of maturation and experience, a general pattern emerges from the evidence to date:

● Brain development continues into adulthood and that is a dynamic process.

● Experience plays a crucial role, since this "sculpts" and "tunes" the brain, a process which continues in adulthood, although some flexibility is lost in later years.

● While "plasticity" of the brain decreases with age, "expertise" in what we have already learned increases.

● We have inbuilt pre-dispositions, but evidence strongly questions the notion that our development is determined from very early on.

● The brain appears to become ready for the development of self-monitoring and goal directed behaviour from late adolescence, with this area of the brain (the dorsolateral parts of the frontal lobes) maturing later.

● There is little evidence to support the existence of "critical periods".

● Sensitive periods appear to exist that are optimal, but not decisive, for the development of particular capacities.

● There are considerable individual differences in brain maturation.

While these general patterns are evident, there have been only limited longitudinal studies to date that can track the changes in the same group of individuals over the years of adolescence and early adulthood. More evidence gained through longitudinal studies is needed to elaborate these general patterns. Also, much knowledge concerning relationships between brain development and behaviour is based on *posthoc* interpretations in that we observe some changes in the brain and subsequently try to connect these changes with what we already know about behaviour. Future studies will need to measure/observe brain and behavioural changes simultaneously to establish more direct evidence.

We have also taken the "social science-led" approach, which reviews what is known, observed and understood about adolescent learning, and considers how "brain research" contributes to that understanding. We have argued that research into the brain can neither confirm nor refute theories of learning, but we can identify some consistencies and inconsistencies between neuroscientific findings and theories based on and informed by the methods of social science, and draw some conclusions about which aspects of which theories it may now be more or less fruitful to pursue.

Neuroscientific findings are consistent with an integrative approach to learning that recognises individual variation in the expansion of human capacities. These capacities are shaped and co-constructed in interaction with the social environment, over a long time frame that extends from childhood to adolescence, into and through adult life. In adolescence, the development of higher cognitive functions, self-monitoring and goal-directed behaviour, and identity formation are of particular significance in this long-term process of learning.

Neuroscience can tell us about the way the brain works. Cognitive science has attempted to model information processing and multiple processes of learning. Other branches of social science shed light on social processes involved in learning and the development of human capacities. Questions about how and why the human capacities have evolved in this way are the subject matter of anthropologists and evolutionary psychologists. It is possible that better integration of these last two disciplines into the analysis could provide some of the links and connections that are missing in our current knowledge. We have already argued, for example, that adaptive mechanisms, in the evolutionary sense, may explain the adolescent risk-taking and experience/challenge seeking behaviours that are not well explained by the other sciences. What is clear is that new interdisciplinary syntheses are needed in addition to advances in the individual disciplines.

The evidence so far is indicating quite strongly that the education and the organisation of teaching and learning need to be much more closely aligned with the way in which human capacities "naturally" develop and mature in interaction with the physical and social environment. Many prior assumptions about "ability" have to be questioned. In the adolescent stage this means more attention to prior learning and experience, acknowledging individual variation, encouraging mastery and challenge, promoting self-monitoring and responsibility when the adolescent is ready for it, and recognising that development of new capacities will go on well beyond the adolescent stage. Bringing new interdisciplinary syntheses to bear on these matters is the biggest challenge for the future.

Karen Evans and Christian Gerlach

B.6. Practitioner's response: I have a dream

I have spent quite a few years teaching, mainly in France but also in other countries, mainly adolescents but also children and adults. I have spent quite some time reading official instructions and attending teacher training sessions. I have also spent quite a few hours looking into the education sciences, social sciences, psycho-sociology, and all the "ologies" one can imagine. I have been dreaming all these years about a different way of looking at learners. I have so many convictions built up around what teaching and schooling should be like, and suddenly this article pops up… extremely clear, incredibly concise pages that not only summarise but also back up, scientifically, most of what I have had in mind for several years about what and how we should teach adolescents.

My reaction to this paper was two-fold: on the one hand it is clearly consistent with my position towards education and learning, yet on the other hand I could only bitterly think, "How long is it going to take before we can actually start working bearing all this in mind?"

So I still have my dream, but I now have the feeling I was not that wrong and, considering what science knows today, I feel more inclined to share my views on the schooling of adolescents. In fact, I believe we need to take a different look at learners, as well as at teaching altogether. We might have to think of defining new goals for schooling, new missions for teachers and new methods of evaluation. I am not saying everything has to change, but we (the education community) have to change the way we see pupils, whatever their age.

B.6.1. Re-considering teaching

Why?

Although we tend to take into consideration the personality of learners much more than we used to do a few decades ago, it seems that we still have some work to do in this direction. Teachers still see in front of them a group of people who have to be prepared for a forthcoming exam and who need to be ready to answer questions (sometimes difficult ones) that test their academic knowledge. It is even clear that, sometimes, students are able to write a paragraph on a topic they have learnt by heart and yet they are totally unable to explain the meaning of it and, subsequently, they are unable to exploit it in a different situation. That's why I strongly believe that we tend to continue feeding them, regardless of what they are ready, or willing, to swallow.

If we consider – as mentioned in the article – that:

1. "learning is a natural process for human beings" (Section B.4 above) we should wonder why adolescents do not like school and how we could reconcile them to the learning process.

2. "learning is much more than content acquisition or the development of cognitive skills" (Section B.4 above), then clearly, teaching is not just a matter of piling up lessons and exercises (even when they are called "activities") according to a heavy curriculum and in order to pass an exam.

Indeed, teaching should not only deal with content but also with meaning. Nobody can properly learn something that has no meaning – and this is all the more true with adolescents. Obviously, curricula are always meaningful and coherent in their progression, but we have to make them transparent and obvious for our adolescents.

Karen Evans and Christian Gerlach agree that "learning always occurs in social contexts through interaction between learners and their environments", which implies that we should always relate what we teach to something concrete, something related to the adolescents' environment and which gradually opens their eyes to the society in which they live: ego, friends and family, school, working environment, society, life.

I would add to this that interaction cannot occur when the issues that students are taught are not clear to them.

Abstract concepts are much better understood when based on concrete matters and explanations. An example of this is the book by Jostein Gaarder "Sophie's World", which provides an easy way of comprehending philosophy.

Similarly, each subject taught at school implies a very specific vocabulary – a jargon – that the students have to learn. In other words, they must, at the same time, learn both the word and the notion that goes with it. Here again, this is not an impossible task when: 1) it is explained first using basic words; and 2) when there are not too many new notions to learn in one day.

I believe that because teachers have to follow an overloaded curricula, they don't have (or take) the time to slow down on complex notions; they explain them once during the class and expect their students to work hard at home, to understand, assimilate and digest the notions on their own. So here is the question teachers should learn to consider everyday: is the average adolescent able/motivated/inclined to take this step on his/her own when there are so many other social experiences to live once the school day is over?

They also probably tend to forget (or minimise?) the importance of the other subjects students have in their curriculum (often as many as 10 or 12, though this varies from country to country). We know that adolescent brains present an enormous learning potential but I do not think we help them make the most of it teaching and grading the way we do.

How?

One point is that we need to work on methodology much more than we do. We often assume that, because they are growing up, "maturing" adolescents are ready to work on their own. But where/when, in their curriculum, do they learn how to learn? I believe it is often explained to them but I think it should be demonstrated to them.

I am not saying we should do the work for them, but why not have some kind of progression, within the curriculum, that would set the various steps a learner should be able to take in terms of learning and reaching autonomy in his/her own learning process?

- First, this means that all teachers should guide the pupils/students into asking and answering the questions "Why am I learning this?"; "Why is this of importance to me?"; "Where does (or will) this fit into my everyday life?"; "When/how/why will this be useful to me?". And if the issue is not linked to something in particular in the student's close, actual environment, how is it connected to another subject? In fact, any reason can be offered, as long as it makes sense to them.

- Next comes the issue of how they should learn such and such a topic. Here again, teachers should guide them into their work and show them (doing it with them over quite a long period) how to constantly look back on what they have done, constantly link it to what came beforehand in the lesson or to personal experience, and to see/feel the natural progression of the whole lesson.

Guiding them into their work also shows them that not everybody can acquire everything at once and that it can take time. How many adolescents are able to work on something they don't understand on their own and persevere in trying to understand on their own? Very few. They need to feel that teachers are working along with them – though not doing the work for them – to reach common goals. Again, teachers must take, give, and be given the time to let knowledge sink in.

At a stage in a person's life where the brain is optimally suited for cognitive learning, I think teachers have a major role to play in guiding its building process.

I am convinced that working on methodology is one of the ways of improving the learning process. Beyond this, if the pupils/students end up understanding how to learn, their teachers will probably have invested their time wisely rather than have wasted it. They will have helped learners to develop the learning skills they will use throughout their life and will probably have helped them to be more self-confident.

Another point that must be taken into account is the one of motivation. We all know that adolescents are more into social interaction with peers than into academic knowledge acquisition. However we cannot just take refuge behind this assertion when facing a refractory teenager. We also know that they are naturally extremely curious. Why do teachers rarely use this curiosity to arouse motivation and to get students to do different work?

I believe time and overloaded curricula are again responsible for this. It would probably be much more interesting and constructive for adolescents to go through the various items that have to be studied via projects in which they would take an active part. Contrary to what they look and seem to be, most adolescents like challenge – as long as it is not too difficult to face – and are (sometimes unconsciously) willing to prove they can manage to do something good.

Whatever the school subject, there is always a way to achieve formal academic objectives via informal and non academic tasks.

Proposing a range of projects or just a framework related to the curriculum, asking students to choose one and to be in charge of it from conception to realisation, is to me a different and necessary way of viewing teaching.

This doesn't mean, of course, that we should put aside basic and fundamental teaching issues, but it would imply having lighter curricula in order for students to have time to learn through doing. As for teachers, they would have time to consider the students as individuals in a constantly developing process, and would thereby allow them to learn via interaction with an environment they would have chosen themselves.

If it was so, learning would come from various sources:

1. Basic learning issues that would come from the teacher in what we could call a "traditional way".

2. Wider issues that would combine the important items set by the curriculum and the personal items introduced by the students.

Everything mentioned in the article by Karen Evans and Christian Gerlach about the brain and learning in adolescence fits the idea of working in such a way. Teachers would not just be leading a bunch of young learners to scale the exam ladder, but they would also be helping them be responsible for projects, take initiatives, build up their personality and be more self-confident and creative.

B.6.2. Should we define new goals for schools and new missions for teachers?

When considering education and learning as an interaction between learners and their environment, we must, at some point, take into account the evolution of the society in which our "learners" live. I don't intend to go deep into this direction but there are two points I'd like to raise:

1. As the means of communication evolve (television, Internet, on line games, etc.) everybody seems to agree that adolescents today have developed other skills and generally have a wider general culture than pupils had a few decades ago. Yet I am not sure the curricula and the ways of teaching have followed this (r)evolution. Besides, between TV, the web and consoles, adolescents tend to be even more passive than they used to be. Hardly anybody forces them into action. They are often viewed as passive consumers unable to make an effort, yet are they given the opportunities to do so at either home or school?

2. Most societies have evolved too and I am not sure parents can take on their role in the same way they used to. In impoverished suburbs we often find parents that cannot cope with what teachers ask their children to do at home. They are often not able to help their children with their homework and cannot afford private lessons. Similarly, in middle-class or upper class society, parents are often tired, stressed and absent because of their jobs and they cannot always be there to help their children out with school. To this list we could also add one-parent families and various other situations...

I sometimes have the feeling that instead of incriminating parents and their way of educating their children, the education community should react and see how it could fill in some of the gaps that today's society has created. I am not saying that school should play the parents' role, but I believe there is something we can do to avoid widening the gap between the children whom can be helped and those who cannot. I especially want to believe that school is the place where the same opportunities are given to everybody.

Bearing in mind the idea of teaching through the realisation of projects, I believe it is also a way of making up for what children can't have at home anymore. School is – to me – the obvious place for helping "moulding" adolescents, "shaping" their brains and developing their curiosity. It should also be the place where they are given the opportunity to put into action the knowledge acquired in and out of school; most of the time they do not even think of doing it. This is directly linked to what is developed in the article in Section B.4.1.

As a consequence of all this, it seems obvious that "changes would have to be made to the organisation of learning in adolescence". The "changes" mentioned in Karen Evans' and Christian Gerlach's article are sufficiently explicit and accurate that they need not be developed further, but they definitely propose that new missions should probably be defined for school and teachers. They should be allowed (or asked) to widen their range of action and be given the means and the time to individualise teaching more often. By working more on motivation, and less on overloaded curricula, they could help develop emotional and cognitive processes in adolescents' brains.

There is one point, however, that doesn't come up in the article and which goes along with teaching: evaluation and grading. If grading can be encouraging and motivating it can also be discouraging and inhibitive. In some school systems, children get marks/grades as early as the age of 4. There is sometimes pressure on children very early in childhood and for those of them who face difficulties at an early stage, one wonders how they manage to develop anything positive and constructive in their learning process.

This is another change we have to go through. I believe we have to re-think evaluation and, again, find a way of evaluating the evolution of competencies as well as of formal knowledge. In a way, the European Language Portfolio developed by the Council of Europe seems interesting, as it allows people to reach different levels of competencies without judging within that competency what is right or wrong. For example, it will attest that you are able to ask simple questions in a foreign language, but it won't say whether you make small grammar mistakes or not, if you have the right accent or not, etc.

Why not think of establishing such an evaluation grid around the projects a student might choose to work on? It would come along with the "traditional" school report and would probably give the opportunity to some students (particularly those who suffer from dyslexia, dyscalculia, ADHD, etc.) to achieve much better results than those they achieve in pure academic testing.

Now I am back to reality, and I feel there's a chance that some doors will soon open to a different way of teaching. However, there is still a lot to do and a long way to go before we can "acknowledge individual variation, encourage mastery and challenge, promote self-monitoring and responsibility when the adolescent is ready for it". I am also aware that all these considerations apply to the "average adolescent", but that they do not take into account those we need to work with on site: psychologists, doctors or social workers.

Sandrine Kelner

References

Alexander, P.A. (2003), "The Development of Expertise: The Journey from Acclimation to Proficiency", *Educational Researcher*, Vol. 32(8), pp. 10-14.

Altenmuller, E.O., W. Gruhn and D. Parlitz *et al.* (1997), "Music Learning Produces Changes in Brain Activation Patterns: A Longitudinal DC-EEG-study Unit", *International Journal of Arts Medicine*, Vol. 5, pp. 28-34.

Baethge, M. (1989), "Individualization as Hope and Disaster", in K. Hurrelmann and U. Engel (eds.), *The Social World of Adolescents*, de Gruyter, Berlin.

Bandura, A. (1977), "Self-efficacy: Toward a Unifying Theory of Behavioural Change", *Psychological Review*, Vol. 84, pp. 191-215.

Bandura, A. (1989), "Regulation of Cognitive Processes through Perceived Self-efficacy", *Developmental Psychology*, Vol. 25, pp. 729-735.

Bandura, A. (1997), *Self-efficacy: The Exercise of Control*, Free Press, New York.

Bechara, A., H. Damasio, D. Tranel and A.R. Damasio (1997), "Deciding Advantageously before Knowing the Advantageous Strategy", *Science*, Vol. 275, pp. 1293-1295.

Blakemore, S.J. and U. Frith (2000), *The Implications of Recent Developments in Neuroscience for Research on Teaching and Learning*, Institute of Cognitive Neuroscience, London.

Bruer, J. (1993), *School for Thought*, MIT Press.

Casey, B.J., J.N. Giedd and K.M. Thomas (2000), "Structural and Functional Brain Development and its Relation to Cognitive Development", *Biological Psychology*, Vol. 54, pp. 241-257.

Casey, B.J., N. Tottenham, C. Liston and S. Durston (2005), "Imaging the Developing Brain: What Have we Learned about Cognitive Development", *Trends in Cognitive Sciences*, Vol. 9, pp. 104-110.

Chugani, H.T. (1998), "A Critical Period of Brain Development: Studies of Cerebral Glucose Utilization with PET", *Preventive Medicine*, Vol. 27, pp. 184-188.

Chugani, H.T. and M.E. Phelps (1986), "Maturational Changes in Cerebral Function in Infants Determined by 18FDG Positron Emission Tomography", *Science*, Vol. 231, pp. 840-843.

Chugani, H.T., M.E. Phelps and J.C. Mazziotta (1987), "Positron Emission Tomography Study of Human Brain Functional Development", *Annals of Neurology*, Vol. 22, pp. 487-497.

Cockram, L. and H. Beloff (1978), "Rehearsing to Be Adult: Personal Development and Needs of Adolescents", National Youth Agency.

Coffield, F. (2004), *Learning Styles and Pedagogy in Post-16 Education*, Learning and Skills Development Agency, London.

Cole, M. and S.R. Cole (2001), *The Development of Children* (4th edition), Worth Publishers, New York.

Coleman, J.C. (1970), "The Study of Adolescent Development Using a Sentence Completion Method", *British Journal of Educational Psychology*, Vol. 40, pp. 27-34.

Coleman, J.S. (1961), *The Adolescent Society*, Free Press, New York.

Csikszentmihalyi, M. (1996), *Creativity: Flow and the Psychology of Discovery and Invention*, Harper Collins, New York.

Damasio, A.R. (1994), *Descartes' Error: Emotion, Reason, and the Human Brain*, G.P. Putnam, New York.

Davies, I.K. (1971), *The Management of Learning*, McGraw-Hill, London.

Devlin, J.T., R.P. Russell, M.H. Davis, C.J. Price, J. Wilson, H.E. Moss, P.M. Matthews and L.K. Tyler (2000), "Susceptibility-induced Loss of Signal: Comparing PET and fMRI on a Semantic Task", *NeuroImage*, Vol. 11, pp. 589-600.

Draganski, B., C. Gaser, V. Busch, G. Schuierer, U. Bogdahn and A. May (2004), "Changes in Grey Matter Induced by Training: Newly Honed Juggling Skills Show up as a Transient Feature on a Brain-imaging Scan", *Nature*, Vol. 472, pp. 111-112.

Durston, S., H.E. Pol, B.J. Casey, J.N. Giedd, J.K. Buitelaar and H. van Engeland (2001), "Anatomical MRI of the Developing Human Brain: What Have we Learned?", *Journal of the American Academy of Child and Adolescent Psychiatry*, Vol. 40, pp. 1012-1020.

Dweck, C.S. and E.L. Leggett (1988), "A Social Cognitive Approach to Motivation and Personality", *Psychological Review*, Vol. 95(2), pp. 256-373.

Elliott, E.S. and C.S. Dweck (1988), "Goals: An Approach to Motivation and Achievement", *Journal of Personality and Social Psychology*, Vol. 54, pp. 5-12.

Entwistle, N. (1984), "Contrasting Perspectives on Learning", in F. Marton, D. Hounsell and N. Entwistle (eds.), *The Experience of Learning*, Scottish Academic Press, Edinburgh, pp. 1-18.

Erikson, E.H. (1968), *Identity, Youth and Crisis*, Norton, New York.

Evans, K. and A. Furlong (1998), "Metaphors of Youth Transitions: Niches, Pathways, Trajectories or Navigations", in J. Bynner, L. Chisholm and A. Furlong (eds), *Youth, Citizenship and Social Change in a European Context*, Avebury, Aldershot.

Evans, K., N. Kersh and S. Kontiainen (2004), "Recognition of Tacit Skills: Sustaining Learning Outcomes in Adult Learning and Work Re-entry", *International Journal of Training and Development*, Vol. 8, No. 1, pp. 54-72.

Evans, K. and W. Heinz (1994), *Becoming Adults in England and Germany*, Anglo-German Foundation for the Study of Industrial Society, London.

Flechsig, P. (1901), "Developmental (myelogenetic) Localisation of the Cerebral Cortex in the Human Subject", Lancet, Oct. 19, pp. 1027-1029.

Fuster, J.M. (2002), "Frontal Lobe and Cognitive Development", *Journal of Neurocytology*, Vol. 31, pp. 373-385.

Gardner, H. (1984), *Multiple Intelligences*, Basic Books, New York.

Giedd, J.N. (2004), "Structural Magnetic Resonance Imaging of the Adolecent Brain", *Annals of the New York Academy of Sciences*, Vol. 1021, pp. 77-85.

Giedd, J. N., J. Blumenthal, N.O. Jeffries, F.X. Castellanos, H. Liu and A. Zijdenbos et al. (1999), "Brain Development during Childhood and Adolescence: A Longitudinal MRI Study", *Nature Neuroscience*, Vol. 2, pp. 861-863.

Hallam, S. (2005), *Learning, Motivation and the Lifespan*, Bedford Way Publications, Institute of Education, London.

Havighurst, R.J. (1953), *Human Development and Education*, Longman's, New York.

Hodkinson, P. and M. Bloomer (2002), "Learning Careers: Conceptualising Lifelong Work-based Learning", in K. Evans, P. Hodkinson and L. Unwin (eds.), *Working to Learn: Transforming Learning in the Workplace*, Routledge, London.

Illeris, K. (2004), *Adult Education and Adult Learning*, Roskilde University Press, Roskilde.

Johnson, M.H. and Y. Munakata (2005), "Processes of Change in Brain and Cognitive Development", *Trends in Cognitive Sciences*, Vol. 9, pp. 152-158.

Keating, D.P. and B.L. Bobbitt (1978), "Individual and Developmental Differences in Cognitive Processing Components of Mental Ability", *Child Development*, Vol. 49, pp. 155-167.

Kohlberg, L. and C. Gilligan (1971), *The Adolescent as Philosopher*, Daedalus, Vol. 100, pp. 1051-1086.

Krawczyk, D.C. (2002), "Contributions of the Prefrontal Cortex to the Neural Basis Human Decision Making", *Neuroscience and Biobehavioral Reviews*, Vol. 26, pp. 631-664.

Mueller, C.M. and C.S. Dweck (1998), "Intelligence Praise can Undermine Motivation and Performance", *Journal of Personality and Social Psychology*, Vol. 75, pp. 33-52.

Paus, T. (2005), "Mapping Brain Maturation and Cognitive Development during Adolescence", *Trends in Cognitive Sciences*, Vol. 9, pp. 60-68.

Paus, T., D.L. Collins, A.C. Evans, G. Leonard, B. Pike and A. Zijdenbos (2001), "Maturation of White Matter in the Human Brain: A Review of Magnetic Resonance Studies", *Brain Research Bulletin*, Vol. 54, pp. 255-266.

Piaget, J. (1967), *Six Psychological Studies*, University of London Press, London.

Pinker, S. (2002), *The Blank Slate: The Modern Denial of Human Nature*, Viking, New York.

Premack, D. and A. Premack (2003), *Original Intelligence: Unlocking the Mystery of Who We Are*, McGraw-Hill, New York.

Rauschecker, J.P. and P. Marler (1987), "What Signals are Responsible for Synaptic Changes in Visual Cortical Plasticity?", in J.P. Rauschecker and P. Marler (eds.), *Imprinting and Cortical Plasticity*, Wiley, New York, pp. 193-200.

Resnick, L.B. (1987), "Learning in School and Out", *Educational Researcher*, Vol. 16, pp. 13-20.

Simos, P.G. and D.L. Molfese (1997), "Electrophysiological Responses from a Temporal Order Continuum in the Newborn Infant", *Neuropsychologia*, Vol. 35, pp. 89-98.

Sowell, E.R., B.S. Peterson, P.M. Thompson, S.E. Welcome, A.L. Henkenius and A.W. Toga (2003), "Mapping Cortical Change across the Human Life Span", *Nature Neuroscience*, Vol. 6, pp. 309-315.

Spear, L.P. (2000), "The Adolescent Brain and Age-related Behavioral Manifestations", *Neuroscience and Biobehavioral Reviews*, Vol. 24, pp. 417-463.

Stevens, B. and R.D. Fields (2000), "Response of Schwann Cells to Action Potentials in Development", *Science*, Vol. 287, pp. 2267-2271.

Sylvester, R. (1995), *A Celebration of Neurons: An Educators' Guide to the Human Brain*, Association for Supervision and Curriculum Development (ASCD), Alexandria.

Twenty-first Century Learning Initiative (2005 draft), "Adolescence: A Critical Evolutionary Adaptation, 21st Century learning Initiative", Bath.

Tyler, L.K., W. Marslen-Wilson and E.A. Stamatakis (2005), "Dissociating Neuro-cognitive Component Processes: Voxel-based Correlational Methodology", *Neuropsychologia*, Vol. 43, pp. 771-778.

Wall, W.D. (1968), *Adolescents in School and Society*, National Foundation for Education Research, Slough.

White, T., N.C. Andreasen and P. Nopoulos (2002), "Brain Volumes and Surface Morphology in Monozygotic Twins", *Cerebral Cortex*, Vol. 12, pp. 486-493.

ISBN 978-92-64-02912-5
Understanding the Brain: The Birth of a Learning Science
© OECD 2007

PART II

Article C

Brain, Cognition and Learning in Adulthood

by

Raja Parasuraman, George Mason University, Fairfax, VA, United States
Rudolf Tippelt, Ludwig-Maximilian-University, München, Germany
Liet Hellwig, TEFL Teacher and Teacher Trainer, Vancouver, Canada

C.1. Introduction

Virtually all societies within the developed world, and an increasing number of the developing countries, are witnessing unprecedented growth in the population of older men and women, and particularly women (Keyfitz, 1990; OECD, 2005). Rapid societal change is also increasingly requiring older adults to acquire and use complex information with new technologies, not just in the workplace but in many aspects of home and everyday life. These requirements can pose considerable challenges to older adults faced with declining sensory, perceptual, and cognitive abilities as they age. Consequently, there are compelling reasons for understanding the effects of aging on adult learning, both from psychological and educational perspectives and from the point of view of the underlying brain mechanisms that support cognition and learning.

Learning and aging cannot be separated from biogenetic, medical, psychological, social, and pedagogic problems. Consequently, understanding brain, cognition and learning in adulthood requires an interdisciplinary approach (Bransford, Brown and Cocking, 2004). On the one side, brain research attempts to understand the operations and functions underlying cognitive, affective, and social behaviour. Brain research and neuroscience therefore provide insights into the development and also the constraints of the learning brain. On the other side, educational and aging research take into consideration that the cultural, economic, and political achievements of modern societies are dependent on the problem-solving abilities of their citizens over their entire life cycle. These abilities are acquired in different institutions of the education system as well as in self-organised, formal and informal learning processes.

C.1.1. What is learning?

It is difficult to provide a consistent definition of the concept "learning" as the term is used in various ways in different disciplines. From a psychological perspective, learning can be defined as a change in the efficiency or use of basic cognitive processes, both conscious and unconscious, that promote more effective problem solving and performance in the tasks of everyday life. In this view, learning and thinking are connected in such a way that cognition is a necessary but not sufficient pre-condition for learning. From an educational point of view "learning" also has to be regarded in its relation to action in the world. Learning therefore not only concerns an expansion of knowledge, but also a change in action patterns.

One has also to distinguish between formal, non-formal, and informal learning processes, especially in adulthood. Formal learning is linked to educational institutions whereas non-formal learning takes place in clubs, associations, or at work. Informal learning, however, concerns the educational processes, in many cases unintended, that occur outside pre-defined learning settings (Tippelt, 2004). Finally, learning can be considered to be a lifelong process. However, partly due to the complexity of the prevailing stages of life and phases of learning, empirical research on childhood, youth, and adult aging still largely co-exist as independent disciplines and rarely merge into a uniform perspective of "lifespan research" (Weinert and Mandl, 1997).

C.1.2. *The human brain in adulthood*

Given that both the psychological and educational perspectives view learning as reflecting, at least in part, the development and expression of different cognitive processes, one can ask what neural mechanisms are involved. From a neurobiological perspective, learning can be viewed as a change in the strength and efficiency of neuronal connections that support cognitive processes (see Spitzer, 2002). It is widely accepted that although the human brain contains all its major structures when a child is born, significant changes in neuron number, connectivity, and functional efficiency continue after birth and throughout infancy. Moreover, the maturation and structural development of the brain continues past childhood and adolescence and well into the early stages of young adulthood, that is, up to about the mid twenties.

If we consider adulthood as the period from the twenties to the eighties, structural and functional brain changes continue to be seen in this age range, though they tend to be less marked and somewhat subtle, except in the case of disorders of the elderly such as dementia, in which case striking changes in brain structure and function are observed. For many years it was believed that because the human brain achieves approximately 90% of its adult size by the age of 6 years, that no further changes take place in adulthood. We now know, however, that the brain undergoes significant changes throughout life. Modern neuroimaging techniques now allow for precise quantification of these alterations. Moreover, these changes reflect not only genetically initiated mechanisms but also represent the brain's response to environmental and lifestyle factors.

Another neural "dogma" that has been overturned in recent years is that no new neurons are found in adulthood. However, the growth of new neurons – or *neurogenesis* – has now been conclusively established to occur in the hippocampus (Eriksson *et al.*, 1998) and perhaps in other brain regions as well. The significance of this finding has been further bolstered by studies showing that adult neurogenesis in the hippocampus is involved in the formation of new memories (Shors *et al.*, 2001). Thus, learning in adulthood is at least partially mediated by structural brain changes, including the formation of new neurons. The challenge is to understand the precise mechanisms by which such changes are associated with age-related variation in cognition and learning in adulthood.

C.1.3. *Overview of paper*

In this paper we focus on neural changes that occur during adulthood, a period that roughly spans the age range of 20 to 80 years. Brain research and educational research are both empirically-based sciences that attempt to provide practitioners and policy makers guidance for their decisions on the basis of validated empirical findings. Today, opinions and ideologies for important policy decisions are not enough, but can we offer empirical knowledge for evidence-based learning and education? Moreover can neuroscientists and educational researchers offer guidance that that does not fall into the trap of naïve "neurologising" – suggesting implications for educational practices on ill-formed and popular accounts of brain function? (Bruer, 1997). We believe so. Accordingly, this paper frames the question from the two perspectives of neuroscience and educational research, consistent with the interdisciplinary, co-operative approach that is necessary for a better understanding of learning (OECD, 2002).

We examine how brain changes relate to the significant behavioral and cognitive changes that also occur during adulthood. The ultimate objective is to consider what implications these age-related changes have for learning, teaching, and education, and for optimal functioning in older adults.

C.2. Adult age-related changes in cognition and learning

C.2.1. Cognitive aging

A large body of research on age-related changes in cognition and learning has been conducted by investigators in the field of cognitive aging (Baltes, 1993; Salthouse, 1996). Most of this work has used cross-sectional designs in which different cohorts of young and older adults are compared. Some studies, however, have used the more powerful longitudinal method in which a cohort of individuals is followed over a period of time as they age and tested on various cognitive tasks. Both cross-sectional and longitudinal studies have led to the conclusion that some perceptual and cognitive functions decline in efficiency with adult aging, whereas others remain stable or become more developed and efficient (Baltes, 1993; Park and Schwarz, 1999). The pattern of age-related change can be summarised very generally as follows. Whereas older adults are generally slower and have poorer memory than the young, they typically display superior general and verbal knowledge, creative problem solving, and what may be termed "wisdom" (Baltes and Staudinger, 2000; Sternberg, 1990). In terms of age-related losses, older adults are slower (Salthouse, 1996; Schaie 2005) and have reduced working memory capacity compared to younger adults (Dobbs and Rule, 1989). There is also convincing evidence of linear declines with age in both peripheral (*e.g.* retinal acuity) and central (*e.g.* motion sensitivity) sensory functioning in both vision and audition (Lindenberger, Scherer and Baltes, 2001).

Another way to characterise age-related changes in cognition is in terms of the distinction between fluid and crystallised intelligence (Cattell, 1963). Fluid intelligence shows a continuous decline from about the fourth decade of life onwards. Crystallised intelligence, on the other hand, remains stable or improves. The superior verbal ability and world knowledge of older adults compared to the young may also allow them to compensate for deficits in processing speed and working memory (Kruse and Rudinger, 1997).

A comprehensive study that examined multiple domains of perception and cognition is illustrative of the kinds of age-related changes that are typically seen in cognitive aging research studies. Park *et al.* (2002) administered several tasks – including ones that tapped processing speed, working memory, long-term memory, and vocabulary – to a sample of about 300 adults aged 20 to 90 years. The use of a relatively large sample enabled normalised (z) performance scores to be computed for each age decade in this age range. Park *et al.* (2002) found roughly linear declines in performance across age decades for processing, working memory, and long-term memory (both free and cued recall), whereas vocabulary (WAIS and Shipley tests) remained stable through middle age and increased from the sixties to the eighties.

A similar pattern of nearly linear age-related declines in processing speed, working memory, and long-term memory was reported by Baltes and Lindenberger (1997) in the Berlin Aging Study from 25 to 103 years. A notable feature of this study is that the age-related declines were shown not to differ significantly as a function of education, social class or income. On the other hand, age-related decline in basic sensory processes – visual and auditory acuity – was a powerful determinant of perceptual and cognitive decline. The authors concluded that as a basic measure of neural integrity uncontaminated by environmental or social factors, sensory functioning provided a fundamental biological mediator of age-related cognitive decline.

Adult aging may also influence the use of task strategies or processing priorities in the performance of cognitive tasks. This may be particularly true when older adults are placed in unfamiliar settings or those requiring them to exhibit new learning. Older adults appear to assign processing priorities differently than the young, perhaps due to the need to re-allocate diminished processing resources such as working memory capacity. This can be demonstrated even for apparently well-learned or "automatic" skills such as walking. For example, Li *et al.* (2001) assessed age effects in a dual-task paradigm in which individuals were trained separately on a memorising task and a walking task and subsequently tested on both tasks together. They found that older adults were quite skilled in walking during the dual walking/memorising task, but at a cost to their memorising performance compared to when memorising alone. The results suggest that the older adults placed greater priority on doing well on the walking task, to the detriment of the memory task, whereas young adults did not need to explicitly adopt a specific strategy and could do well on both tasks. Compared to the young, older adults are also more likely to depend on "top-down" information when it is available when asked to perform complex attention tasks. For example, searching for targets among distractors in a cluttered visual scene is dependent both on "bottom-up" factors – the saliency of the target and its similarity to distractors – as well as top-down factors such as knowledge of target location or form. Greenwood and Parasuraman (1994, 1999) had young and older participants search a large visual array for a particular target specified by a conjunction of form and color. Prior to the presentation of the search array, participants were given prior knowledge of the likely spatial location of the target by cues that varied in the precision of localisation. These cues had a greater effect in speeding up the time to find the target in the older adults than in the young. The use of different processing priorities may also be interpreted as protective strategy given the simultaneous decline in older adults in processing capacity and increased knowledge (Park *et al.*, 2002). For example, Hedden, Lautenschlager and Park (2005) reported that older adults relied more on their (superior) verbal knowledge in a paired-associates memory task than did younger adults, who tended to rely more on speed of processing and working memory capacity.

In sum, there is substantial evidence for age-related decline in sensory functions, speed of processing, working memory, and long-term memory. At the same time, vocabulary, semantic and world knowledge, and wisdom tend to improve with age. These are key cognitive functions that contribute individually and collectively to the success of learning. Aging also seems to be accompanied by changes in task priorities and strategies, so that older adults appear to be more dependent on top-down factors and so can be penalised when such top-down support is not available. Each of these changes in cognitive functioning can negatively impact on new learning in older adults. At the same time, however, older adults may be able to compensate for their decline in basic cognitive functioning by bringing their considerably greater verbal and world knowledge to bear upon the solution of a given problem. Exploiting the gains of adult aging to offset the losses would appear to be a possible strategy for educational and training interventions aimed at enhancing learning in older adults.

C.2.2. Lifelong learning: a perspective from adult education

The lifespan developmental perspective provides a complementary approach to the cognitive aging framework described in the previous section. This view acknowledges the fundamental contribution of Erik Erikson (1966) that ontogenetic development is a lifelong process. No specific age has a monopoly on human development and learning is continual and cumulative. At the same time, surprising and discontinual learning processes can occur throughout life. Whereas Erikson was mainly concerned with how individuals cope with crises throughout life, modern educational research concerning learning, development, and education emphasises the multi-directionality of ontogenetic changes. As an example, recent studies have established the interplay of autonomy and dependence, so that a dynamic between growth, maintenance, and the regulation of loss can be observed amongst the elderly in the course of life (Baltes, 1993; Lehr, 1991). Therefore learning and developing do not always signify an increase in capacity or an increase in a sense of higher efficiency. Whereas in the first stage of one's life the main aim of learning is to reach a high amount of autonomy and empathy, it becomes increasingly important as one grows older to handle the loss of physical independence and to harness social support networks creatively and productively.

Prevention of dependency is one of the most important learning goals of higher adulthood, because the maintenance of competence ensures that the ability to perform and personal growth act together (Alterskommission, 2005). The course of development is strongly dependent on social background and the individual living situation. Sociologists, on the other hand, emphasise the historical learning context, i.e., that lifelong learning is not solely constrained by biological and cognitive pre-conditions but also by social and cultural conditions associated with historical epochs. In this view, age-related development is associated with collective cohort experiences related to economic crisis or welfare, cultural values and basic political experiences. For example, in modern societies adults have a higher average level of education and a greater familiarity with educational opportunities than earlier generations. Increased life expectancy and technical innovation contain the risk of systems of knowledge quickly becoming outdated. The focus of educational processes only on early ages in life is not sufficient anymore; moreover, from the perspective of individuals and society, work must be linked to lifelong learning. On the one hand, such learning focuses socially on the improvement of economic ability of being competitive and the support of individual ability of occupation and also the strengthening of social cohesion in modern plural and individualised societies. On the other hand, lifelong learning, especially individually, aims at the independent unfolding of personality and the maintenance of independence in adulthood. Such perspectives must be kept in mind when considering the relationship of brain and learning in adulthood.

In research concerning learning and education as well as in gerontology one distinguishes between normal, optimal, and pathological forms of growing older (Thomae, 1970; Kruse, 1997; Lehr, 1991). The aging processes of individuals are generally very different to one another. According to this, learning strategies and approaches of research also have to differentiate and take into account individual factors. We consider individual differences from cognitive, genetic, and neural perspectives in subsequent sections of this paper.

The normal aging process and the optimal forms of learning in adulthood focus on the improvement of pedagogical, medical, psychological and social basic conditions and service offers. The approach assumes that early learning commitments in the family or at school have a positive effect on the further personal development and the active learning in the adulthood of the individual (see Feinstein et al., 2003). This approach analyses the aging process with the help of an optimistic competence model, replacing the more limited deficit model which simply equates aging with cognitive, psychological and social losses (see Kruse and Rudinger, 1997).

Independent of the deficit model, a second path of research has to analyse the given pathological concomitants of aging. This concerns the different dementias and the various diseases acting in combination which often occur in advanced adulthood. The prevention, starting in youth or in early adulthood, contains the chance of a long life with good health, independence and share of responsibility. There is a vision of reducing illnesses and mental impairment – at a further increase in life expectancy – to the last years of one's life (Baltes, 2003) in order for this compression to make a preferably long active and explorative aging possible. These issues raise important questions concerning brain research: What do we know about brain imaging studies in older adults? Do inter-individual differences affect brain functions and can cortical information processes be compensated? How are the different forms of cognitive and emotional processes represented in the brain? Finally, how can we exploit the growing literature on the cognitive neuroscience of aging to achieve optimal aging and enhance learning opportunities in adulthood? We address some of these questions in the following sections.

C.3. Aging and brain function: structural neuroimaging

The aging brain is associated with a variety of structural changes at multiple levels of neural organisation, from intra-cellular, to neuronal, to inter-cortical. At a gross level, *postmortem* studies have revealed that aging is accompanied by an approximately 2% decline in brain weight and volume per decade (Kemper, 1994). Computed tomography (CT) and magnetic resonance imaging (MRI) studies have confirmed that global brain volume shows a reduction that is systematically negatively correlated with age (Raz *et al.*, 2005). Of these two techniques, MRI has the greater sensitivity and is particularly useful for distinguishing between gray matter (neurons) and white matter (axons). With high-resolution MRI, one cannot only differentiate gray and white matter volume changes, but also quantify volume changes in specific cortical and subcortical structures of the brain, as well as in the ventricles.

MRI data indicate that aging is accompanied by reduced gray matter volume (Resnick *et al.*, 2003; Sowell *et al.*, 2003). The reduction in volume can be observed as early as the thirties (Courchesne *et al.*, 2000) but is typically most reliably detected in older adults aged 50 and over. As can be seen from a recent study in which a wide age range of 15-90 years was assessed (Walhovd *et al.*, 2005), cortical gray matter loss shows a steady decline throughout life, but the decrease is more evident after middle age. However, it should be noted that the age-related reduction in gray matter volume has not been shown to reflect a reduction in neuronal numbers. (Thus the old adage that one loses brain cells as one ages is not necessarily true.) While some evidence of neuronal attrition has been reported (Kemper, 1994), this remains controversial, and others have suggested that the gray matter volume loss may reflect neuronal shrinkage rather than a loss of neurons.

In addition to gray matter changes, aging is also associated with white matter alterations, although at present the evidence is mixed with respect to the extent of such changes. Some studies have reported no age-related changes in total white matter volume (Good *et al.*, 2001) while others have found that white matter volume (Guttman *et al.*, 1998; Jernigan *et al.*, 2001) is reduced with age. Walhovd *et al.* (2005) did observe an overall reduction, although the pattern of deline was not consistent. White matter abnormalities – "hyperintensities" that may represent either local axonal or vascular degradation – have also been found in older adults (Guttmann *et al.*, 1998).

C.4. Aging and brain function: functional neuroimaging

In addition to structural changes, several positron emission tomography (PET) and functional MRI (fMRI) studies in young and older adults have revealed age-related differences in regional brain activation patterns during the performance of perceptual and cognitive tasks. In a PET study, Grady *et al.* (1994) showed that compared to the young, older adults showed decreased activation of occipital cortex in a face-matching task. In general, older adults have been found to show reduced activation of modality-specific cortical regions devoted to primary perceptual processing, *e.g.* of occipital and temporal cortices during visual detection and recognition tasks.

At the same time that aging appears to be associated with a reduction in activation of perceptual-specific cortical processing regions, evidence of activation of other brain regions not seen in young adults has also been reported. In particular, several studies have shown that older adults show increased activation of prefrontal cortex (PFC), including bilateral activation, in tasks of lexical decision (Madden *et al.*, 1996), visual search (Madden *et al.*, 2004), and problem solving (Rypma and D'Esposito, 2000).

Several explanations have been offered for these age-related differences in activation patterns. One theory is that the additional activation, particularly of PFC in older adults, may compensate for a neural system with declining processing capacity (Park *et al.*, 2002; Rosen *et al.*, 2002). For example, Gutchess *et al.* (2005) found that older adults had greater activation of medial PFC than the young while encoding pictures in a memory task, whereas the young adults showed more hippocampal activation. This suggests that older adults compensated for hippocampal processing deficits by recruiting additional resources from the frontal areas, a neural theory that is similar to the cognitive view discussed earlier of older adults using different processing strategies, including their superior verbal knowledge, to compensate for declining processing capacity.

Another view is that the findings reflect reduced cerebral lateralisation in old adults, with old people showing bilateral activation during tasks which in young are left-hemisphere lateralised (*e.g.* episodic encoding) or right-hemisphere lateralised (*e.g.* visual attention) (Cabeza, 2002). Also consistent with the idea that old people require bilateral processing for the successful completion of tasks that are unilaterally processed in young people is evidence from transcranial magnetic stimulation (TMS). Applying TMS only to the right dorsolateral PFC in young people interfered with retrieval while in old people applying TMS to either hemisphere interfered (Rossi *et al.*, 2004).

The results of these neuroimaging studies suggest that older adults may compensate for decline in processing capacity by recruiting different and/or additional neural areas, particular PFC. If this view is correct, it would suggest a flexibility indicating not only that plasticity is also a feature of the older adult brain, but that it continues into late adulthood. However, not all neuroimaging results can be easily accommodated by the "compensation" hypothesis of additional recruitment of cortical regions by older adults. For example, compared to the young, old people showed weaker hippocampal activation in a range of tasks, but stronger parahippocampal activation during episodic retrieval (Grady, McIntosh and Craik, 2003). Furthermore, Colcombe *et al.* (2005) recently reported that while some older adults did show additional PFC activation in an inhibitory-control (flanker) task compared to the task, this pattern was seen only in those who scored in the bottom half of performance on the flanker task. Those who had good performance scores showed the same pattern of PFC activation as did the young group.

In summary, functional neuroimaging studies have consistently revealed age-related differences in regional brain activation during the performance of perceptual and cognitive tasks. These differences are especially marked in the PFC, which is known to be important for higher-order "executive" functions of the brain and which may be particularly sensitive to aging. However, at present there is not a consensus on the theoretical significance of age-related changes in brain activation. The recent study by Colcombe et al. (2005) offers the possibility that a synthesis may emerge, given that these authors showed that "abnormal" activation patterns in older adults could be linked to lowered task performance, whereas good-performing individuals showed a "normal" pattern. Given that task performance in older adults can be improved by training and other interventions, one possibility is that the degree of "normalisation" of brain activation patterns can be used as a marker for evaluating the success of training interventions in the older population.

C.5. Individual differences in age-related brain and cognitive changes

In the previous sections we have reviewed evidence indicating that as a whole, older adults show a variety of changes in different aspects of cognitive functioning. Aging is also accompanied by both global and regional changes in both gray and white matter volume. It is possible that these changes are correlated, so that the changes in brain structure are causally linked to age-related changes in cognition. However, any such linkage must be able to account for individual differences in the age-related changes that have been observed. The extent of age-related cognitive decline varies substantially between individuals. Some persons exhibit a precipitous decline in cognitive efficiency as they age, whereas others show only modest losses, and a few maintain cognitive functioning at a near-constant level throughout their life. Furthermore, when higher cognitive abilities are assessed, a substantial fraction of older individuals show only modest losses or maintain functioning with age (Wilson et al., 2002).

Given that older adults exhibit deficiencies in processing speed, working memory, and executive functioning compared to younger adults, and that adults are accompanied by structural brain changes and alterations in the regional pattern of brain activation during cognitive task performance, one asks whether the cognitive and brain changes are associated. There is certainly evidence of parallel patterns in cognitive changes with age and the possible neural mediators of these cognitive changes (Cabeza, Nyberg and Park, 2005). Both gray and white matter volume shrink with age (Bartzokis et al., 2003; Resnick et al., 2003). However, as with cognitive decline, there is considerable individual variation (Raz et al., 2005).

What factors underlie these individual differences in brain and cognitive integrity in aging? It is tempting to attribute both normal and age-related cognitive variation to brain volume changes, and several studies have shown an association between prefrontal cortical volume loss and reductions in cognitive functions tapping "executive" and inhibitory processes (for a review, see Raz et al., 2005). With the advent of new neuroimaging techniques such as diffusion tensor imaging for assessing axonal integrity in the living human brain, associations have also been reported between cognitive changes and white matter volume (Bartzokis et al., 2003). Such reported associations are significant because of the general finding that higher-order executive functions are strongly sensitive to age-related change, and that such functions are critical to learning and the expression of fluid intelligence (Cabeza, Nyberg and Park, 2005). However, associations between cortical or white matter volume and age-related cognitive changes are not consistently strong. For example, a recent meta-analysis of hippocampal volume and memory functioning in middle-aged and older adults found the relationship to be weak (van Petten, 2004).

The variable pattern of brain volume/cognition associations in older adults suggests that additional moderating variables need to be studied. Environmental factors include opportunities for new leaning and social interaction, training, exercise, mental stimulation, etc. The effects of some of these environmental modifiers on brain and cognitive aging have been identified (e.g. Raz et al., 2005).

In addition to environmental factors, genes also play a major role. Twin studies have shown that genetic factors contribute substantially to normal variation in general cognitive ability, or g (Plomin, DeFries, McClearn and McGuffin, 2001). The heritability of g increases over the lifespan, reaching .62 in those aged over 80 years (McClearn et al., 1997). Both high g (Schmand et al., 1997; Whalley et al., 2000) and high cognitive functioning (Snowdon et al., 1996) early in life protect against Alzheimer Disease (AD) late in life. The high heritability of g compellingly suggests that genetics must also play a role in the variation between individuals in the extent of age-related cognitive changes.

C.6. Genetics and individual differences in cognition

Much of what we know about the genetics of cognition has come from twin studies in which identical and fraternal twins are compared to assess the heritability of a trait. This paradigm has been widely used in behavioral genetics research for over a century. For example, the method has been used to show that general intelligence, or g, is highly heritable (Plomin and Crabbe, 2000). However, this approach cannot identify the *particular genes* involved in intelligence or the cognitive components of g. Recent advances in molecular genetics now allow a different, complementary approach to behavioral genetics, that of *allelic association*. This method has been recently applied to the study of individual differences in cognition in healthy individuals, revealing evidence of modulation of cognitive task performance by specific genes (Fan, Fossella, Sommer, Wu and Posner, 2003); Greenwood et al., 2000; Parasuraman, Greenwood and Sunderland, 2002; see Greenwood and Parasuraman, 2003, for a review).

In the allelic association method *normal* variations in candidate genes – those deemed likely to influence a given cognitive ability due to the functional role of each gene's protein product in the brain – are identified and examined for possible association with cognitive functions. More than 99% of individual DNA sequences in the human genome do not differ between individuals and hence are of limited interest in investigating individual differences in normal cognition. However, a small proportion of DNA base pairs (bp) occur in different forms or alleles. Allelic variation is due to slight differences in the chain of nucleic acids making up the gene – commonly the result of a substitution of one nucleotide for another – a single nucleotide polymorphism (SNP). Consequently, the protein whose production is directed by that gene is correspondingly altered (see Parasuraman and Greenwood, 2003).

If the neurotransmitter innervation of brain networks underlying a particular cognitive function is known, then in principle one can link SNPs that influence neurotransmitter function to cognitive function. For example, with respect to attention and working memory, a growing body of evidence from lesion, electrophysiological, neuroimaging, and pharmacological studies points to the role of cholinergically-mediated posterior brain networks in spatial attention and dopaminergically-rich PFC networks in working memory and executive control processes (Everitt and Robbins, 1997). Dopaminergic receptor genes are likely candidates for genetic effects on working memory and executive control processes due to the importance of dopaminergic innervation for these functions.

One candidate gene that may be linked to individual differences in working memory is the dopamine beta hydroxylase (DBH) gene, which is involved in converting dopamine to norepinephrine in adrenergic vesicles of neurons. A polymorphism in the DBH gene, a G to A substitution at 444, exon 2 (G444A) on chromosome 9q34, has been associated with familial cases of attention deficit hyperactivity disorder. There are three genotypes associated with this SNP, AA, AG, and GG. Parasuraman et al. (2005) examined the role of DBH in attention and working memory in a group of healthy adults aged 18-68 years. The working memory task they used involved maintaining a representation of up to three spatial locations (black dots) over a period of three seconds. At the end of the delay, a single red test dot appeared alone, either at the same location as one of the target dot(s) (match) or at a different location (non-match). Participants had 2 seconds to decide whether the test dot location matched one of the target dots. Matching accuracy decreased as the number of locations to be maintained in working memory increased, demonstrating the sensitivity of the task to variations in memory load. Accuracy was equivalent for all three DBH genotypes at the lowest memory load, but increased with higher "gene dose" of the G allele, particularly for the highest (3 target) load. At this highest load, memory accuracy for the GG allele (G gene dose = 2) was significantly greater than that for both the AG (G gene dose = 1) and AA alleles (G gene dose = 0), with an effect size of .25, which is a "moderate" size effect in Cohen's (1988) terminology. Parasuraman et al. (2005) also administered a visuospatial attention task with little or no working memory component to the same group of participants. Individual differences in performance of this task were not significantly related to allelic variation in the DBH gene. Furthermore, working memory accuracy at the highest memory load was not correlated with performance on the attention task. In sum, these findings point to a substantial association between the DBH gene and working memory performance.

Parasuraman et al. (2005) found that increasing gene dose of the G allele of the DBH was associated with better working memory performance. This effect was most apparent when the number of target locations to be retained was high. Thus, the association between the DBH gene and working memory was particularly marked under conditions that most taxed the working memory system. These results as well as others (see Parasuraman and Greenwood, 2003), indicate that molecular genetic analyses can be used to identify the genetic contribution to individual differences in cognition. Moreover, these findings have been extended to middle-aged and older adults, in whom the interactive effects of neurotransmitter genes such as DBH with neuronal repair genes like apolipoprotein E (APOE) have been demonstrated (Greenwood et al., 2005; Espeseth et al., 2007). The finding that individual differences in working memory can be associated with the DBH gene is interesting because working memory capacity has been linked to the efficiency of learning, decision making, problem solving, and many other complex cognitive tasks that draw upon executive functions of the brain. Individual differences in working memory capacity are well documented (Conway and Engle, 1996) and known to be highly reliable (Klein and Fiss, 1999). High working memory capacity is associated with better ability to filter out interference in the Stroop task (Kane and Engle, 2003). Thus older individuals with higher working memory capacity may be better able to adapt to new learning situations and show less need for task-induced top-down control in compensating for age-related decline in processing speed, as discussed previously. Furthermore, as discussed in a subsequent section, cognitive training methods could also be used to enhance working memory capacity in selected older adults.

C.7. Training and aging

C.7.1. Cognitive training

There is some evidence that "cognitive stimulation", either self-initiated or by family members, may play a role in maintaining cognitive function in older adults who might otherwise show significant age-related decline (Karp *et al.*, 2004). Wilson *et al.* (2002) asked older adults to rate their level of involvement in a variety of cognitively demanding activities such as listening to the radio, reading, playing games, and going to theatre. Individuals who self reported greater involvement in these activities performed better on a cognitive test battery than those reported less involvement. Importantly, longitudinal analysis indicated that such cognitive stimulation also led to a lower risk of developing Alzheimer disease in these individuals.

If certain lifestyle choices are successful in maintaining cognitive function in older adults, can related benefits be demonstrated through specific forms of cognitive training? Given the well-accepted finding that older adults exhibit declines in basic cognitive functions such as processing speed, working memory, and long-term memory, there has been considerable interest in the use of cognitive training programmes aimed at enhancing the efficiency of these functions. Typically, older adults are trained on a specific cognitive task in which they exhibit age-related decline and evidence is sought for potential improvement on related tasks.

C.7.2. Training – the developmental perspective

A developmental perspective on training indicates that an individual's life – the transition from school to work, finding a partner, parenthood, different career stages, dealing with crisis, and retirement – can be divided into successive "development tasks" (Lehr, 1986; Kruse, 1999). In this view, every age can be characterised by certain development tasks involving the interplay of: 1) biological-psychological maturation; 2) social expectations and demands of that age; and 3) individual interests and learning opportunities. Moreover, experiences in earlier stages of life and the current living situation are jointly responsible for the realisation of this developmental potential (see Tippelt, 2002).

Recent large-scale longitudinal data sets on 1 958 individuals aged in the thirties and fourties have pointed to the wider benefits of participating in any form of education (Schuller *et al.*, 2004). Adult literacy and numeracy had positive effects on health behaviour (smoking, drinking, level of exercise, body mass index), well-being (life satisfaction, depression, general health), and political involvement (political interest, voting, civic membership). Participation in education has even been found to foster racial tolerance, at least in men (see Bynner, Schuller and Feinstein, 2003). Higher education is clearly correlated with membership of voluntary organisations in later life, so graduates in their later middle age, *i.e.* are more likely to engage in the local community than others. Participation in learning, such as active attendance of courses in adult education had positive effects on health behaviour, social tolerance and active citizenship. On the other hand, participation in leisure courses had no general preventive effect on depression. The conclusion of these longitudinal data is clear: education is not so much an option for government but an absolute prerequisite for the promotion of personal well-being and a cohesive society (see Feinstein *et al.*, 2003).

These findings from studies of adults in their middle age also extend to older adults. Further education experience has indicated the importance to older adults of education and prior knowledge for many types of further learning (see Becker, Veelken and Wallraven, 2000). There is a current *zeitgeist* of "successful aging" which has, to some extent (but not completely), overcome the "deficit" model of prior decades. The view now is that competence and the ability to perform can be maintained until late adulthood. While learning processes do change as one grows older, learning capacity remains (see Schaie, 2005; Baltes and Staudinger, 2000). Thus, despite age-related decline in basic sensory and cognitive processing operations (as described in Section C.2) and in the functioning of the supporting brain structures (as described in Sections C.3 and C.4), knowledge acquired in earlier stages of life is well retrievable and can be put to use for new learning. Learning acquired at an early stage of life, whether created in formal (education) or informal settings (family, school, work and social environments), can be used to good effect to provide the pre-knowledge necessary for effective learning strategies in adulthood. The outcome of such early learning experiences continue to have effects up to late adulthood (Kruse, 1999).

In addition, self-awareness and identity are important components of adult development, particularly the experience and emotional context of remembrances (memories related to the own self). Such memories provide the frame for an integrated view of development including life span, brain development, learning and social environment, and genetic predisposition. Surprisingly, very little research on the emotional quality of memories has been conducted by educational research studies (see Welzer and Markowitsch, 2001, p. 212).

A person's need for developing not only concerns childhood and youth but extends over the entire course of life. Adults, however, are to a greater extent responsible for engaging in and the contents and forms of learning (see Tippelt, 2000). In psychological research on motivation this interest in learning appears in the concept of "flow" (Csikszentmihalyi, 1982) or the pleasure of "being lost in time" while being apparently effortlessly engaged in a challenging activity. This concept has also variously been termed by White (1959) as "the feeling of efficacy", by deCharms (1976) as the "feeling of one's own efficiency" and of "self-determination" and by Heckhausen (1989) as "the matching of action and goal of action". Moreover, positive convictions of self-effectiveness and "internal attribution" (locus of control) have a supportive effect of cognitive ability to perform amongst all learners (see Jennings and Darwin, 2003).

A special approach is the concept of wisdom, which is considered as an ideal endpoint of human development, although high levels of wisdom-related knowledge are rare. The period of late adolescence and early adulthood is the primary age window for wisdom-related knowledge to emerge. The foundation of wisdom lies in the orchestration of mind and virtue towards the personal and public good. The most powerful predictors of wisdom-related knowledge are not cognitive factors such as intelligence (*cf.* Sternberg, 1990). Specific life experiences (*e.g.* practising in a field concerned with complex life problems) and personally-related factors, such as openness to experience, creativity and a preference for comparing, evaluating and judging information are better predictors (see Baltes, Glück and Kunzmann, 2002; Baltes and Staudinger, 2000).

From a development-oriented perspective, the competences of older adults include numerous abilities, skills and interests that do more than allow the goal of maintaining one's independence. By competence one understands a person's ability to maintain or re-establish a dependent, task-related and meaningful life in an encouraging, supportive environment which promotes the active and conscious confrontation with tasks and strains (see Kruse, 1999, p. 584). The unfolding of competence therefore is always linked to positive characteristics of the social and institutional environment. Disabilities that may occur therefore require an unrestricted and technically supportive environment as well as the support by other people and organisations.

Competence can also be linked to *human capital* similar to other resources in an economy, *e.g.* competences of the over 50-year-old create a wealth of human capital in the working sector that the service and knowledge society is currently not making full use of. Older employees have the reputation of being less flexible and less stable with regard to their health, but not only persons responsible for staff in companies increasingly stress their important working experiences, mental stamina, operational loyalty and reliability, thorough ability to make decisions and act as well as their social and communicative competence (see Lahn, 2003; Karmel and Woods, 2004; Williamson, 1997; Wrenn and Maurer, 2004). It could be shown that health problems of older employees do not occur in general but increasingly in cases of inconvenient working places with no learning possibilities which do not allow unfolding (see Baethge and Baethge-Kinsky, 2004; Feinstein *et al.*, 2003). Education programmes for older people have preventative functions and serve the maintenance of cognitive abilities as well as the physical and mental health (see Lehr, 1991; Alterskommission, 2005). But we need more output (short time effects) and outcome (long term consequences) evaluations in longitudinal studies to discuss on a better empirical basis the effects of different trainings.

Good examples of such educational measures are the programmes "Coming back 45 plus" which especially appeal to women and which react to the very low employment numbers of over 50-year-olds in many countries (Eurostat, 2003). In the European Union, Sweden has the highest employment rate of over 50-year-olds at 75% and Belgium has the lowest one at 42%. If one assumes that the small number of over 50-year-old women and men in the employment sector is contradictory to the working potentials of older employees, the following elements in educational measures proved empirically make sense concerning re-entering work – if one makes use of the special competences of older employees and allows their weaknesses (see Kruse, 2005):

● Social-communicative techniques: conversations in the form of dialogues and in groups, co-operation and team work, application training, negotiation training.

● Cognitive training: learning and memory training, the application of familiar cognitive strategies, the acquisition of new problem-solving competences, synthetic and conceptual thinking, training of the planning behaviour.

● Knowledge of information and communication technology: active search for relevant information, exchange and storage of knowledge.

● Deepening of practical experiences and internships: knowledge transfer, increasing of motivation and confidence.

● General knowledge concerning the labour market and the new working role: strategies of the re-entry, job perspective in the second half of one's life and the double role in family and job.

In relevant courses, cognitive and social effects were reached *e.g.* the concentration abilities, the speed of handling things and the conversation competences of the participants were improved. Health promoting effects can be seen in that aspects of neuroticism i.e. fear, irritability, depressive moods and vulnerability decreased significantly so that daily situations of conflict, crisis and stress were dealt with in a better way due to higher mental robustness. The evaluation of such measures is therefore encouraging but it also shows that especially socially exalted milieus are being reacted to.

Overall, it can be observed – and this is shown by representative adult-education studies (see Barz and Tippelt, 2004) – that in lifelong learning, especially concerning professional development, significant inequalities exist as the degree of education, job-related qualifications, working status, sex, nationality, age, but also lifestyles have a serious effect – so that specific and target group-oriented learning environments are necessary. Barriers of learning and education which are often being named in empirical studies *do not have a lot to do with the principal learning capability* of older adults, but in comparison they strongly discuss the "structure of offers and opportunities" of further education. As people grow older, the participation in further education and especially in work-related further education decreases strongly. From a learning-theoretical point of view this age-related decrease in the participation in further education cannot be justified – an exception are older academics who maintain and even increase their participation in further education. The still widespread reality of company strategies of not operationally integrating older adults through further education remains contradictory to the insights of research of development and learning.

C.8. Creating positive learning environments for adults

Naturally, the learning-related environmental conditions and the concrete didactic learning settings must take learning-theoretical, adult-pedagogical and gerontological insights into consideration. But what does this signify for adult-adequate learning? Three modern learning concepts are especially important: competence-based, constructivist and situated learning.

C.8.1. *Competence-based learning: prepare for solving problems*

The term "competence" was developed in further education out of a debate concerning key qualifications in vocational training (see Achatz and Tippelt, 2001), but is since then established in the field of school education and further training. Especially in the international comparison of education (OECD, 2004; OECD PISA survey) and concerning the efforts to develop national education standard, this term was brought in and was made the definition of educational goals. Here, a definition by Weinert (2001) is helpful, who understands competences as "*cognitive abilities and skills in order to solve certain problems as well as being able to make use of the motivational, volitional and social willingness and abilities related to this in a successful and responsible way in variable situations*" (ibid., p. 27f). Competences therefore are being introduced as a yardstick for successful learning which makes in reverse the creation of competences also the goal of educational events. Reaching this goal at first requires teachers and learners to change their understanding of roles which allows more activity, self-drive and own responsibility for the learners and primarily allocates the teachers supporting functions such as company, support and reflection (see Achatz and Tippelt 2001, p. 124f). In order to create competence it can be recommended to take different learning principals into consideration in class. Learning with the help of meaningful activities and practical problems as well as connecting learning contents to relevant application contexts belong to these. Over and above, the independency of the learning individual and the co-operation of the learners have to be promoted and one has to pay attention to methodical diversity. In this context, project-related learning which also finds expression in enterprise-operational education models, proves to be especially suitable.

C.8.2. Constructivist learning: making use of subjective experiences

If one follows the theoretical discussions about learning processes of the last decades two neurobiologists are viewed as the founders of modern constructivist learning philosophy: Maturana and Varela (see Siebert, 1998). Their central insights are the foundation-stone of constructivist learning theory and they are also being confirmed in current brain research studies (e.g. Spitzer, 2002; Siebert and Roth, 2003).

In conclusion, from a constructivist point of view, learning always occurs individually and in an experience-based way. New knowledge always resumes already existent knowledge and can lead to its transformation and differentiation. This expanding of knowledge occurs through new experiences or through the critical reflection of one's own cognitive constructs in the confrontation with others. Specific requirements concerning the organisation of learning opportunities can be derived from this (see Tippelt and Schmidt, 2005):

● The creation of knowledge cannot be solely initiated by the teachers but is always the responsibility of the learners as well. Teachers are responsible for providing learning resources and organising stimulating learning surroundings (expository teaching is still an important factor, but not dominant).

● Social exchange favours learning processes and should therefore be promoted.

● Likewise, problem-oriented learning is desirable and contributes to the creation of application-oriented knowledge.

● New learning contents should always resume individual previous knowledge. A class which is organised according to constructivist principles could, for example, encourage the learning people in advance to expatiate their experiences, opinions and views related to a subject.

Neuroscientific results have increasingly come into the pedagogic field of view in the past few years, and have to be regarded additionally and complementary to these insights – and not as fashionable (see Stern, 2004; Pauen, 2004). Neurobiological studies show that the structuring of our brain mainly occurs in childhood and youth. Nevertheless, structural and functional changes continue into adulthood and throughout the lifespan, although they tend to be less marked than in early life (except in the case of dementing diseases of the elderly). Because of age-related changes in processing efficiency, as one grows older the learning of new material takes longer whilst already existent knowledge is being more and more differentiated and made more precise (see Spitzer, 2002, and Sections C.4 and C.5 of this paper). The significant roles that attention, motivation and emotion play in the learning success, which brain researchers have pointed out (see Singer, 2002), validate pedagogical findings.

It is generally true that learning is related to the creation of meaning. Learning happens through the interpretation of sensory impressions. The meanings constructed in this way enable the creation of new synaptic links between the neurons of the brain and therefore learning (Roth, 2004). The results of neurobiological research replenish and support the central statements of constructivist learning research and likewise draw attention to the meaning of the organisation of suitable learning surroundings. This also is the central job of teachers. The provision of learning-promoting and target group-adequate basic conditions and the teachers' activation with the aid of suitable didactic principles are more important than forms and techniques of presentation.

C.8.3. Situated learning: organising learning environments

Based on the constructivist learning ideas, different approaches developed in which the displayed ideas on learning and the construction of knowledge were made more precise. Due to its learning-theoretical foundations and because of it having been converted into practical learning concepts many times already, here an approach is being chosen, which emphasises the meaning of the context in which learning takes place.

The approach of situated learning is based on the insight that knowledge always is acquired in a certain context, i.e. the application of knowledge is not independent on the situation it was being learned in. The more similar the learning and application contexts are, the more securely can knowledge be converted into successful action. But adults – unlike children and teenagers – do mainly not learn in school but in the companies or institutions in which they are supposed to apply the learning contents. This is an advantage of the learning of adults and corresponds to the recommendations concerning situated learning. It is important to create a vicinity to possible application contexts. The possibilities concerning this matter expand from the multi-perspective examination of problems (Cognitive Flexibility), over a step-by-step transition to independent problem solving (Cognitive Apprenticeship) to the inclusion of contents into complex problems (Anchored Instruction). The two latter strategies will be briefly sketched out in the following (see Tippelt and Schmidt, 2005).

The Cognitive-Apprenticeship-approach is strongly linked to the traditional training in trade as it leads the learner step-by-step from instructional orders towards the independent solving of problems. The learners are confronted with complex tasks from the beginning, the handling of which is being modelled by experts for them. Only in the second step, the learner is supposed to solve problems him/herself in which the expert, or rather the teacher instructs and supports him. The presence of the expert decreases step-by-step in the course of the learning process so that s/he merely has the role of an observer during the last learning stage. This methodical course of action is not least of all based on the articulation of problem solving strategies of the learners. They are being encouraged to comment on their own course of action and to expatiate the applied problem solving strategies during the learning process.

The Anchored-Instruction-approach has also proved itself. This approach involves learning contents into authentic and complex problems in order to promote the learners independent analysis of problems in the sense of explorative learning. Problems based on the world of everyday life and experiences are authentic, and complex signifies that the problems are not being restricted to the details relevant for the solution but that the filtering of exactly this relevant information out of a whole pool of information also belongs to this task. While doing so, it becomes evident that this is not about teacher-centred lessons but that a self-driven way of working is required from the adult as well as the younger learners, whilst the teachers moderate and specifically give the support, sometimes also clearly by giving instructions (see Tippelt and Schmidt, 2005). At a repeated application, the learners also create general problem-solving strategies by categorising different types of tasks and solutions and therefore developing comprehensive patterns of solution.

In sum, there are adequate learning strategies for adult education, but we have to take into consideration that aging seems to be accompanied by changes in task priorities and learning strategies, that older adults compensate for their decline in basic cognitive functioning by bringing in greater verbal and experience based world knowledge to solve problems, that cognitive changes and learning show enormous individual differences – and most important for learning of adults that plasticity is a feature of adults brain and this continues into late adulthood.

C.9. Future agenda

Informed by the distinction between social science-based and neuroscience-based knowledge of learning, it is obvious that age differences in cognition exist, that historical and ontogenetic plasticity of intellectual performance is given and that a valid and meaningful strategy of learning needs both, brain research and educational science. Nevertheless neuroscience and educational or even broader social science have their own tasks, different perspectives and languages, but in the future competence-based learning should be understood in a deeper and meaningful way – taking into consideration both perspectives.

Promising for collaboration is especially the analysis of constraints and of learning problems of certain groups of learners in the aging society: physical, mental and social well-being of the older learners are an enormous challenge for learning research. Brain research is able to show that prior learning can improve learning in later years, going beyond explanations from educational science and psychology. Analysing abilities and constraints of the brain, brain research and educational science can better explain why special learning environments in adulthood work and why others fail.

From an optimistic point of view of societies and individuals, the question arises: in what way the potentials of aging and of age can be of use for future generations? The promotion and use of older people's potentials in a society is therefore based on the contact of the generations as learning only makes sense in a society oriented on the model of the solidarity of generations (see Tippelt, 2000). For example, the advantages of working and learning teams including people of all ages and a balanced staff structure in companies, have to be made more visible by research. An increased effort to make the transition from the working life to the postworking life more flexible is required because of the individually very different cognitive and motivational potentials of the elderly. This has to be analysed in more detail by research. And surely, a differentiated learning which is competence-promoting and which takes the lifespan into consideration, is necessary and will lastingly challenge brain research and educational science.

Raja Parasuraman and Rudolf Tippelt

C.10. Practitioner's response

Neuroscience and educational research are spheres that are far removed from the world of classroom teaching. The article "Brain, Cognition and Learning in Adulthood" by Parasuraman and Tippelt does not represent bedtime reading for most teachers, as it focuses on adult learning from the point of view of brain and educational research. Nevertheless, it is a great relief for a teacher like me to read about the empirical findings from these fields, because they reinforce my own classroom observations and practices.

A practising teacher is most aware of the input and output in the formal educational setting, but can only have some intuitions about what happens in between inside the skull of the learner. In other words, a good teacher has a fairly substantial understanding of the teaching process and is usually also in a position to observe and evaluate any evidence of subsequent learning. However, the practising teacher is not aware of the actual processes in the brain that constitute the actual learning itself, and can at best only make some guesses about the mystery of brain mechanisms. It is therefore reassuring that brain and educational research validates such guesses and that the article also complements the insights which have come out of research into linguistics and language teaching. In particular, Section C.8 of the article dealing with creating positive learning environments for adults supports the classroom practices that I am familiar with.

Admittedly, my professional experience is much narrower than the general field of learning addressed by Parasuraman and Tippelt. I am concerned with English Language Teaching (ELT) to the adult learner who did not acquire any English in early childhood. My comments about their article need to be assessed against the backdrop of ELT.

Firstly, such teaching refers to the two categories of EFL and ESL. Adult EFL (= English as a Foreign Language) learners live in countries where they do not use English on a daily basis. My EFL professional activities have taken place in Mexico, France, Jordan, Palestine, Indonesia and the Netherlands. In contrast, adult ESL (= English as a Second Language) learners are those who have come to stay as immigrants or international students in an English-speaking environment (Canada and Great Britain in my teaching career) and whose daily needs therefore require mastery of English.

Secondly, my ELT work has also included teacher education courses in several countries (Jordan, Mexico, Indonesia, Canada and China) preparing people to become English language teachers in EFL/ESL classrooms. These have included courses which either prepare people without any previous experience in language teaching to become English language teachers (pre-service courses), or alternatively help already practising English teachers to upgrade their professional skills (in-service courses).

One red thread runs through the work in these different countries during different periods: plentiful evidence of adult learning. The myth that the adult brain can no longer learn (the "deficit model" mentioned in the article) got refuted again and again by the successful learning that I witnessed around me. The growth of new neurons must have taken place continually, whether with adults in their early twenties or those who were twice or three times that age. Clearly, that corroborates what is argued in the article.

However, one issue of adult learning and training has not been addressed in the article, but it is of utmost importance to me. The article deals with "how" an adult learns – and thus learns best – given specific parameters such as age, genetics and opportunities. In a general overview of learning from the fields of empirical neuroscience and educational research, the question of "what" is actually learned seems irrelevant. The subject to be learned could be anything from art history to information technology to aerobics. Presumably, the principal medium for the teaching and learning of the given subject is the language that the teacher and the learner share, be it Japanese, Dutch or any other language. Therefore the "how" and the "what" are distinct entities and totally separate from each other. In ELT this used to be – and regrettably in many parts of the world still is – the case, since the English language was taught through the medium of the learner's mother tongue, with courses which required a great deal of lecture-mode presenting from the teacher and which were top-heavy on grammatical analysis and translation. This educational tradition comes from a long-standing and historically determined approach.

In practical terms the results of this approach were, and still are, disastrous: young adults who in their adolescence received between 1 000 and 2 000 hours of English teaching (excluding homework), and who start travelling to places where the use of the English language is needed, frequently discover that their years of hard work did not pay off with any practical benefits. More often than not they cannot understand large stretches produced by an English speaker, are incapable of having a simple conversation of moderate length, do not have the reading skills to tackle and enjoy even a short story, cannot write a 1-page letter without numerous mistakes and cause constant confusion to their English listener due to their incomprehensible pronunciation. This shows that the way "how" the teaching and learning took place has not led to the desired outcome of "what" was supposed to be learned. Furthermore, it is painful evidence of a waste of time, money, energy, mental effort, infrastructure and human resources.

Fortunately, over the last few decades increasingly more ELT (and in general FLT = Foreign Language Teaching) practitioners have diverted from this traditional approach. The realisation that the English language is an important tool for communication has slowly pushed classroom practices into a radically different direction. The learner's mother tongue is no longer the means to an end, i.e. the goal of knowing the English language. Instead, the "what" has also become the "how", so that the distinction between medium and message has become blurred. A learner whose aim is to gain a level of proficiency in English uses English as the very tool to achieve that aim. The learner's limited English competence constitutes the springboard to reach a higher language level. In turn, once that level is sufficiently established, the learner moves on to a yet higher level of English. Only partly does the learning process entail an expansion of knowledge, such as vocabulary, grammar rules, and spelling and pronunciation rules. A more substantial part of classroom practice is the fostering of skills to manage communication, viz. speaking, listening, reading and writing. The traditional accumulation of factual knowledge makes way for the development of language skills.

Such an approach calls for a skillful and well-trained teacher. The teacher needs to steer the class carefully from one stage to the next, constantly building on previous skills and knowledge, facilitating not only the progress of purely linguistic skills (such as accuracy and fluency in the language), but also increasing the learner's confidence and maintaining motivation for continued learning.

Obviously, the need for such teachers requires carefully designed teacher preparation courses: TEFL (Teaching English as a Foreign Language), TESL (Teaching English as a Second Language) or more generally TESOL (Teaching English to Speakers of Other Languages). During their vocational preparation aspiring teachers are no longer mainly concerned with the academic study of their subject – linguistics, education – but should get extensive training in an educational approach where they use the English language as the communicative vehicle for teaching that English language for communication.

I have been privileged in that, for more than two decades, my ELT work has included a great deal of teacher education/teacher training. Although the student teachers from vastly different parts of the world often exhibited huge variation as adult learners owing to factors such as their English language competence, mother tongue background, previous learning experience, cultural and national characteristics or individual motivation, there was always the red thread of learning in all these courses. Invariably, learning occurred in the student teachers and, more importantly, with it came their understanding and appreciation of the underlying ELT principles.

What should then be the common denominator for teacher education programmes in ELT (or by extension in Foreign Language Teaching)? What principles should any teacher preparation adhere to? What traditional components are no longer deemed desirable and necessary, but should be scrapped as superfluous and useless? What elements are absolute prerequisites in TESOL and must be part and parcel of any course? In my view there are seven elements that deserve serious consideration.

Knowledge about English, education and pedagogy still has its place, as it has always had, because a teaching professional must know the subject-matter. However, all this knowledge can no longer be divorced from any implementation in the real ELT classroom. While many teacher preparation programmes used to contain a great deal of theory about education and linguistics, such a syllabus is no longer regarded as satisfactory. Knowledge and theory should only be included if they bear some – direct or indirect – relation to what the future teacher needs in the working environment. If this means that long-standing courses in medieval literature or Old English linguistics must be axed, so be it.

This leads to another principle, namely that all the theory must be integrated with the practice of the teaching/learning world. In a TESOL programme, there is no place for ivory-tower theorising that does not make the link with the needs of the teaching professional. Any inclusion of theory should be justified in the light of practical ELT application. If the gap between a given theoretical model and its relevance to practical reality cannot be bridged within the TESOL syllabus, then that theory ought to be taken out of the curriculum.

It is imperative that a major part should be devoted to teaching methods and techniques, as well as their underlying rationale. This demands a great deal of practice teaching from the student teacher. The corollary is that it also necessitates a great deal of feedback from the teacher educators about the student teacher's practical performance. Such a process is so labour-intensive and logistically complex that it calls for meticulous planning and extensive human resources. Practical tasks in various forms should be included right from the beginning to the end of the teacher education programme, whether it extends over merely 4 weeks (as with many of my Canadian courses) or 4 years (as with the TEFL university degree in my Mexican experience).

Furthering these principles means that the actual way in which the TESOL programme is delivered by the teacher educator must reflect what is advocated by the theory. First of all, the teacher educator is preferably an EFL/ESL teacher with considerable experience from the ELT classroom. Therefore, any course content of the teacher education programme (the "what" of the syllabus) has to be reflected in the way that the programme is presented. The implication is that all the techniques described in theoretical models are actually practised in sessions that impart such models. For instance, if it is argued that pairwork discussions – or games or role plays – are extremely useful in the EFL/ESL classroom, then the same techniques of pairwork, games or role play ought to be used in preparing teachers during their teacher education course. The commitment to the methodological content of the course is measured through the methodology of the delivery. Hence, a teacher educator who merely lectures in traditional fashion is definitely unacceptable. "Practise what you preach" is a quintessential principle in teacher education programmes.

One highly successful – and to me indispensable – component to ensure that student teachers understand many different aspects of teaching communicatively is the introduction of an unfamiliar foreign language (UFL) into a TESOL programme where no student teacher speaks that language. The student teacher is put in the shoes of the beginner learner who acquires a foreign language that is totally new and unfamiliar, for instance Serbo-Croat, Arabic, Spanish or Hebrew. The student teacher invariably develops, through such experiential learning, a high degree of empathy with the learner and an understanding of what is required of the language teacher. More specifically, the student teacher realises how strenuous and intimidating it is from the learner's perspective to embark on a new foreign language; how rewarding and exciting it is to accomplish positive results even at beginner level; how important it is for the teacher to be clear and inventive with visuals, body language, mime and facial expression; how vital the teacher's patience and encouragement are; and how it is perfectly feasible and manageable for a teacher to use the new foreign language exclusively, *i.e.* without ever resorting to the learner's mother tongue. This foreign language learning experience may be given in as short as a half-hour lesson or as long as a 10-week course of 2 hours per week, with varying tasks of reflection and analysis (*e.g.* group discussions, questionnaires about emotional responses, journal writing). My professional observations of a UFL in widely different settings and locations (Mexico, Canada, Indonesia) show that the insights gained by student teachers from being put in the situation of beginning foreign language learners are invaluable. It is particularly valid for aspiring teachers who want to teach English and whose mother tongue is also English, so who have no personal experience with acquiring English in the formal classroom setting.

One added set of skills that student teachers may need to acquire – depending on their formerly acquired learning habits – is study skills. Student teachers need to realise what successful *versus* unhelpful learning strategies are. If they do not understand this fully, for instance due to bad learning habits fostered in their own school career (*e.g.* short-term cramming before an exam, reluctance to try out the language, or walking up and down to learn text by rote), then they first need to "learn how to learn". This includes note-taking, library research skills, analytical thinking, expressing an opinion, problem solving, etc. It is unrealistic to expect ELT practitioners to teach appropriate study skills without ever having implemented these themselves.

Last but not least, in a teacher education programme for student teachers with a mother tongue other than English, it is important to upgrade their own English language skills. In pre-service courses student teachers can often perceive the value of improving their own language levels and are happy to follow the entire TESOL programme in English, thus making considerable progress in English. Therefore very few, if any, courses should be delivered in another language, regardless of the question if the teacher educator shares the student teachers' mother tongue. However, sometimes during in-service training, resistance to English upgrading is expressed by allegedly experienced English language teachers. In fact, such teachers often demonstrate – but do not perceive in themselves – a great need for English language upgrading, despite their previous experience in ELT jobs. Further improvement of their English skills is usually recommended, if not absolutely necessary.

Summarising, the most important principles to programmes that train and prepare adults for ELT are:

1. Any theory relates to the application in the real world of teaching.

2. Theory and practice are integrated.

3. Practical teaching tasks are organised all through the course.

4. Teacher educators practise what they preach.

5. A course in an unfamiliar foreign language is included.

6. Study skills are taught if necessary.

7. The upgrading of English language skills is included if necessary.

Programme developers responsible for TESOL courses have the professional obligation to ensure that their adult learners are able to learn and be trained in the best possible manner. Once these adults are later involved with their professional activities in ELT, they in turn will become the models in their classes and facilitate the learning process for their own learners preparing for the future. With the help of empirical research findings we can invest in our human resources by creating the most efficient and effective learning environments for adults. It is high time we do just that.

Liet Hellwig

References

Achatz, M. and R. Tippelt (2001), "Wandel von Erwerbsarbeit und Begründungen kompetenzorientierten Lernens im internationalen Kontext", in A. Bolder, W. Heinz and G. Kutscha (eds.), *Deregulierung der Arbeit – Pluralisierung der Bildung?*, Leske and Budrich, Opladen, pp. 111-127.

Achtenhagen, F. and W. Lempert (eds.) (2000), *Lebenslanges Lernen im Beruf – Seine Grundlegung im Kindes- und Jugendalter*, Bd. 1-5, Leske and Budrich, Opladen.

Alterskommission (2005), "Zusammenfassung wesentlicher Thesen des Fünften Altersberichts", Berlin.

Baethge, M. and V. Baethge-Kinsky (2004), *Der ungleiche Kampf um das lebenslange Lernen*, Waxmann, Münster/New York/München and Berlin.

Ball, K., D.B. Berch, K. Helmers, J. Jobe, M. Leveck, M. Marsiske, J. Morris, G.W. Rebok, D.M. Smith, S.L. Tennstedt, F. Unverzagt and S. Willis (2002), "Effects of Cognitive Training Interventions with Older Adults. A Randomized Controlled Trial", *JAMA* 288, pp. 2271-2281.

Baltes, P.B. (1993), "The Aging Mind: Potential and Limits", *The Gerontologist*, Vol. 33/5, pp. 580-594.

Baltes, P.B. (2003), "Das hohe Alter – mehr Bürde als Würde?", MaxPlanckForschung, Vol. 2, pp. 15-19.

Baltes, P.B., J. Glück and U. Kunzmann (2002), "Wisdom: Its Structure and Function in Successful Lifespan Development", C.R. Snyder and S.J. Lopez (eds.), *Handbook of Positive Psychology*, Oxford University Press, New York, pp. 327-350.

Baltes, P.B. and U. Lindenberger (1997), "Emergence of a Powerful Connection between Sensory and Cognitive Functions across the Adult Life Span: A New Window to the Study of Cognitive Aging?", *Psychology and Aging*, Vol. 12(1), pp. 12-21.

Baltes, P.B. and U.M. Staudinger (2000), "Wisdom: A Metaheuristic (pragmatic) to Orchestrate Mind and Virtue toward Excellence", *American Psychologist*, Vol. 55, pp. 122-136.

Bartzokis, G., J.L. Cummings, D. Sultzer, V.W. Henderson, K.H. Nuechterlein and J. Mintz (2003), "White Matter Structural Integrity in Healthy Aging Adults and Patients with Alzheimer Disease: A Magnetic Resonance Imaging Study", *Archives of Neurology*, Vol. 60(3), pp. 393-398.

Barz, H. and R. Tippelt (eds.) (2004), *Weiterbildung und soziale Milieus in Deutschland*, Bd.1 u.2, Bertelsmann, Bielefeld.

Becker, S., L. Veelken and K.P. Wallraven (eds.) (2000), *Handbuch Altenbildung. Theorien und Konzepte für Gegenwart und Zukunft*, Leske and Budrich, Opladen.

Bransford, J.D., A.L. Brown and R.R. Cocking (2004), *How People Learn: Brain, Mind, Experience, and School*, Expanded Edition, National Academy Press, Washington DC.

Bruer, J.T. (1997), "Education and the Brain: A Bridge too Far", *Educational Researcher*, Vol. 26(8), pp. 4-16.

Bynner, J., T. Schuller and L. Feinstein (2003), "Wider Benefits of Education: Skills, Higher Education and Civic Engagement", *Z.f.Päd.*, 49.Jg., Vol. 3, pp. 341-361.

Cabeza, R. (2002), "Hemispheric Asymmetry Reduction in Older Adults: The HAROLD Model", *Psychology and Aging*, Vol. 17(1), pp. 85-100.

Cabeza, R., L. Nyberg and D.C. Park (2005), *Cognitive Neuroscience of Aging*, Oxford University Press, New York.

Cattell, R.B. (1963), "Theory of Fluid and Crystallized Intelligence: A Critical Experiment", *Journal of Educational Psychology*, Vol. 54, pp. 1-22.

Cohen, J. (1988), *Statistical Power Analysis for the Behavioral Sciences* (2nd ed.), Lawrence Erlbaum Associates, Hillsdale, New Jersey.

Colcombe, S.J. and A. Kramer (2002), "Fitness Effects on the Cognitive Function of Older Adults: A Meta Analytic Study", *Psychological Science*, Vol. 14, pp. 125-130.

Colcombe, S.J., A. Kramer, K.I. Erickson and P. Scalf (2005), "The Implications of Cortical Recruitment and Brain Morphology for Individual Differences in Inhibitory Function in Aging Humans", *Psychology and Aging*, Vol. 20, pp. 363-375.

Courchesne, E., H.J. Chisum, J. Townsend, A. Cowles, J. Covington, B. Egaas *et al.* (2000), "Normal Brain Development and Aging: Quantitative Analysis at In Vivo MR Imaging in Healthy Volunteers", *Radiology*, Vol. 216, pp. 672-681.

Csikszentmihalyi, M. (1982), "Towards a Psychology of Optimal Experience", R. Gross (ed.), *Invitation to Life-long Learning*, Fowlett, New York, pp. 167-187.

DeCharms, R. (1976), *Enhancing Motivation: Change in the Classroom*, Irvington, New York.

Deci, E.L. and R.M. Ryan (1985), *Intrinsic Motivation and Self-determination in Human Behavior*, Plenum Press, New York.

Dobbs, A.R. and B.G. Rule (1989), "Adult Age Differences in Working Memory", *Psychology and Aging*, Vol. 4(4), pp. 500-503.

Erikson, E. (1966), *Identität und Lebenszyklus*, Frankfurt.

Eriksson, P.S., E. Perfilieva, T. Björk-Eriksson, A.M. Alborn, C. Nordborg, D.A. Peterson and F.H. Gage (1998), "Neurogenesis in the Adult Human Hippocampus", *Nature Medicine*, Vol. 4, pp. 1313-1317.

Espeseth, T., P.M. Greenwood, I. Reinvang, A.M. Fjell, K.B. Walhovd, L.T. Westlye, E. Wehling, A. Astri Lundervold, H. Rootwelt and R. Parasuraman (2007), "Interactive Effects of APOE and CHRNA4 on Attention and White Matter Volume in Healthy Middle-aged and Older Adults", *Cognitive, Behavioral, and Affective Neuroscience*.

Feinstein, L., C. Hammond, L. Woods, J. Preston and J. Bynner (2003), "The Contribution of Adult Learning to Health and Social Capital. Wider Benefits of Learning Research", Report 8, Center of Research on the Wider Benefits of Learning, London.

Good, C.D., I.S. Johnsrude, J. Ashburner, R.N.A. Henson, K.J. Friston and R.S.J. Frackowiak (2001), "A Voxel-based Morphometric Study of Aging in 465 Normal Adult Human Brains", *Neuroimage*, Vol. 14, pp. 21-36.

Grady, C.L., J.M. Maisog, B. Horwitz, L.G. Ungerleider, M.J. Mentis and J.A. Salerna (1994), "Age-related Changes in Cortical Blood Flow Activation during Visual Processing of Faces and Location", *Journal of Neuroscience*, Vol. 14, pp. 1450-1462.

Grady, C.L., A.R. McIntosh and F.I. Craik (2003), "Age-related Differences in the Functional Connectivity of the Hippocampus during Memory Encoding", *Hippocampus*, Vol. 13(5), pp. 572-586.

Greenwood, P.M. and R. Parasuraman (1994), "Attentional Disengagement Deficit in Nondemented Elderly over 75 Years of Age", *Aging and Cognition*, Vol. 1(3), pp. 188-202.

Greenwood, P.M. and R. Parasuraman (1999), "Scale of Attentional Focus in Visual Search", *Perception and Psychophysics*, Vol. 61, pp. 837-859.

Greenwood, P. and R. Parasuraman (2003), "Normal Genetic Variation, Cognition, and Aging", *Behavioral and Cognitive Neuroscience Reviews*, Vol. 2, pp. 278-306.

Greenwood, P.M., C. Lampert, T. Sunderland and R. Parasuraman (2005), "Effects of Apolipoprotein E Genotype on Spatial Attention, Working Memory, and their Interaction in Healthy, Middle-aged Adults: Results from the National Institute of Mental Health's BIOCARD Study", *Neuropsychology*, Vol. 19(2), pp. 199-211.

Greenwood, P.M., T. Sunderland, J. Friz and R. Parasuraman (2000), "Genetics and Visual Attention: Selective Deficits in Healthy Adult Carriers of the 4 Allele of the Apolipoprotein E Gene", *Proceedings of the National Academy of Science*, Vol. 97, pp. 1661-1666.

Gropengießer, H. (2003), "Lernen und Lehren: Thesen und Empfehlungen zu einem professionellen Verständnis", Report 3/2003, Literatur und Forschungsreport Weiterbildung, Gehirn und Lernen, pp. 29-39.

Guttmann, C.R., F.A. Jolesz, R. Kikinis, R.J. Killiany, M.B. Moss and T. Sandor (1998), "White Matter Changes with Normal Aging", *Neurology*, Vol. 50, pp. 972-981.

de Haan, M. and M. Johnson (eds.) (2003), "The Cognitive Neuroscience of Development", *Psychology Press*, Hove.

Heckhausen, H. (1989), *Motivation und Handeln*, Springer, Heidelberg.

Hedden, T., G.J. Lautenschlager and D.C. Park (2005), "Contributions of Processing Ability and Knowledge to Verbal Memory Tasks across the Adult Lifespan", *Quarterly Journal of Experimental Psychology*.

Jennings, J.M. and A.L. Darwin (2003), "Efficiacy Beliefs. Everyday Behavior, and Memory Performance among Elderly Adults", *Educational Gerontology*, Vol. 29, pp. 34-42.

Jernigan, T.L., S.L. Archibald, C. Fennema-Notestine, A.C. Gamst, J.C. Stout and J. Bonner (2001), "Effects of Age on Tissues and Regions of the Cerebrum and Cerebellum", *Neurobiology of Aging*, Vol. 22, pp. 581-594.

Johnson, M., Y. Munakata and R.O. Gilmore (eds.) (2002), *Brain Development and Cognition – A Reader*, Blackwell, Oxford.

Karmel, T. and D. Woods (2004), "Lifelong Learning and Older Workers", NCVER, Adelaide.

Kemper, T. (1994), "Neuroanatomical and Neuropathological Changes in Normal Aging and in Dementia", in M.L. Albert (ed.), *Clinical Neurology of Aging* (2nd edition), Oxford University Press, New York.

Keyfitz, N. (1990), *World Population Growth and Aging: Demographic Trends in the Late Twentieth Century*, University of Chicago Press, Chicago, IL.

Kruse, A. (1997), "Bildung und Bildungsmotivation im Erwachsenenalter", in F.E. Weinert and H. Mandl (eds.), *Psychologie der Erwachsenenbildung*, Hogrefe, Göttingen, pp. 115-178.

Kruse, A. (1999), "Bildung im höheren Lebensalter. Ein aufgaben-, kompetenz- und motivationstheoretischer Ansatz", in R. Tippelt (ed.), *Handbuch der Erwachsenenbildung/Weiterbildung*, Leske and Budrich, Opladen, pp. 581-588.

Kruse, A. (2005), "Qualifizierungsmaßnahmen für Wiedereinsteigerungen in den Beruf", Heidelberg (unveröffentl. Manuskript).

Kruse, A. and G. Rudinger (1997), "Lernen und Leistung im Erwachsenenalter", in F.E. Weinert and H. Mandl (eds.), *Psychologie der Erwachsenenbildung*, Hogrefe, Göttingen, pp. 46-85.

Lahn, L.C. (2003), "Competence and Learning in Late Career", *European Educational Research Journal*, Vol. 2/1, pp. 126-140.

Lchr, U. (1986), "Aging as Fate and Challenge", in H. Häfner, G. Moschel and N. Sartorius (eds.), *Mental Health in the Elderly*, Heidelberg, pp. 57-77.

Lehr, U. (1991), *Psychologie des Alterns* (7th edition), Heidelberg.

Leibniz-Gemeinschaft (2005), "Wie wir altern: Megathema Alternsforschung", *Journal der Leibniz-Gemeinschaft*, pp. 6-13.

Lerner, R.M. (2002), "Concepts and Theories of Human Development", Lawrence Erlbaum, Malwah, NJ.

Li, K.Z., U. Lindenberger, A.M. Freund and P.B. Baltes (2001), "Walking while Memorizing: Age-related Differences in Compensatory Behavior", *Psychological Science*, Vol. 12(3), pp. 230-237.

Lindenberger, U., H. Scherer and P.B. Baltes (2001), "The Strong Connection between Sensory and Cognitive Performance in Old Age: Not Due to Sensory Acuity Reductions Operating during Cognitive Assessment", *Psychology and Aging*, Vol. 16(2), pp. 196-205.

Madden, D.J., T.G. Turkington, R.E. Coleman, J.M. Provenzale, T.R. DeGrado and J.M. Hoffman (1996), "Adult Age Differences in Regional Cerebral Blood Flow during Visual Work Identification: Evidence from H2 15O PET", *Neuroimage*, Vol. 3, pp. 127-142.

Madden, D.J., W.L. Whiting, J.M. Provenzale and S.A. Huettel (2004), "Age-related Changes in Neural Activity during Visual Target Detection Measured by fMRI", *Cerebral Cortex*, Vol. 14(2), pp. 143-155.

Nelson, C.A. and M. Luciana (eds.) (2001), *Handbook of Developmental Cognitive Neuroscience*, MIT Press, Cambridge, MA.

OECD (2002), *Understanding the Brain: Towards a New Learning Science*, OECD, Paris.

OECD (2004), *Education at a Glance: OECD Indicators*, OECD, Paris.

OECD (2005), *Aging and Employment Policies: United States*, OECD, Paris.

Parasuraman, R., P.M. Greenwood, R. Kumar and J. Fossella (2005), "Beyond Heritability: Neurotransmitter Genes differentially Modulate Visuospatial Attention and Working Memory", *Psychological Science*, Vol. 16(3), pp. 200-207.

Parasuraman, R., P.M. Greenwood and T. Sunderland (2002), "The Apolipoprotein E Gene, Attention, and Brain Function", *Neuropsychology*, Vol. 16, pp. 254-274.

Park, D.C., G. Lautenschlager, T. Hedden, N. Davidson, A.D. Smith and P. Smith (2002), "Models of Visuospatial and Verbal Memory across the Adult Life Span", *Psychology and Aging*, Vol. 17(2), pp. 299-320.

Park, D. and N. Schwarz (1999), *Cognitive Aging: A Primer*, Psychology Press, Hove.

Pate, G., J. Du and B. Havard (2004), "Instructional Design – Considering the Cognitive Learning Needs of Older Learners", *International Journal of Instructional Technology and Distance Learning*, Vol. 1/5, pp. 3-8.

Pauen, S. (2004), "Zeitfenster der Gehirn- und Verhaltensforschung: Modethema oder Klassiker?", *Z.f.Päd.*, 50Jg., Vol. 4, pp. 521-530.

Pillay, H., G. Boulton-Lewis, L. Wilss and C. Lankshear (2003), "Conceptions of Work and Learning at Work: Impressions from Older Workers", *Studies in Continuing Education*, Vol. 25/1, pp. 95-111.

Raz, N., U. Lindenberger, K.M. Rodriue, K.M. Kennedy, D. Head and A. Williamson (2005), "Regional Brain Changes in Aging Healthy Adults: General Trends, Individual Differences and Modifiers", *Cerebral Cortex*, Vol. 15(11), pp. 1676-1689.

Resnick, S.M., D.L. Pham, M.A. Kraut, A.B. Zonderman and C. Davatzikos (2003), "Longitudinal Magnetic Resonance Imaging Studies of Older Adults: A Shrinking Brain", *Journal of Neuroscience*, Vol. 23(8), pp. 3295-3301.

Rosen, A.C., M.W. Prull, R. O'Hara, E.A. Race, J.E. Desmond, G.H. Glover, J.A. Yesavage and J.D.E. Gabrieli (2002), "Variable Effects of Aging on Frontal Lobe Contributions to Memory", *Neuroreport*, Vol. 13, pp. 2425-2428.

Rossi, S., C. Miniussi, P. Pasqualetti, C. Babiloni, P.M. Rossini and S.F. Cappa (2004), "Age-related Functional Changes of Prefrontal Cortex in Long-term Memory: A Repetitive Transcranial Magnetic Stimulation Study", *Journal of Neuroscience*, Vol. 24(36), pp. 7939-7944.

Roth, G. (2004), "Warum sind Lehren und Lernen so schwierig?", *Z.f.Päd.*, 50. Jg., Vol. 4, pp. 496-506.

Rypma, B. and M. D'Esposito (2000), "Isolating the Neural Mechanisms of Age-related Changes in Human Working Memory", *Nature Neuroscience*, Vol. 3(5), pp. 509-515.

Saczynski, J.S., S.L. Willis and K.W. Schaie (2002), "Strategy Use in Reasoning Training with Older Adults", *Aging Neuropsychology and Cognition*, Vol. 9/1, pp. 48-60.

Salthouse, T.A. (1996), "The Processing-speed Theory of Adult Age Differences in Cognition", *Psychological Review*, Vol. 103, pp. 403-428.

Schaie, K.W. (2005), "Developmental Influences on Adult Intelligence. The Seattle Longitudinal Study", University Press, Oxford.

Schuller, T., J. Preston, C. Hammond, A. Brassett-Grundy and J. Bynner (eds.) (2004), *The Benefits of Learning. The Impact of Education on Health, Family Life and Social Capital*, Routledge Farmer, London.

Shors, T.J., G. Miesegaes, A. Beylin, M. Zhao, T. Rydel and E. Gould (2001), "Neurogenesis in the Adult is Involved in the Formation of Trace Memories", *Nature*, Vol. 410, pp. 372-376.

Siebert, H. (1998), *Konstruktivismus: Konsequenzen für Bildungsmanagement und Seminargestaltung*, Schneider, Frankfurt a.M.

Siebert, H. and G. Roth (2003), "Gespräch über Forschungskonzepte und Forschungsergebnisse der Gehirnforschung und Anregungen für die Bildungsarbeit", *Report 3, Literatur- und Forschungsreport Weiterbildung*, Gehirn und Lernen, pp. 14-19.

Singer, W. (2002), *Der Beobachter im Gehirn*, Suhrkamp, Frankfurt a.M.

Snowdon, D.A., S.J. Kemper, J.A. Mortimer, L.H. Greiner, D.R. Wekstein and W.R. Markesbery (1996), "Linguistic Ability in Early Life and Cognitive Function and Alzheimer's Disease in Late Life. Findings from the Nun Study", *Journal of the American Medical Association*, Vol. 275, pp. 528-532.

Sowell, E.R., B.S. Peterson, P.M. Thompson, S.E. Welcome, A.L. Henkenius and A.W. Toga (2003), "Mapping Cortical Change across the Human Life Span", *Nature Neuroscience*, Vol. 6(3), pp. 309-315.

Spitzer, M. (2000), "Geist, Gehirn and Nervenheilkunde. Grenzgänge zwischen Neurobiologie", *Psychopathologie und Gesellschaft*, Schattauer, New York.

Spitzer, M. (2002), *Lernen. Gehirnforschung und die Schule des Lebens*, Spektrum Akademischer Verlag, Heidelberg.

Spitzer, M. (2004), *Selbstbestimmen. Gehirnforschung und die Frage: Was sollen wir tun?*, Spektrum Akademischer Verlag, Heidelberg.

Stern, E. (2004), "Wie viel Hirn braucht die Schule? Chancen und Grenzen einer neuropsychologischen Lehr-Lern-Forschung", *Z.f.Päd*, 50.Jg., Vol. 4, pp. 531-538.

Stern, E., R. Grabner, R. Schumacher, C. Neuper and H. Saalbach (2005), "Lehr-Lern-Forschung und Neurowissenschaften: Erwartungen, Befunde und Forschungsperspektiven", Bildungsreform Bd.13, BMBF, Berlin.

Sternberg, R.J. (1990), *Wisdom: Its Nature, Origin, and Development*, Cambridge University Press, New York.

Thomae, H. (1970), "Theory of Aging and Cognitive Theory of Personality", *Human Development 13*, pp. 1-16.

Tippelt, R. (1999) (ed.), *Handbuch Erwachsenenbildung/Weiterbildung*, Leske and Budrich, Opladen.

Tippelt, R. (2000), "Bildungsprozesse und Lernen im Erwachsenenalter. Soziale Integration und Partizipation durch lebenslanges Lernen", D. Benner and H.-E. Tenorth (eds.), *Bildungsprozesse und Erziehungsverhältnisse im 20. Jahrhundert, Z.f.Päd.*, Vol. 42, Beiheft, Beltz, Weinheim, pp. 69-90.

Tippelt, R. (2002) (ed.), *Handbuch Bildungsforschung*, Verlag für Sozialwissenschaften, Wiesbaden.

Tippelt, R. (2004), "Lernen ist für Pädagogen keine Blackbox: Basiselemente einer pädagogisch konzipierten Lerntheorie", *Grundlagen der Weiterbildung*, Vol. 3/15, pp. 108-110.

Tippelt, R. and B. Schmidt (2005), "Was wissen wir über Lernen im Unterricht?", *Pädagogik*, Vol. 3, pp. 6-11.

Walhovd, K.B., A.M. Fjell, I. Reinvang, A. Lundervold, A.M. Dale and D.E. Eilertsen (2005), "Effects of Age on Volumes of Cortex, White Matter and Subcortical Structures", *Neurobiology of Aging*, Vol. 26(9), pp. 1261-1270.

Wechsler, D. (1939), *The Measurement and Appraisal of Adult Intelligence*, Williams and Wilkens, Baltimore.

Weinert, F.E. (2001), "Vergleichende Leistungsmessung in Schulen – eine umstrittene Selbstverständlichkeit", *Leistungsmessungen in Schulen*, Beltz, Weinheim, pp. 17-31.

Weinert, F.E. and H. Mandl (1997) (ed.), *Psychologie der Erwachsenenbildung*, Hogrefe, Göttingen.

Welzer, H. and H.J. Markowitsch (2001), "Umrisse einer interdisziplinären Gedächtnisforschung", *Psychologische Rundschau*, Vol. 52(4), pp. 205-214.

White, R.W. (1959), "Motivation Reconsidered: The Concept of Competence", *Psychological Review*, Vol. 66, pp. 297-333.

WHO (2003), *Gender, Health, and Aging*, World Health Organization, Geneva, Switzerland.

Williamson, A. (1997), "You Are Never too Old to Learn! Third-Age Perspectives on Lifelong Learning", *International Journal of Lifelong Education*, Vol. 16/3, pp. 173-184.

Wilson, R.S., C.F. Mendes De Leon, L.L. Barnes, J.A. Schneider, J.L Bienias, D.A. Evans et al. (2002), "Participation in Cognitively Stimulating Activities and Risk of Incident Alzheimer Disease", *JAMA*, Vol. 287(6), pp. 742-748.

Wrenn, K.A. and T.J. Maurer (2004), "Beliefs about older Workers Learning and Development Behaviour in Relation to Beliefs about Malleability of Skills, Age-Related Decline, and Control", *Journal of Applied Social Psychology*, Vol. 34/2, pp. 223-242.

ISBN 978-92-64-02912-5
Understanding the Brain: The Birth of a Learning Science
© OECD 2007

ANNEX A

Fora

The OECD/CERI "Teach-the-Brain" on-line forum discussions (*www.ceri-forums.org/forums*) played a vital role in linking two important communities together – teachers and scientific experts. The forum provided an excellent opportunity for teachers to ask questions about how the brain learns, gain advice on how to incorporate brain-based learning into their teaching methods, obtain justification from neuroscience as to how and why some of their teaching practices and intuitions work, and offer insights from a practitioner's perspective. Scientific experts were available to respond and offer research. Also, experts and OECD staff were available to clarify misconceptions and terminology.

The results of this two-year venture proved to be very successful both in terms of the turnout (over 160 threads created and over 2 000 replies) but also in its ability to continue growing (where the average number of visits had increased from 300 per month in 2005 to 600 per month during the first 6 months of 2006).

These forums were able to capture the pressing questions teachers currently have (could an underperformance in math be related to a brain problem?), the common misconceptions about the brain teachers often acquire (is Attention Deficit Hyperactivity Disorder characterised by a lack of motivation?), and the hunger educational specialists have regarding what information from neuroscience can apply and work in the classroom setting (how to incorporate in teaching various neural developments).

The following are a selection of extracts on varying topics of discussion, from "Emotions and Learning", "Literacy and the Brain" to "Dyscalculia" and "Brain Science and Education". Meet our members, who hail from Solana Beach, California to a farm in East York, United Kingdom, from mid-west America to the outback hills of Australia, and even beyond OECD member countries to include members from countries like India and Nigeria. They are teachers, education counselors, neuroscientists and OECD experts. They go by aliases such as "segarama", "puppet-maker", "just me" "the foreign brain" and "4th grade teacher". Some are even retired, well in their 60s, still lit by the fire to educate, by the driving force of learning, and to find out the latest about what neuroscientists are discovering and what can be possibly applied in the future.

As you discover and follow along in their stories, they will serve as a reminder of where all communities of practice find themselves during these exciting times – in the giant gap between neuroscience and education – between information exchange and application. Share in their experiences as we all move forward.

All entries have been directly extracted from the teach-the-brain forum, the content and personal styles have not been edited so as to preserve the individuality and authenticity of these exchanges.

On Dyslexia...

Dear All,

I would like to find out what causes a student with dyslexia to not be able to hear the distinctions between some sounds. Why do they have difficulties rhyming, segmenting, blending? What is the exact cause of the phonological processing deficit? Could it be damage from an ear infection as an infant or toddler? Or an injury or whack on the head? Is it even damage as opposed to being underdeveloped for some reason? I've always been interested in the cause of disabilities. I would be interested in your thoughts on causes of the phonological processing problems that plague dyslexia.

justme

Hi justme,

You can learn more about dyslexia from our dyslexia primer on the OECD *Brain and Learning* website: *www.oecd.org/document/51/0,2340,en_2649_14935397_35149043_1_1_1_1,00.html*. I would also highly recommend *Overcoming Dyslexia* by Sally Shaywitz (2003). Best wishes,

Christina Hinton
Expert, OECD consultant

Dear All,

I would like to hear comments on the view that dyslexia is not a "special" condition of difficulty with reading, but that poor readers are so because they lack an ability to segment and blend sounds: an ability which is independent of intelligence in the same way as colour blindness is independent of intelligence. This view was presented in a recent UK television programme but I regret that I have forgotten the names of the proponents.

Deborah

Dear Deborah,

Re: the television program and Professor Elliot's position. Where any Child with a "learning difficulty", must be officially given a "Label" for their difficulty before they will receive direct assistance to address it. No label, no assistance. Though, the problem is that Labels are generalised terms, whereas very few children/adults actually fit the precise definition. This creates a further problem, what happens after having been given a label? Remediation is then adopted, which addresses what is defined under the label. Which fails to recognise the variety of different potential causations of a difficulty with reading/writing/ Dyslexia. Whereas Prof Elliot states; "It is a catch all label". What Elliot is suggesting, is that we need to move beyond the labels, and directly focus on the causations so that it becomes an "individualised approach".

I would suggest that we need to move beyond labels towards the development of the general public's understanding of our Brain's Processes?

Geoff.

Dear Forum,

Dyslexia in these days is seen as a problem. I hope one day people will recognise dyslexia as a different way of thinking, which could lead to new ideas or new ways of solving a problem. If it is looked at this way dyslexia can be an extra... it just depends on how you look at it...

frulle

On Mathematics...

Dear All,

I know a very smart child who has excellent grades in all his courses except in math. His parents say that he makes an effort to study math as much as for other courses. He is just 8 years old. Is this normal? Do you think that his math underperfomance could be related to a brain problem?
Thanks,

The foreign brain

Dear teachers,

Last weekend I read a great article about math difficulties on this website. Sorry I can't tell you just how I find it. There is so much good information here that I can spend hours reading and forget where I've been. Anyway if I am correct, there may be a problem with number sense. Underlying the problem comes from the child not understanding the basis of what are numbers and what they represent. The article explains that simultaneously a child must understand that the number, for example five, must recognize the symbol 5, the spelling five, and the concept of five. I hope that you can find the article and will report back to this forum what you learn. I am very interested.

Cathy Trinh

It sounds as though the child you are talking about may be dyscalculic. Dyscalculia is the equivalent of dyslexia for maths. However, unlike dyslexia, it is very understudied. Dyscalculia definitions and tests vary from country to country. The terminology used even differs, *e.g.* in the US, it is known as "mathematical disabilities". In England, there is a national test, the "Dyscalculia Screener" by Brian Butterworth. In France, speech therapists perform their own custom tests. In the USA, your local school psychologist will test the child and decide if he or she qualifies as having a specific learning disability in math.

Cathy, you probably read an article about the symposium on remediation software for dyscalculia (which OECD is funding the development of), you can find the article again by going to "Brain and Learning" from the main page and scrolling down.

If you want to learn more about dyscalculia, you can visit Brian Butterworth's webpage at *www.mathematicalbrain.com*.

And the OECD team will shortly be putting up more information about dyscalculia on this website, so stay tuned!

Anna Wilson

On Assessment...

Hi Christina,

Thanks for your reply.

In regard to the implications that this could have for teaching, I would return to our recent discussion of assessment/evaluation.

Where this highlights the value of Formative assessment as a means of moving beyond the vague generalised Summative assessment of Subjects, which simply indicates a Student being "good or bad" at a Subject and provides no indication of where precise intervention is required.

Whereas a Formative assessment of a Subject could address the various neural processes that are utilised within the learning of subject. This could help identify the precise neural process/es that need developmental assistance. Though this comes back to the issue of what Teachers need to learn about Neuroscience and Learning?

Geoff.

Dear Geoff,

Your idea of directly linking neuroscience and assessment is thought-provoking. Might there be a future in which a student's assessment portfolio would include a neuroscience component, such as an fMRI scan?
Take good care,

Christina Hinton
Expert, OECD consultant

Dear Christina,

I suspect it may be controversial? Though it would be of considerable value, in that it would highlight a "problem" very early, for targeted intervention, rather than waiting for a few years of low grades before recognising a problem, by which time a Student has fallen a long way behind. Therefore it could be of notable benefit?

Geoff.

On Emotions and Learning...

Dear Teachers,

Do you feel that emotions can be separated from cognitive processes in a classroom context?

Christina Hinton
Expert, OECD consultant

Dear All,

In a word, no.
In an ideal situation, teachers would have students that learn everything they are taught as soon as they are given the information, and would be able to produce evidence of understanding on their own. This is the image many of us have of education. If this isn't happening, then something is wrong with the teacher or the student. I believe there is nothing wrong with either except a lack of understanding of their emotions.

One example of this is in learning basic math facts. In fourth grade, students who can "automatically" give answers to times facts learn more difficult computation problems more quickly than those who don't. (Of course there are many students who are poor in timings and good at more complex computation, but when dealing with large numbers of children, teachers tend to look at the big picture.) So, many teachers have timings on these facts. Long ago, I realized that doing this as a group activity in the classroom was too destructive to continue. Students who excelled were ecstatic. Everyone who didn't get the most felt bad. For a few, the competition made them work harder. For more than a few, the competition reinforced the image of themselves as being bad in math. Every year I have parents tell me their children hate being timed because for one reason or another they can't think quickly enough. Therefore, they say, math facts are not important, so please don't give my child timings. I say, let's give it a couple weeks and see how it goes. Now I have parents do individual timings in the hall. When students are by themselves, no one sees how they work, and every one of my students makes progress. The kids are happy, the parents are happy, and some of them even thank me.

I'm reading with interest the comments on this forum regarding the involvement of holistic learning, motor skills, physics, philosophy, and biological underpinnings. In order to understand the content of each message, I try to envision the person writing and their background to find their intent. Having an emotional understanding of the person helps to put into context their reasoning and use of words. This is another example of what teachers do in teaching literacy. All people have "voice" in their writing, no matter what they are writing. Voice is characterized by the writer's personality, which is based on the person's emotions.

So, separating emotions from cognitive processes does not seem do-able to me. In fact, I think it is quite necessary to develop a certain level of passion in learning. However, focusing, changing, containing in other words, dealing with emotions, is very do-able, and is, in fact, what teachers spend most of their time doing. Perhaps it depends on what you mean by separating.

What does neuroscience say about this?

4th grade teacher

Dear 4th grade teacher,

Thanks very much for this.

Emotive and cognitive processes operate seamlessly in the brain. This is because emotion and cognition are categorical concepts that do not reflect brain organisation. The brain is organized into assemblies of neurons with specialized properties and functions, a principle termed modularity. These assemblies regulate very specific functions, such as spatial perception or tonal discrimination. A stimulus elicits a network response of various assemblies to produce a certain experience. Particular components of this experience can usefully be labelled cognitive or emotive, but the distinction among the two is categorical and not based in brain function. Therefore, from my perspective, the emotion that your students feel when learning mathematics and the cognitive processes they engage in cannot be separated. It seems that you view the experience in a holistic way as well.

All the best,

Christina Hinton
Expert, OECD consultant

On Educational Neuroscience...

Dear Teachers,

View the quotations by an educator and from a renowned neuroscientist present at our *Lifelong Learning Network* meeting which was held in Tokyo in January 2005. We would be interested in any comments you wish to make on these reflections in this thread.

"Engineers don't really rely upon physics to build their bridges, nor do they wait for physics to come up with principles that will tell them how to do so. Rather they study physics and adapt the principles in that field to the practical decisions that need to be made that include the cultural values governing vehicular travel that will eventually move over the bridge."

Michael Posner, Neuroscientist

"We are all talking in metaphors because there is no real field – education and the brain – to relate to. All we can do is refer to the concept as a relation to something that social groups will recognize."

Frank Coffield

Dear All,

Posner is exactly right but I recommend the book "Why Buildings Fall Down" by Matthys Levy and Mario Salvadori. The reality is that progress will not be made without failure and no one would say that the advances of the architects should not have been attempted. We have to move forward with some ignorance because there is no other choice.
So who will do the pioneering?

Karl

ANNEX B

Brain Imaging Technologies

Neuroscientific research techniques vary and can include invasive procedures, including neurosurgery. However, the now most well-known and used tools are non-invasive brain-imaging technologies. Brain-imaging tools can be divided into two general categories, those that provide high-resolution spatial information and those that provide high-resolution temporal information about brain activity. Among those tools that provide high-resolution spatial information about brain activity, the best known are Positron Emission Tomography (PET) and functional Magnetic Resonance Imaging (fMRI). PET techniques, using radioisotopes, detect brain activity by monitoring changes in oxygen utilisation, glucose utilisation, and cerebral blood flow changes. fMRI, with the use of radio frequencies and magnets, identifies changes in the concentration of deoxygenated haemoglobin (see Box A). Both techniques require subjects to remain motionless for accurate imaging.

One thing to note from this is that this signal that is measured – the change in oxygen – is produced naturally by the body – no contrast needs to be injected. This means it is a non invasive procedure, unlike other forms of brain imaging like PET scanning which requires an injection of radioactive materials. This means fMRI can safely be used to scan children's brains and it can also be performed multiple times in the same person – which means it is possible to now look at the effects of training, interventions, etc. The second thing to note from this is that this signal measured – while it is inferred to be representative of neuronal activity – is an INDIRECT measure of brain activity. That is an important limitation to remember about the method – at this point the primary way of measuring neuronal activity directly is still limited to animal studies.

Because PET and fMRI provide spatial resolution in the millimetre range, but temporal resolution only in seconds, these techniques are useful for measuring changes in brain activity during relatively prolonged cognitive activity. Another technique, Transcranial Magnetic Stimulation (TMS) is used to create a temporary disruption of brain function (a few seconds) in order to help locate brain activity in a circumscribed region of the brain. Nonetheless, processes such as performing mathematical calculations or reading involve many processes that occur over the course of a few hundred milliseconds. For that reason, PET and fMRI are able to localise brain regions involved in reading or mathematical activity, but cannot illuminate the dynamic interactions among mental processes during these activities.

Another set of tools provides accurate temporal resolution in the millisecond range, but their spatial resolution is coarse, providing data only in centimetres. These techniques measure electric or magnetic fields at the scalp surface during mental activity. Among these tools are electroencephalography (EEG), event-related potentials (ERP), and

Box A. **What is fMRI?**

fMRI, which allows us to see brain function is a variant of standard MRI. Standard MRI, such as is used in hospitals to look at knees and backs and brains, gives us a picture of soft tissue. This is as opposed to x-rays, for example, which let us see bone and calcifications. So, MRI allows us to view ligaments and tendons and other soft tissues inside the body, including the brain, which is entirely soft tissue. Since fMRI is a variant of MRI it has many of the same benefits and limitations. Some limitations are the same as any MRI: 1) you have to remain extremely still; 2) if you are claustrophobic it is very difficult to tolerate the very small tube; and 3) it is very loud. Anyone who has a standard MRI has experienced these qualities and it is the same with fMRI. Standard MRI works because the different tissues in our body have slightly different magnetic properties and the magnets in the MRI make them respond slightly differently and then computers help make that different magnetic response into an image. With fMRI, we take advantage of the fact that oxygen and its carrier, hemoglobin, have magnetic properties and respond to the magnetic field. This gives us a way to measure brain function, since neurons that are more active use more oxygen and have different magnetic signal than areas which are not active.

To summarise this technique briefly

A person has a thought or idea or does some cognitive or perceptual task this leads to an increase in neural activity – in a specific region or regions of the brain – a FOCAL increase in neural activity. This in turn leads to a FOCAL increased blood flow to that same brain region which leads to increased oxygen delivery to that region – (or more accurately a change in the ratio of deoxy and oxy hemoglobin). It is this change in oxygen that gives us our signal in fMRI (see Figure A).

Figure A. **Functional magnetic resonance imaging**

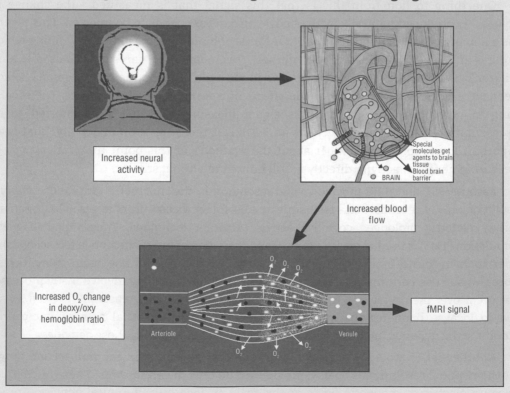

Increased neural activity

Special molecules get agents to brain tissue
Blood brain barrier
BRAIN

Increased blood flow

Increased O$_2$ change in deoxy/oxy hemoglobin ratio

Arteriole

Venule

fMRI signal

Source: Courtesy of Elize Temple, image by Cassandra Davis.

magnetoencephalography (MEG). EEG and ERP use electrodes placed on particular areas on the scalp. Because of their ease of use, these techniques are often used successfully with children. MEG uses super-conducting quantum interference devices (SQUIDs) at liquid helium temperature. Using these tools, accurate measures in the milliseconds of changes in brain activity during cognitive tasks can be obtained.

A new method for non-invasive brain function imaging is optical topography (OT), which was developed by using near infrared spectroscopy (NIRS) (Box B). Unlike conventional methodologies, it can be used for behavioural studies because the flexible optical fibres allow a subject to move, and a light and compact system can be built. This method can be applied to infants as well as adults. The observation of early development on a monthly time-scale will provide information about the architecture of the neuronal processing system in the brain. Optical topography may bring important implications for learning and education.[1]

Box B. **Near-infrared optical topography (NIR-OT) for learning sciences and brain research**

Near-infrared optical topography (NIR-OT) is a new non-invasive methodology for higher-order brain-function analyses, which captures local area brain activation among multiple subjects under natural conditions, such as learning at home or in a classroom. This new methodology helps to assess individual and integrated brain activities in cross-sectional longitudinal studies.

Other brain scanning methodologies such as functional magnetic resonance imaging (fMRI) and magneto-encephalography (MEG) have been used since the early 1990s. However, those methodologies have exhibited various limitations because a subject has to be rigidly fixed to the machine during the entire process of measurement. Although this point has not been so important in conventional neurology in which the subjects were primarily patients, there has been an increasing interest in enabling higher-order brain-function imaging under natural conditions. Indeed, the NIR-OT methodology provides brain images of multiple subjects without any restrictions, which is an essential aspect in analysing the interaction between learners and teachers.

NIR-OT relies on a near-infrared light carried by optical fiber, which is radiated on the scalp, some of that light will reach a depth of approximately 30 mm for adults. The cerebral cortex will then reflect the light and pass it back through the scalp. This reflected and scattered light will later be detected by another optical fiber situated about 30 mm from the point of irradiation. The near-infrared light is completely non-invasive, equivalent to the irradiation power of the sun on a cloudy day in winter. The NIR-OT analytical device is a mobile semiconductor that may in the future become a small integrated circuit of hands-on size that would significantly enable enlargement of the environmental measurement conditions.

Various preliminary studies related to learning sciences and brain research have already been performed over the years using the NIR-OT methodology. They include developmental studies with healthy infants, language studies on working memory. NIR-OT has also been applied to other research areas such as changes in brain function during cooking and driving.

Hideaki Koizumi, Fellow, Hitachi, Ltd.

---

1. Koizumi, H. *et al.* (1999), "Higher-order Brain Function Analysis by Trans-cranial Dynamic Nearinfrared Spectroscopy Imaging", *Journal Biomed*, Opt., Vol. 4.

Effective research in cognitive neuroscience requires a combination of these techniques in order to provide information on both spatial location and temporal changes in brain activity associated with learning. In making the link with learning processes, it is important for the neuroscientist to have fine-grained elementary cognitive operations and analyses in order to make powerful use of brain-imaging tools. Among those disciplines associated with learning, such fine range and analyses are most typically available from studies in cognitive science or cognitive psychology, and, to date, typically in studies of visual processing, memory, language, reading, mathematics and problem-solving.

Other research options available to neuroscientists include examining brains during autopsy (for example, to measure synaptic density) and in some rare cases, working with certain medical populations, such as those suffering from epilepsy (to learn about brain processes from people who have suffered brain damage or brain lesions due to disease or injury). Some neuroscientists study children suffering from fetal alcohol syndrome or Fragile X syndrome, and others study the cognitive decay prevalent during the onset of Alzheimer's disease or senile depression. Still others study the brains of primates or of other animals, such as rats or mice, in order to better understand how human mammalian brains function. In the past, without brain-imaging techniques available, it has been difficult to collect direct neuroscientific evidence of learning in the general, healthy human population.

A further limitation is presented by the fact that no single set of well-understood developmental learning tasks has been applied to normal human populations across the lifespan. Much work has been carried out in regards to early childhood learning, but less regarding adolescent learning and even less again regarding adult learning. Without a baseline of normal cognitive development, it is difficult to understand any pathological occurrences in learning.

Understanding both the power and limitations of brain-imaging technology and the necessity of conducting rigorous cognitive protocols is the first step in trying to understand how cognitive neuroscience can guide education eventually in the formation of brain-based curricula. Recent findings are beginning to show that eventually education will emerge at the crossroads of cognitive neuroscience and cognitive psychology along with sophisticated and well-defined pedagogical analysis. In the future, education will be trans-disciplinary, with an intersection of different fields merging to produce a new generation of researchers and educational specialists adept at asking educationally significant questions at the right grain size.

Current research methods in cognitive neuroscience necessarily limit the types of questions that are addressed. For example, questions such as "How do individuals learn to recognise written words?" are more tractable than "How do individuals compare the themes of different stories?". This is because the first question leads to studies where the stimuli and responses can be easily controlled and contrasted with another task. As such, it becomes understandable in reference to known cognitive models. The second question involves too many factors that cannot be successfully separated during experimental testing. For this reason, the type of educational tasks favoured by society will remain more complex than the ones that might suit cognitive neuroscience.

Researchers also stress the methodological necessity of testing for learning not only immediately after some educational intervention (which is typical of current practice), but also at certain intervals thereafter, especially in the case of age-related comparisons. These longitudinal studies take the research projects out of the laboratory and into real-life situations, which places limits on when the results can be interpreted and available for education use.

When attempting to understand and analyse scientific data, it is important to retain critical standards when judging claims about cognitive neuroscience and its educational implications. Some points to consider:

- the original study and its primary purpose;
- if the study is a single study or a series of studies;
- if the study involved a learning outcome;
- the population used.[2]

The importance of developing an informed critical community for the progress of science (that comes to consensus, over time, on the evidentiary and inferential basis of purported scientific claims) has recently been re-emphasised.[3]

The development of such a community (composed of educators, cognitive psychologists, cognitive neuroscientists, and policy makers, etc.) around the emerging sciences of learning is crucial. In order for that community to develop, an appropriately critical judgement in matters of "brain-based" claims about learning and teaching is necessary. Integrated into this community, education policy makers will more successfully enter into appropriate brain-based curricula if there is a recognition of the following:

a) the popularity of a neuroscientific claim does not necessarily imply its validity;

b) the methodology and technology of cognitive neuroscience is still a work in progress;

c) learning is not completely under conscious or volitional control;

d) the brain undergoes natural developmental changes over the lifespan;

e) much cognitive neuroscience research has been directed at understanding or addressing brain-related pathologies or diseases;

f) a satisfactory science of learning considers emotional and social factors in addition to cognitive ones; and

g) although a science of learning and brain-based education are just beginning, important gains are already being made.

There are ample data at the psychological level (drawn primarily from well designed studies in cognitive psychology) from which to draw lessons for learning and teaching. Data from cognitive neuroscience can help by refining hypotheses, disambiguating claims, and suggesting directions for research. In other words, a major contribution of cognitive neuroscience to an emerging science of learning may be to imbue the discipline with a scientific scepticism toward unfettered claims and unexamined advocacy about how to improve teaching and learning.

But scepticism toward some current claims about the neuroscientific basis for learning should not breed cynicism about the potential benefits of cognitive neuroscience for education. Indeed, the emerging data about brain plasticity are encouraging. The evidence for claims about learning is unlikely to come from neuroscientific studies alone, however. In the future, improved brain imaging technologies and more sophisticated learning protocols may allow us to further illuminate this question.

---

2. Whether using human primates or non-human primates, questioning the representativeness of the sample, and asking to what population the claims do apply, is of utmost importance.
3. In a US National Research Council report on Scientific Inquiry in Education.

Glossary

Acalculia. See dyscaculia.

Accumbens area. See nucleus accumbens.

Action potential. This occurs when a neuron is activated and temporarily reverses the electrical state of its interior membrane from negative to positive. This electrical charge travels along the axon to the neuron's terminal where it triggers the release of an excitatory or inhibatory neurotransmitter.

Activation study. Study performed with imaging techniques (see also PET and fMRI).

ADHD (Attention Deficit Hyperactivity Disorder). A syndrome of learning and behavioural problems characterised by difficulty sustaining attention, impulsive behaviour (as in speaking out of turn), and hyperactivity.

Alzheimer's disease. A progressive degenerative disease of the brain associated with ageing, characterised by diffuse atrophy throughout the brain with distinctive lesions called senile plaques and clumps of fibrils called neurofibrillary tangles. Cognitive processes of memory and attention are affected (see also neurodegenerative diseases).

Amygdala. A part of the brain involved in emotions and memory. Each hemisphere contains an amygdale ("shaped like an almond") and located deep in the brain, near the inner surface of each temporal lobe.

Angular gyrus. An area of the cortex in the parietal lobe involved in processing the sound structure of language and in reading.

Anhedonia. Recognised as one of the key symptoms of the mood disorder depression. Patients with anhedonia are unable to experience pleasure from normally pleasurable life events such as eating, exercise, and social/sexual interactions.

Anterior cingulated cortex. Frontal part of the cingulate cortex. It plays a role in a wide variety of autonomic functions, such as regulating heart rate and blood pressure, and is vital to cognitive functions, such as reward anticipation, decision-making, empathy, and emotions.

Aphasia. Disturbance in language comprehension or production.

Apolipoprotein E (or "apoE"). Has been studied for many years for its involvement in cardiovascular diseases. It has only recently been found that one allele (gene factor) of the aopE gene (E4) is a risk factor for Alzheimer's disease.

Artificial intelligence (AI). A field of computer science which attempts to develop machines that behave "intelligently".

Attention. Attention is the cognitive process of selectively concentrating on one task while ignoring other tasks. Imaging studies have been able to show the distinct networks of neural areas which carry out the various functions of attention such as maintaining the alert state, orienting to sensory information and resolving conflict among competing thoughts or feelings.

Auditory cortex. The region of the brain that is responsible for processing of auditory (sound) information.

Auditory nerve. A bundle of nerve fibers extending from the cochlea of the ear to the brain, which contains two branches: the cochlear nerve that transmits sound information and the vestibular nerve that relays information related to balance.

Autism/autistic spectrum disorders. A spectrum of neurodevelopmental conditions, characterised by difficulties in the development of social relationships, communication skills, repetitive behaviour, and learning difficulties.

Axon. The fiberlike extension of a neuron by which the cell sends information to target cells.

Basal ganglia. Clusters of neurons, which include the caudate nucleus, putamen, globus pallidus and substantia nigra, that are located deep in the brain and play an important role in movement. Cell death in the substantia nigra contributes to Parkinsonian signs.

Bipolar disorder. Otherwise known as manic depression. Bipolar disorder involves extreme swings of mood from mania (a form of euphoria) to deep depression. There is no simple cause, although there is strong evidence that it is associated with internal chemical changes to various natural transmitters of mood to the brain, but the precise way in which this happens is not yet known. The disorder can be triggered by the stresses and strains of everyday life, or a traumatic event or, in rare cases, physical trauma such as a head injury.

Brainstem. The major route by which the forebrain sends information to and receives information from the spinal cord and peripheral nerves. It controls, among other things, respiration and regulation of heart rhythms.

Broca's area. The brain region located in the frontal lobe of the left hemisphere, involved in the production of speech.

Caudate or caudale nucleus. A telencephalic nucleus located within the basal ganglia in the brain. The caudate is an important part of the brain's learning and memory system.

Cerebellum. A part of the brain located at the back and below the principal hemispheres, involved in the regulation of movement.

Cerebral hemispheres. The two specialised halves of the brain. The left hemisphere is specialised for speech, writing, language and calculation; the right hemisphere is specialised for spatial abilities, face recognition in vision and some aspects of music perception and production.

Cerebrospinal fluid. A liquid found within the ventricles of the brain and the central canal of the spinal cord.

Cerebrum. Otherwise known by more technical term telencephalon. Refers to cerebral hemispheres and other, smaller structures within the brain, and is composed of the following sub-regions: limbic system, cerebral cortex, basal ganglia, and olfactory bulb.

Circadian clock/rhythm. A cycle of behavior or physiological change lasting approximately 24 hours.

Classical conditioning. Learning in which a stimulus that naturally produces a specific response (unconditioned stimulus) is repeatedly paired with a neutral stimulus (conditioned stimulus). As a result, the conditioned stimulus can become able to evoke a response similar to that of the unconditioned stimulus.

Cochlea. A snail-shaped, fluid-filled organ of the inner ear responsible for transducing motion into neurotransmission to produce an auditory sensation.

Cognition. Set of operations of the mind which includes all aspects of perceiving, thinking, learning, and remembering.

Cognitive maps. Mental representations of objects and places as located in the environment.

Cognitive networks. Networks in the brain involved in processes such as memory, attention, perception, action, problem solving and mental imagery. This term is also used for artificial networks as in artificial intelligence.

Cognitive neuroscience. Study and development of mind and brain research aimed at investigating the psychological, computational, and neuroscientific bases of cognition.

Cognitive science. Study of the mind. An interdisciplinary science that draws upon many fields including neuroscience, psychology, philosophy, computer science, artificial intelligence, and linguistics. The purpose of cognitive science is to develop models that help explain human cognition – perception, thinking, and learning.

Cognitive training. Teaching methods and training to remediate cognitive deficits.

Cohort study. A type of longitudinal study used in medicine and social sciences that compares a cohort, or group of people who share a common characteristic or experience, to an outside group.

Competences. Referring to student ability. The mental capacity to perform particular tasks.

Constructivism. A learning theory whereby individuals actively construct understanding from their experiences.

Corpus callosum. The large bundle of nerve fibers linking the left and right cerebral hemispheres.

(cerebral) Cortex. Outer layer of the brain.

Cortisol. A hormone manufactured by the adrenal cortex. In humans, it is secreted in greatest quantities before dawn, readying the body for the activities of the coming day.

Critical period. Concept referring to certain periods when the brain's capacity for adjustment in response to experience is substantially greater than during other periods. In humans, critical periods only exist during prenatal development. Sensitive periods, however, are known to occur in childhood (see sensitive period).

Cross-sectional study. A type of descriptive study that measures the frequency and characteristics of a population at a particular point in time.

CT (Computed Tomography). Originally known as computed axial tomography (CAT or CT scan) and body section roentgenography. A medical imaging method employing tomography where digital geometry processing is used to generate a three-dimensional image of the internals of an object from a large series of two-dimensional X-ray images taken around a single axis of rotation.

Decoding. An elementary process in learning to read alphabetic writing systems (for example, English, Spanish, German or Italian) in which unfamiliar words are deciphered by associating the letters of words with corresponding speech sounds.

(senile) Dementia. A condition of deteriorated mentality that is characterised by marked decline from the individual's former intellectual level and often by emotional apathy. Alzheimer's disease is one form of dementia.

Dendrite. A tree-like extension of the neuron cell body. It receives information from other neurons.

Depression. A lowering of vitality of functional activity: the state of being below normal in physical or mental vitality. Senile depression refers to depression in later life which may be dominated by agitation and hypochondria. Whether this form of depression is distinct from depression during earlier life is not clear.

Development. Progressive change that occurs in human beings as they age. Biological inclinations interact with experience to guide development throughout life.

DNA (Deoxyribonucleic acid). DNA is a long polymer of nucleotides (a polynucleotide) that encodes the sequence of amino acid residues in proteins, using the genetic code.

Dopamine. A catecholamine neurotransmitter known to have multiple functions depending on where it acts. Dopamine-containing neurons in the substantia nigra of the brainstem project to the caudate nucleus and are destroyed in Parkinson's victims. Dopamine is thought to regulate emotional responses, and play a role in schizophrenia and cocaine abuse.

DTI (Diffusion Tensor Imaging). A magnetic resonance imaging (MRI) technique that enables the measurement of the restricted diffusion of water in tissue. It allows the observation of molecular diffusion in tissues in vivo and therefore the molecular organisation in tissues.

Dyscalculia. Impairment of the ability to perform simple arithmetical computations, despite conventional instruction, adequate intelligence and socio-cultural opportunity.

Dyslexia. A disorder manifested by difficulty in learning to read despite conventional instruction, adequate intelligence, and socio-cultural opportunity.

Dyspraxia. Motor co-ordination difficulties in carrying out any complex sequence.

ECG (Electrocardiogram). A recording of the electrical voltage in the heart in the form of a continuous strip graph.

EEG (Electroencephalogram). A measurement of the brain's electrical activity via electrodes. EEG is derived from sensors placed in various spots on the scalp, which are sensitive to the summed activity of populations of neurons in a particular region of the brain.

Electrochemical signals. These signals are the means by which neurons communicate with one another.

Emotional intelligence. Sometimes referred to as emotional quotient ("EQ"). Individuals with emotional intelligence are able to relate to others with compassion and empathy, have well-developed social skills, and use this emotional awareness to direct their actions and behaviour. The term was coined in 1990.

Emotional regulation. Ability to regulate and appropriately temper emotions.

Emotions. There is no single universally accepted definition. The neurobiological explanation of human emotion is that emotion is a pleasant or unpleasant mental state organised mostly in the limbic system of the mammalian brain.

Endocrine organ. An organ that secretes a hormone directly into the bloodstream to regulate cellular activity of certain other organs.

Endorphins. Neurotransmitters produced in the brain that generate cellular and behavioral effects similar to those of morphine.

Epigenetic. Changes in gene function, often elicited by environmental factors.

Epilepsy. A chronic nervous disorder in humans which produces convulsions of greater or lesser severity with clouding of consciousness; it involves changes in the state of consciousness and of motion due to either an inborn defect of a lesion of the brain produced by tumour, injury, toxic agents, or glandular disturbances.

ERP (Event-related potentials). Electric signals are first recorded with an EEG. Data from this technology is then time locked to the repeated presentation of a stimulus to the subject, in order to see the brain in action. The resulting brain activation (or event-related potentials) can then be related to the stimulus event.

Evoked potentials. A measure of the brain's electrical activity in response to sensory stimuli. This is obtained by placing electrodes on the surface of the scalp (or more rarely, inside the head), repeatedly administering a stimulus, and then using a computer to average the results.

Excitation. A change in the electrical state of a neuron that is associated with an enhanced probability of action potentials.

Excitatory synapses. Synapses where neurotransmitters decrease the potential difference across neuron membranes.

Experience-dependent. A property of a functional neural system in which variations in experience lead to variations in function, a property that might persist throughout the life-span.

Experience-expectant. A property of a functional neural system in which the development of the system has evolved to critically depend on stable environmental inputs that are roughly the same for all members of species (*i.e.* stimulation of both eyes in new-borns during development of ocular dominance columns). This property is thought to operate early in life.

Explicit memory. Memories that can be retrieved by a conscious act, as in recall, and can be verbalised, in contrast to implicit or procedural memories, which are less verbally explicit.

Fatty acids. The human body can produce all but two (linoleic acid and alpha-linolenic acid) of the fatty acids it needs which the brain is made up of. Since they cannot be made in the body from other substrates and must be supplied in food (namely in plant and fish oils) they are called essential fatty acids. (See also Omega and HUFA).

Fear/fear conditioning. Fear conditioning is a form of classical conditioning (a type of associative learning pioneered on animals by Ivan Pavlov in the 1920s) involving the repeated pairing of a harmless stimulus such as a light, called the conditioned stimulus, with a noxious stimulus such as a mild shock, called the unconditioned stimulus, until the animal shows a fear response not just to the shock but to the light alone, called a conditioned response. Fear conditioning is thought to depend upon the amygdala. Blocking the amygdala can prevent the expression of fear.

fMRI (Functional Magnetic Resonance Imaging). Use of an MRI scanner to view neural activity indirectly through changes in blood chemistry (such as the level of oxygen) and investigate increases in activity within brain areas that are associated with various forms of stimuli and mental tasks (see MRI).

Forebrain. The largest division of the brain, which includes the cerebral cortex and basal ganglia. It is credited with the highest intellectual functions.

Frontal lobe. One of the four divisions (parietal, temporal, occipital) of each hemisphere of the cerebral cortex. It has a role in controlling movement and associating the functions of other cortical areas, believed to be involved in planning and higher order thinking.

Functional imaging. Represents a range of measurement techniques in which the aim is to extract quantitative information about physiological function.

Fusiform gyrus. A cortical region running along the ventral (bottom) surface of the occipital-temporal lobes associated with visual processes. Functional activity suggests that this area is specialised for visual face processing and visual word forms.

Gene. A gene is the unit of heredity in living organisms. Genes influence the physical development and behaviour of the organism. See also genetics.

Genetics. The science of genes, heredity, and the variation of organisms. **Classical genetics** consists of the techniques and methodologies of genetics predating molecular biology. **Molecular genetics** builds upon the foundation of classical genetics but focuses on the structure and function of genes at a molecular level. **Behavioral genetics** studies the influence of varying genetics on animal behaviour, and the causes and effects of human disorders.

Glia/glial cells. Specialised cells that nourish and support neurons.

Graphemes. The smallest unit of written language, including letters, Chinese characters, numerals, and punctuation marks.

Grey matter/gray matter. Gray matter consists of neurons' cell bodies and dendrites.

Gyrus/gyri. The circular convolutions of the cortex of which each has been given an identifying name: middle frontal gyrus, superior frontal gyrus, inferior frontal gyrus, left interior frontal gyrus, posterior middle gyrus, postcentral gyrus, supermaginal gyrus, angular gyrus, left angular gyrus, left fusiform gyrus, cingulated gyrus.

Hard-wired. Meaning "not changeable". In contrast to concept of plasticity in which brain is malleable to change.

(cerebral) Hemisphere. One of two sides of the brain classified as "left" and "right".

Hippocampus. A seahorse-shaped structure located within the brain and considered an important part of the limbic system. It functions in learning, memory and emotions.

Hormones. Chemical messengers secreted by endocrine glands to regulate the activity of target cells. They play a role in sexual development, calcium and bone metabolism, growth and many other activities.

HUFA (Highly unsaturated fatty acids).

Hypothalamus. A complex brain structure composed of many nuclei with various functions. These include regulating the activities of internal organs, monitoring information from the autonomic nervous system and controlling the pituitary gland.

Immune system. The combination of cells, organs and tissues which work together to protect the body from infection.

Implicit memory/learning. Memories that cannot be retrieved consciously but are activated as part of particular skills or action, and reflect learning a procedure of a pattern, which might be difficult to explicitly verbalise or consciously reflect upon (i.e. memory that allows you to engage in a procedure faster the second time, such as tying a shoe).

Information-processing. An analysis of human cognition into a set of steps whereby abstract information is processed.

Inhibition. In reference to neurons, this is a synaptic message that prevents the recipient cell from firing.

Insomnia. Inability to remain asleep for a reasonable period.

Intelligence. Characeristic of the mind lacking a scientific definition. Can be fluid or crystallised intelligence (see also multiple intelligences, IQ).

Interference theory. A theory of forgetting in which other memories interfere with the retention of the target memory.

Ions. Electrically charged atoms.

IQ. A number held to express the relative intelligence of a person originally determined by dividing mental by chronological age and multiplying by 100.

Left-brained thinking. A lay term based on the misconception that higher level thought processes are strictly divided into roles that occur independently in different halves of the brain. Thought to be based on exaggerations of specific findings of left hemisphere speclialisations, such as the neural systems that control speaking.

Limbic system. Also known as the "emotional brain". It borders the thalamus and hyphothalamus and is made up of many of the deep brain structures – including the amygdala, hippocampus, septum and basal ganglia – that work to help regulate emotion, memory and certain aspects of movement.

Lobe. Gross areas of the brain sectioned by function (see also occipital, temporal, parietal and frontal).

Long-term memory. The final phase of memory in which information storage may last from hours to a lifetime.

Longitudinal study. Studies that track the development of individuals over an extended period of time.

Long-term potentiation (LTP). The increase in neuron responsiveness as a function of past stimulation.

MEG (Magnetoencephalography). A non-invasive functional brain imaging technique sensitive to rapid changes in brain activity. Recording devices ("SQUIDs" for *Superconducting Quantum Interference Devices*) placed near the head are sensitive to small magnetic fluctuations associated with neural activity in the cortex. Responses to events can be traced out on a millisecond time scale with good spatial resolution for those generators to which the technique is sensitive.

Melatonin. Produced from serotonin, melatonin is released by the pineal gland into the bloodstream. It affects physiological changes related to time and lighting cycles.

Memory. Working memory/or short-term memory refers to structures and processes used for temporarily storing and manipulating information. **Long-term memory** stores memory as meaning. Short-term memory can become long-term memory through the process of rehearsal and meaningful association.

Memory consolidation. The physical and psychological changes that take place as the brain organises and restructures information in order to make it a part of memory.

Memory span. The amount of information that can be perfectly remembered in an immediate test of memory.

Mental imagery. Also known as visualisation. Mental images are created by the brain from memories, imagination, or a combination of both. It is hypothesised that brain areas responsible for perception are also implicated during mental imagery.

Mental images. Internal representations consisting of visual and spatial information.

Metabolism. The sum of all physical and chemical changes that take place within an organism and all energy transformations that occur within living cells.

Meta-cognition. Conscious awareness of one's own cognitive and learning processes. In short, "thinking about thinking".

Micro-array. A tool for analysing gene expression that consists of a glass slide or other solid support with the sequences of many different genes attached at fixed locations. By using an array containing many DNA samples, scientists can determine the expression levels of hundreds or thousands of genes within a cell in a single experiment.

Micro-genetics. A method of tracking change during development. The micro–genetic method stresses that change is continual and occurs at many different points aside from the major stage changes. Tracking these ongoing changes can help researchers understand how children learn.

Mind. The mind is what the brain does, it includes intellect and consciousness.

Mirror neurons. A neuron which fires both when a human performs an action and when a human observes the same action performed by another. Mirror neurons therefore "mirror" behaviours as if the observer himself was performing the action.

Mnemonic technique. A technique which enhances memory performance.

Morphology. In linguistics, morphology is the study of word structure.

Motivation. Can be defined as whatever causes to act. Motivation reflects states in which the organism is prepared to act physically and mentally in a focussed manner, that is, in states characterised by raised levels of arousal. Accordingly, motivation is intimately related to emotions as emotions constitute the brain's way of evaluating whether things should be acted upon. **Intrinsic motivation** is evident when people engage in an activity for its own sake, without some obvious external incentive present, as opposed to **external/extrinsic motivation** which is reward-driven.

Motor cortex. Regions of the cerebral cortex involved in the planning, control, and execution of voluntary motor functions.

Motor neuron. A neuron that carries information from the central nervous system to the muscle.

MRI (Magnetic Resonance Imaging). A non-invasive technique used to create images of the structures within a living human brain, through the combination of a strong magnetic field and radio frequency impulses.

Multiple intelligences. Theory that each individual has multiple, partially distinct, intelligences, including: linguistic, logical-mathematical, spatial, bodily-kinesthetic, musical, interpersonal, and intrapersonal.

Multiple sclerosis/MS. A chronic, inflammatory disease affecting the central nervous system.

Multi-tasking. Simultaneous performance of two or more tasks.

Myelin/myelination. Compact fatty material that surrounds and insulates axons of some neurons. Process by which nerves are covered by a protective fatty substance. The sheath (myelin) around the nerve fibres acts electrically as a conduit in an electrical system, increasing the speed at which messages can be sent.

Myth of three. Also known as the "Myth of the Early Years". This assumption states that only the first three years really matter in altering brain activity and after that the brain is insensitive to change. This extreme "critical period" viewpoint is not accurate. In fact, the brain is responsive to change throughout the lifespan.

Neurobiology. The study of cells and systems of the nervous system.

Neurodegenerative diseases. Disorders of the brain and nervous system leading to brain dysfunction and degeneration including Alzheimer's diseases, Parkinson's disease and other neurodegenerative disorders that frequently occur with advancing age.

Neurogenesis. The birth of new neurons.

Neuromyth. Misconception generated by a misunderstanding, a misreading or misquoting of facts scientifically established (by brain research) to make a case for use of brain research, in education and other contexts.

Neuron. Nerve cell. It is specialised for the transmission of information and characterised by long fibrous projections called axons, and shorter, branch-like projections called dendrites. Basic building block of the nervous system; specialised cell for integration and transmission of information.

Neurotransmitter. A chemical released by neurons at a synapse for the purpose of relaying information via receptors.

NIRS (Near Infrared Spectroscopy). Non-invasive imaging method which allows measures of the concentrations of deoxygenated haemoglobin in the brain by near-infrared absorption (near-infrared light at a wavelength between 700 nm and 900 nm can partially penetrate through human tissues).

Nucleus accumbens. The nucleus accumbens (also known as the accumbens nucleus or nucleus accumbens septi) is a collection of neurons located where the head of the caudate and the anterior portion of the putamen meet just lateral to the septum pellucidum. The nucleus accumbens, the ventral olfactory tubercle, and ventral caudate and putamen collectively form the ventral striatum. This nucleus is thought to play an important role in reward, pleasure, and addiction.

Nurture. The process of caring for and teaching a child as the child grows.

Occipital lobe. Posterior region of the cerebral cortex receiving visual information.

Occipito-temporal cortex. Also known as Brodman's area is part of the temporal cortex in the human brain.

Omega fatty acids. Polyunsaturated fatty acids which cannot be synthesised in the body.

Ontogenesis. The developmental history of an individual.

Orthography. The set of rules about how to write correctly in the writing system of a language.

OT (Optical Topography). Non-invasive trans-cranial imaging method for higher-order brain functions. This method, based on near-infrared spectroscopy, is robust to motion, so that a subject can be tested under natural conditions.

Oxytocin. Also known as the "love hormone". Oxytocin is involved in social recognition and bonding, and might be involved in the formation of trust between people.

Pallidum/globus pallidus. A sub-cortical structure of the brain.

Parasympathetic nervous system. A branch of the autonomic nervous system concerned with the conservation of the body's energy and resources during relaxed states.

Parietal lobe. One of the four subdivisions of the cerebral cortex. It plays a role in sensory processes, attention and language. Involved in many functions such as processing spatial information, body image, orienting to locations, etc. Can be subdivided into superior parietal lobule and inferior parietal lobule. The precuneus, postcentral gyrus, supramaginal gyrus and angular gyrus make up the parietal lobe.

Parkinson's disease. A degenerative disorder of the central nervous system that affects the control of muscles, and so may affect movement, speech and posture (see also neurodegenarative disorders).

Peripheral nervous system. A division of the nervous system consisting of all nerves which are not part of the brain or spinal cord.

Perisylvian areas. Cortical regions that are adjacent to the sylvian fissure – major fissure on the lateral surface of the brain running along the temporal lobe.

PET (Positron Emission Tomography). A variety of techniques that use positron emitting radionucleides to create an image of brain activity; often blood flow or metabolic activity. PET produces three-dimensional, coloured images of chemicals or substances functioning within the brain.

Phonemes. Basic units of oral speech that make up words.

Phylogenic development. The process of evolution which favors those genetic behavioural traits in both genders that best assure survival of the species.

Pineal gland. An endocrine organ found in the brain. In some animals, it seems to serve as a light influenced biological clock.

Pituitary gland. An endocrine organ closely linked with the hypothalamus. In humans, it is composed of two lobes and secretes a number of hormones that regulate the activity of other endocrine organs in the body.

Plasticity. Also "brain plasticity". The phenomenon of how the brain changes and learns in response to experience. See also experience-expectant/experience-dependent plasticity.

Precuneus. Structure in the brain positioned above the cuneus and located in the parietal lobe.

Prefrontal cortex. The region in front of the frontal cortex which is involved in planning and other higher-level cognition.

Primary motor cortex. Works in association with pre-motor areas to plan and execute movements.

Primary visual cortex. The region of the occipital cortex where most visual information first arrives.

Pruning/synaptic pruning. The natural process of eliminating weak synaptic contacts.

Putamen. A component of the limbic system. This part is responsible for familiar motor skills.

Qualia. A term for subjective sensations. In "Phantoms In The Brain", Professor Ramachandran describes the riddle of qualia like this. How can the flux of ions and electrical currents in little specks of jelly, which are the neurons in my brain, generate the whole subjective world of sensations like red, warmth, cold or pain? By what magic is matter transmuted into the invisible fabric of feelings and sensations?

Reasoning. The act of using reason to derive a conclusion from certain premises using a given methodology. Two most commonly used explicit methods to reach a conclusion are **deductive reasoning** in which the conclusion derived from previously known facts, and **inductive reasoning**, in which the premises of an argument are believed to support the conclusion but do not ensure it.

REM (Rapid eye movement sleep). The stage of sleep characterised by rapid movements of the eyes, when the activity of the brain's neurons is quite similar to that during waking hours.

"Reptilian" brain (so-called). Refers to the brain stem which is the oldest region in the evolving human brain.

Right-brained thinking. A lay term based on the misconception that higher level thought processes are strictly divided into roles that occur independently in different halves of the brain. Thought to be based in exaggerations of specific findings of right hemisphere specialisation in some limited domains.

Schizophrenia. A mental disorder characterised by impairments in the perception or expression of reality and/or by significant social or occupational dysfunction.

Science of learning. Term that attempts to provide a label for the type of research possible when cognitive neuroscience and other relevant disciplines research joins with educational research and practice.

Second messengers. Recently recognised substances that trigger communications between different parts of a neuron. These chemicals are thought to play a role in the manufacture and release of neurotransmitters, intracellular movements, carbohydrate metabolism and, possibly, even processes of growth and development. Their direct effects on the genetic material of cells may lead to long-term alterations of behavior, such as memory.

Sensitive period. Time frame in which a particular biological event is likely to occur best. Scientists have documented sensitive periods for certain types of sensory stimuli (such as vision and speech sounds), and for certain emotional and cognitive experiences (such as attachment and language exposure). However, there are many mental skills, such as reading, vocabulary size, and the ability to see colour, which to do not appear to pass through tight sensitive periods in the development.

Serotonin. A monoamine neurotransmitter believed to play many roles including, but not limited to, temperature regulation, sensory perception and the onset of sleep. Neurons using serotonin as a transmitter are found in the brain and in the gut. A number of antidepressant drugs are targeted to brain serotonin systems.

Short-term memory. A phase of memory in which a limited amount of information may be held for several seconds to minutes.

SPECT. Functional imaging using single photon emission computerised tomography.

Stimulus. An environmental event capable of being detected by sensory receptors.

Stress. The physical and mental responses to anything that causes a real or imagined experiences and changes in life. Persistent and/or excessive stress may lead to depressive (withdrawal) behaviour.

Striatum. A subcortical part of the telencephalon, best known for its role in the planning and modulation of movement pathways but is also involved in a variety of other cognitive processes involving executive function.

Stroop task. A psychological test for mental vitality and flexibility. E.g. If a word is printed or displayed in a color different from the color it actually names; for example, if the word "green" is written in blue ink , a delay occurs in the processing of the word's color, leading to slower test reaction times and an increase in mistakes.

Sulcus/sulci. A furrow of convuluted brain surface. While gyri protrude from the surface, sulci recede, forming valleys between gyri.

Sympathetic nervous system. A branch of the autonomic nervous system responsible for mobilising the body's energy and resources during times of stress and arousal.

Synapse. A gap between two neurons that functions as the site of information transfer from one neuron to another (called "target cell" or "postsynaptic neuron").

Synaptic density. Refers to the number of synapses associated with one neuron. More synapses per neuron are thought to indicate a richer ability of representation and adaption.

Synaptic pruning. Process in brain development whereby unused synapses (connections among neurons) are shed. Experience determines which synapses will be shed and which will be preserved.

Synaptogenesis. Formation of a synapse.

Temporal lobe. One of the four major subdivisions of each hemisphere of the cerebral cortex. It functions in auditory perception, speech and complex visual perceptions.

Terminal/Axon terminal. A specialised structure at the end of the axon that is used to release neurotransmitter chemicals and communicate with target neurons.

Thalamus. A structure consisting of two egg-shaped masses of nerve tissue, each about the size of a walnut, deep within the brain. It is the key relay station for sensory information flowing into the brain, filtering out only information of particular importance from the mass of signals entering the brain.

TMS (Transcranial magnetic stimulation). A procedure in which electrical activity in the brain is influenced by a pulsed magnetic field. Recently, TMS has been used to investigate aspects of cortical processing, including sensory and cognitive functions.

Trans-disciplinarity. Term used to explain the concept of fusing completely different disciplines resulting in a new discipline with its own conceptual structure, known to extend the borders of the original sciences and disciplines included in its formation.

Ventricles. Of the four ventricles, comparatively large spaces filled with cerebrospinal fluid, three are located in the brain and one in the brainstem. The lateral ventricles, the two largest, are symmetrically placed above the brainstem, one in each hemisphere.

Visual cortex. Located in the occipital lobe; involved in detection of visual stimuli.

Wernicke's area. A brain region involved in the comprehension of language and the production of meaningful speech.

White matter. White matter consists of myelinated axons that connect various grey matter areas of the brain.

Also Available in the CERI Collection

Demand-Sensitive Schooling? Evidence and Issues
146 pages • November 2006 • ISBN: 978-92-64-02840-1

Think Scenarios, Rethink Education
200 pages • April 2006 • ISBN: 978-92-64-02363-5

Personalising Education
128 pages • February 2006 • ISBN: 978-92-64-03659-8

Students with Disabilities, Learning Difficulties and Disadvantages – Statistics and Indicators
152 pages • October 2005 • ISBN: 978-92-64-00980-6

E-learning in Tertiary Education: Where do We Stand?
290 pages • June 2005 • ISBN: 978-92-64-00920-2

Formative Assessment – Improving Learning in Secondary Classrooms
280 pages • February 2005 • ISBN: 978-92-64-00739-0

Quality and Recognition in Higher Education: The Cross-border Challenge
205 pages • October 2004 • ISBN: 978-92-64-01508-1

Internationalisation and Trade in Higher Education – Opportunities and Challenges
250 pages • June 2004 • ISBN: 978-92-64-01504-3

Innovation in the Knowledge Economy – Implications for Education and Learning
Knowledge Management series
96 pages • May 2004 • ISBN: 978-92-64-10560-7

www.oecdbookshop.org

OECD PUBLICATIONS, 2, rue André-Pascal, 75775 PARIS CEDEX 16
PRINTED IN FRANCE
(96 2007 01 1 P) ISBN 978-92-64-02912-5 – No. 55383 2007